John Butler Johnson

Chemistry of the Materials of the Machine and Building Industries

First Edition

John Butler Johnson

Chemistry of the Materials of the Machine and Building Industries
First Edition

ISBN/EAN: 9783337214616

Printed in Europe, USA, Canada, Australia, Japan

Cover: Foto ©berggeist007 / pixelio.de

More available books at **www.hansebooks.com**

CHEMISTRY OF MATERIALS

OF THE

MACHINE AND BUILDING INDUSTRIES

BY

ROBERT B. LEIGHOU, Sc. B.

ASSOCIATE PROFESSOR OF CHEMISTRY, IN THE CARNEGIE
INSTITUTE OF TECHNOLOGY

FIRST EDITION

McGRAW-HILL BOOK COMPANY, Inc.
239 WEST 39TH STREET. NEW YORK

LONDON: HILL PUBLISHING CO., Ltd.
6 & 8 BOUVERIE ST., E. C.
1917

THE MAPLE PRESS YORK PA

PREFACE

The preparation of this text has been brought about by the pressure of circumstances. The author has for nine years been conducting a course in Chemistry of Materials in the School of Applied Industries, of the Carnegie Institute of Technology, and has been considerably inconvenienced by the lack of a text. The particular object of the work is to supply information concerning the chemical properties of the materials employed in the various courses in Building Construction and Equipment, and in Machinery Construction and Operation, so that these materials might be used more intelligently and, therefore, to a better advantage. Although there are many texts on Industrial Chemistry covering the methods by which these materials are manufactured, nothing has appeared emphasizing their properties from the standpoint of the user. This is, then, the special purpose of the present text. It will naturally be understood that it is not possible to divorce entirely the discussion of the properties of any material from the discussion of the methods of its manufacture, but the latter have not been emphasized or presented in detail.

In preparing the material, the aim has been to avoid complexity of treatment in so far as possible, without becoming superficial. Although the treatment is by no means exhaustive, it is sufficiently complete to enable the student to gain a reasonably thorough knowledge of the various materials. Some theories, such as those that account for corrosion, the hardening of cements, the effect of the heat-treatment of steels, the electro-chemical action in primary and secondary cells, etc., have been introduced; and although the nature of the work has made a full consideration of them inadvisable, it is believed that they have been developed in sufficient detail to avoid inexactness.

A prior course in Elementary General Chemistry has been assumed; indeed, this is considered essential not only for the study of this text, but for the understanding of any literature dealing with these or similar subjects. No special effort has been made to avoid the use of technical terms. It is considered one of the

v

purposes of the text to explain such terms, and render them sufficiently familiar so that students whose training otherwise might be insufficient for the demand, may be encouraged to make use of reference works, scientific magazines, journals, and similar literature in which technical terms are commonly employed. In fact, the author feels that if the text accomplish no other object than to point the way to such sources of knowledge, it will have performed a valuable service.

The references to literature, from which a considerable amount of the material of the text has been drawn, have been indicated in the footnotes and listed at the ends of the chapters in order that the student may have recourse to them to extend the study of the subject when this is desired. An effort has been made in all cases to acknowledge the source of information obtained from these writings, and if any such acknowledgment has been omitted, it has not been done designedly.

The author wishes to acknowledge his indebtedness to Mr. C. A. Coulter for his aid in preparing certain illustrations. Also, he wishes to thank the following publishers for permission to use illustrations: Charles Griffin and Co., Ltd., for photomicrographs from their treatises on Alloys by Gulliver and Law, Sauveur and Boylston for photomicrographs from Sauveur's "Metallography of Iron and Steel," McGraw-Hill Book Co., Inc., for photomicrographs from Gardner's "Paint Technology and Tests;" and the following manufacturers for photographs and electrotypes: Harbison-Walker Refractories Co. for photomicrographs and illustrations of metallurgical furnaces and coke ovens, Pratt and Lambert for photographs relating to varnish and varnish materials, The Edison Storage Battery Co., and the Electric Storage Battery Co. for electrotypes showing their products.

R. B. L.

Pittsburgh, Pennsylvania,
January, 1917.

CONTENTS

CHAPTER VI

CHAPTER VII

CHAPTER VIII

CHAPTER IX

CHAPTER X

CHAPTER XI

CHAPTER XII

CHAPTER XIII

CHAPTER XIV

CHAPTER XV

CHAPTER XVI

CHAPTER XVII

CHAPTER XX

CHEMISTRY OF MATERIALS

CHAPTER I

WATER FOR STEAM GENERATION

Because of the intimate connection of steam with the development of power, and because of its wide application in other ways, the whole subject of steam generation is of considerable importance. But, since the conversion of water from the liquid to the vapor phase is essentially such a simple process, the need for intelligent supervision in the development of steam from natural waters on an industrial scale is apt to be overlooked. However, there are many real problems connected with the process, and it is the purpose of the present chapter to briefly discuss the more important of these. The difficulties that attend the successful operation of steam boilers are due in large measure to the impurities in the waters available for use, and it will be necessary, therefore, to consider these first.

Impurities in Natural Waters.—In a chemical sense, all natural waters are to some degree impure. Even rain water, which is the purest form, contains solid matter, both organic and mineral, obtained by washing out soot and wind-raised dust from the air. In fact, the generally accepted theory now is that the production of rain depends upon the presence in the air of dust particles, which serve as nuclei about which the drops may form. All waters obtained from the earth, whether from the surface or lower depths, contain dissolved substances taken up from the rocks and soils with which they have been in contact. Water from very low levels, as from deeply bored wells, is likely to contain larger quantities of dissolved substances than surface water, because of the great mass of rock through which it has percolated. In general, water from regions of granite, sandstone, and clay formations, contains much less dissolved mineral substances than that from limestone regions. Further, water from rocky regions is purer than that from regions where the

1

rock has been disintegrated to form soil, since rocks are generally
less soluble than soils. Mountain waters are relatively pure
because they usually come into contact with little soil. Beside
the dissolved substances, water may carry a great deal of suspended matter, as silica, clay, organic material, etc. This is
especially true of surface waters, as from rivers and other
streams.

Water that contains dissolved calcium, magnesium, and iron
compounds, generally as the sulfates and bicarbonates, is known
as "hard water." The application of this term should be confined to water containing these substances, but it is used loosely
in popular language to apply to acidulated water, or water
containing any substance that will destroy soap.

Temporary and Permanent Hardness.—The hardness of
water is generally recognized as being of two kinds. That which
is removable by boiling is said to be "temporary," that which
persists after boiling is said to be "permanent."

Temporary hardness is caused by the presence of bicarbonates,
especially of calcium, but also of magnesium and iron. The
normal carbonates of these metals are practically insoluble in
water alone, but in water that contains carbon dioxide gas they
are converted into their corresponding bicarbonates and these
are a great deal more soluble. Consequently, water coming
from limestone regions where much organic matter also exists,
will likely contain much of this sort of hardness. The organic
matter, decaying, produces carbon dioxide, and this in the
presence of water produces carbonic acid which acts on the
limestone as follows:

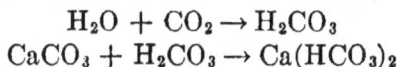

$$H_2O + CO_2 \rightarrow H_2CO_3$$
$$CaCO_3 + H_2CO_3 \rightarrow Ca(HCO_3)_2$$

Marl is a mixture of calcium carbonate and organic matter,
hence water from marl deposits contains much calcium
bicarbonate.

Permanent hardness is caused by the sulfates of calcium,
magnesium, and (ferric) iron, especially the sulfate of calcium.
These sulfates are not removed by boiling, although they are
less soluble in boiling water than in water at atmospheric
temperatures.

If present in sufficient amount the bicarbonates and sulfates are very objectionable in boiler water. To aid in forming an estimate of the suitability of such water for steam generation, the following table quoted by Benson[1] is shown:

RATING FOR WATER CONTAINING TEMPORARY HARDNESS
(GEBHARDT)

Less than 8 grains per gallon	Very good
12 to 15 grains per gallon	Good
15 to 20 grains per gallon	Fair
20 to 30 grains per gallon	Bad
Over 30 grains per gallon	Very bad

With permanently hard water, for the same ratings, the number of grains shown in the preceding table should be divided by four.

The results of chemical analyses are often stated in parts per million. Grains per U. S. gallon may be converted into parts per million by multiplying the number of grains by 17.1; or conversely, parts per million may be converted into grains per U. S. gallon, by multiplying parts per million by 0.0584.

Effect of Impurities on Boiler Service.—The objectionable features that result from the presence of impurities in boiler water may be classified under the following heads: production of scale, corrosion, and foaming.

Production of Scale.—During the conversion of water into steam, there is deposited within the boiler both the dissolved and suspended matter that the water carried. The deposition is due to concentration brought about by evaporation of water, to a lessening of the solubility of the dissolved substance by the increased heat and pressure, or to reactions that produce insoluble substances from others previously soluble. The deposit may be in the form of a loose sediment, sludge, or hard scale, depending upon the substance carried by the water, the temperature and pressure within the boiler, and other factors. As an example of the effect of the increased pressure and temperature, the following table[2] is given in which is shown the lessened solubility of calcium sulfate as the temperature and pressure rise:

[1] Industrial Chemistry, p. 74.

[2] Quoted from *Engineering*, Dec. 25, 1903, by BENSON: Industrial Chemistry, p. 66.

SOLUBILITIES OF CALCIUM SULFATE IN GRAINS PER U. S. GALLON

Temperature, degrees F.	Corresponding steam pressure in pounds	Calcium sulfate, grains per gallon
68.0	....	140.6
212.0	0	125.9
284.0	37	45.6
323.6	79	32.7
356.0	131	15.7
464.0	484	10.5

The effect on solubility produced by the conversion of the bicarbonates into normal carbonates, as the carbon dioxide gas is eliminated by heat, is shown by the following figures also taken from a table quoted by Benson:[1]

SOLUBILITIES IN PARTS PER MILLION

Substance	In water at 60°F.	In water containing carbon dioxide at 60°F.	In water at 212°F.
Calcium carbonate........	12	1,100	21
Magnesium carbonate.....	385	27,500	nearly 0

Effects of Deposits.—The substances deposited in the various ways shown in the preceding paragraphs, collect on the flues, in the tubes, or other parts of the boiler and act as heat insulators, so that heat that would otherwise be available cannot pass into the water. The degree to which the deposit interferes with the transmission of heat, depends largely upon whether it is in the form of a loose sediment or a hard compact scale. The latter form is the most objectionable in this respect. Collet[2] gives the following figures to show the comparative resistance of iron and certain other substances, the last two of which, or substances of the same composition, enter into the formation of boiler scale. The resistance of wrought iron being taken as 1, that of copper is 0.4; of slate, 9.5; of brick, 16; of chalk, 17; and of sulfate of lime, 48. Because of the poor transmission when

---

[1] *Loc. cit.*, p. 67.
[2] *Water Softening*, p. 17.

the metal is coated with scale, it becomes overheated—may even become red hot—so that it is soft and subject to deformation. Beside the actual insulating effect, the scale is objectionable in other ways. As the tubes become clogged, the surface area of water exposed to the heat is lessened and the heating efficiency falls. Also the boilers must be cleaned, which is an expensive process, especially if the scale is hard and closely adherent.

Composition and Physical Character of Scales.—Whether the scale is soft and loose, or hard and closely adherent, depends upon various factors, but no doubt it is determined chiefly by the composition of the scale. If the amount of calcium sulfate and magnesium compounds in the water is low, or if the quantity of suspended matter is high, the scale will be soft and loose, so that it may be removed from the boiler in the form of a sludge. But if the water is clear, having but little suspended matter, and the amount of calcium sulfate and magnesium compounds is high, the scale will be hard, dense and difficult to remove. Thus, hard scales are more objectionable than soft ones, both because of their greater insulating effect and because of the greater expense in removing them.

The location in the boiler has much to do with the composition of the scale at that point. Near the feed pipe the carbonates are found, these being deposited where the water first comes into contact with the heat and gives up its carbon dioxide. The calcium sulfate remains dissolved until it reaches the hottest part of the boiler.

Effect of Grease.—In respect to its heat insulating effect, a film of grease on the heating surface of the boiler is far worse than many times its thickness of scale. Booth says[1] a mere film of grease will cause overheating and collapse. Oils, especially heavy oils and animal fats, should not be introduced into boilers for the purpose of removing scale,[2] for although they may accomplish this result, the final effect is more harmful than good.

Foaming and Priming.—The collection of a mass of bubbles in the steam space above the water is designated as foaming. It is closely connected with priming, which is the passing from

[1] Water Softening and Treatment, p. 33.
[2] CHRISTIE: Purification of Water, pp. 165 and 172.

the boiler of steam mixed with water.[1]   Foaming is caused when the concentration of dissolved salts, especially the salts of sodium and potassium, becomes high.   The dissolved salts increase the surface tension and thereby lessen the readiness with which the steam bubbles break.   These salts may be naturally present in the water or they may be introduced by the methods of softening with soda ash, as described under "Water Treatment."   They consist largely of chlorides, sulfates, and carbonates, the last mentioned being particularly objectionable.   As the water is evaporated the concentration increases, and when it reaches about 100 grains per gallon, foaming is apt to occur.[2]

However, foaming is not due alone to dissolved salts.   Suspended matter, such as mud, calcium and magnesium salts, oil, and especially organic matter from sewage and other sources, which rises to the surface and forms a scum, may cause it.   Or it may be caused by a combination of dissolved salts and suspended matter, when neither alone is present in sufficient amount to bring it about.   In addition, the type of boiler and operating conditions have been shown to be important factors.   The remedy lies in frequent "blowing off," or in changing the water within the boiler at stated intervals.

Corrosion.—The corrosion of boilers is dependent upon the same general conditions and factors as corrosion elsewhere (see Chapter V).   The corrosion of boilers and boiler tubes usually manifests itself in a form known as "pitting," this being due to local action, as explained on page 131.   The presence of fatty oils in the boiler is objectionable in this connection, because under the influence of the pressure and temperature existing there, they are hydrolyzed (see page 339) to form glycerine and fatty acids, which exert a marked corrosive action.   Also, some inorganic salts are more corrosive than others.   For example, it is generally held that magnesium chloride reacts with the water to produce magnesium hydroxide, thus splitting off hydrochloric acid, as:

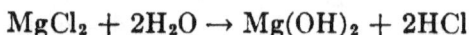

$$MgCl_2 + 2H_2O \rightarrow Mg(OH)_2 + 2HCl$$

The hydrochloric acid then becomes an active corroding agent.

---

[1] See ROGERS and AUBERT: Industrial Chemistry, p. 38.

[2] BENSON: Industrial Chemistry, p. 72.

This is largely the reason why sea water, which contains considerable magnesium chloride, is so objectionable for boiler use.

Ost[1] holds that magnesium chloride exerts its corrosive action in a somewhat different manner than that just shown. In his opinion, ferrous hydroxide is first formed by the reaction of iron with water, this being removed by the magnesium chloride, acting as follows:

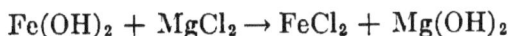

$$Fe(OH)_2 + MgCl_2 \rightarrow FeCl_2 + Mg(OH)_2$$

As the ferrous hydroxide is removed, more is formed. Ost's remedy is based upon the fact that at the higher pressures magnesium chloride and calcium carbonate react, forming calcium chloride, which is practically inactive in this manner. The reaction between the magnesium chloride and calcium carbonate is never complete, but it is estimated that only one-quarter as much calcium carbonate as magnesium chloride is needed to prevent the corrosive action of the latter.

TREATMENT OF BOILER WATERS

The scale forming tendencies of water may be overcome by suitable treatment. As a preliminary measure, the suspended matter may be removed by some purely mechanical device. Coarse substances may be taken out by screens, while the more finely divided material may be removed by filtering through sand, coke, wood fiber, or similar material. Or to lessen the work of the filters, a great deal of the suspended matter may be gotten rid of by allowing the water to stand in sedimentation tanks or basins.

The dissolved solids are less easily removed. Two general methods are employed, namely, cold-water softening, in which the water is treated with chemical reagents in the cold followed by sedimentation; and hot-water softening, which consists of heating with or without the use of chemical reagents.[2]

Cold-water Softening.—In this process the temporary hardness is removed by converting the soluble bicarbonates into the insoluble normal carbonates by the use of lime, while the per-

[1] *Engineering,* **74,** 482 (1902).

[2] For descriptions of commercial softeners of various types, see CHRISTIE: Water Purification and Its Use in the Industries, and *Jour. Ind. Eng. Chem.,* **3,** 326 (May, 1911).

manent hardness is cared for by converting the sulfates and chlorides also into the normal carbonates by the use of "soda ash" (sodium carbonate). In order that the bicarbonates may be converted into the normal form it is necessary to neutralize the carbonic acid present. The essential reactions, then, are as follows:

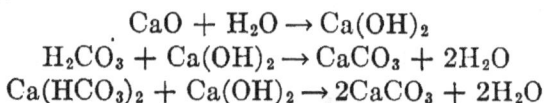

$$CaO + H_2O \rightarrow Ca(OH)_2$$
$$H_2CO_3 + Ca(OH)_2 \rightarrow CaCO_3 + 2H_2O$$
$$Ca(HCO_3)_2 + Ca(OH)_2 \rightarrow 2CaCO_3 + 2H_2O$$

The bicarbonates of magnesium and iron are acted upon in the same manner.

For permanent hardness, the reaction is as follows:

$$CaSO_4 + Na_2CO_3 \rightarrow CaCO_3 + Na_2SO_4$$

In a similar manner the magnesium sulfate and the chlorides of calcium and magnesium are converted into carbonates with the formation of the corresponding sodium salt. The normal carbonates of calcium, magnesium, and iron, formed by these reactions are insoluble, and to remove them it is necessary only to allow the water to stand for a time and they will settle out. As they settle, suspended matter, as silica, clay, and organic matter is carried down with them. The clear water is then drawn off and used. Sometimes, when the water has become partly clarified by settling, the residual suspended material is removed by rapid filtration through thin beds of coke, excelsior, or similar substances. The sodium salts formed by the use of sodium carbonate, being soluble, remain in the water, and although they do not contribute directly to the formation of scale, they may become objectionable because of their tendency to cause foaming. If the amount of permanent hardness is very great, it is sometimes not practicable to neutralize all of it, because of the large amounts of sodium salts that would be left in the water.

In determining the amount of calcium hydroxide and sodium carbonate necessary to correct the hardness, an analysis of the water is required. The softening cannot possibly be properly carried out without it; and, since the amount of hardness may vary widely within short periods, on account of the prevalence or absence of rains, and other varying factors, the amount present should be determined often, at least once a day.[1] If the degree of hardness is not accurately known, then the chances are very great that either too little of the reagents will be added, and the hardness will be incompletely corrected; or too much will be used, with the result that not only will material be wasted, but the excess in the water will be objectionable. Calcium hydroxide itself will cause hardness, and by secondary reactions will produce scale, while the sodium carbonate causes foaming as stated before.

Hot-water Softening.—The process of hot-water softening is carried out in feed-water heaters. These are adjuncts to the boiler, designed to conserve heat that would be wasted otherwise. The saving of heat is their most important function, but because of the heating, bicarbonates are decomposed, and so they effect a certain amount of softening as well. In many cases the heater is specially designed to take advantage of this softening effect. The heat conserved may be either from flue gases, or from exhaust steam: if from the former source, the device is generally known as an economizer; if from the latter, a feed-water heater.

Economizers consist especially of water tubes set in the flues leading from the furnaces.[2] They are so arranged that scale may be cleaned from the inside, and soot from the outside of the tubes. Because of the source of heat, much higher temperatures can be obtained in the economizer than in the feed-water heater; in fact, the conditions of pressure in the economizer very closely approximate the conditions in the boiler itself.

Feed-water heaters which use steam as a source of heat are of two kinds, open and closed. Open heaters operate at atmospheric pressure, and are more serviceable in softening water than those of the closed type, because the gases can escape and the bicarbonates are to a greater extent removed, although the sulfates are not much affected. As has been explained, the sulfates tend to produce hard, and the carbonates soft scale; consequently boiler scale from water that has been merely preheated will be harder than that from raw water, although it will not be so great in amount.

[1] For methods of analysis, see MASON: Examination of Water.

[2] ROGERS and AUBERT: Industrial Chemistry, p. 60.

In closed heaters, the water is heated by pipes which pass through them carrying steam under high pressure. Or these positions may be reversed; the water may pass through the pipes and the steam, under pressure, about them. Suitable arrangements are provided for cleaning the scale from the tubes.

In order that the sulfates also may be removed from the water, it is necessary to treat it with sodium carbonate in much the same way as in the cold process, but because of the heat the sulfates are converted into the corresponding carbonates more readily. After the precipitate has been formed, the water is filtered.

"Boiler Compounds."—These are substances added either to the boiler direct, or to the water just as it enters the boiler, so that whatever softening or other action that the substance is capable of bringing about, takes place within the boiler itself. Although this process of treating the water while within the boiler is widely used, it is not to be recommended, since much more satisfactory results are produced by the use of a separate purifying apparatus of some one of the types previously mentioned. There are, perhaps, certain exceptions; for example, in such cases where it is not possible to treat the water beforehand, or when the supply is of such quality that it needs but little treatment. Boiler compounds can in no manner lessen the amount of scale-forming ingredients; in fact, they may increase it. The only function they can have is to convert the scale-producing constituents into some form in which they will produce a deposit less objectionable. For example, soda ash will convert calcium sulfate into calcium carbonate which will form a deposit much more easily removable. Sodium phosphate acts in a similar manner, and is used to some extent as a boiler compound. Beside its use as a precipitant, soda ash neutralizes free acids, and thus aids in lessening the corrosion of the boiler. On the other hand, it increases the tendency of the water to foam.

Aside from the sodium carbonate and phosphate, whose action is quite well understood, there is another class of boiler compounds that includes a great variety of heterogeneous substances, that are apparently effective, but the manner of whose action is less easily explained. In this class are included tannin from bark or spent tan liquors, glue, starches, sugars, petroleum products, graphite, etc., all of which, and many similar substances beside, have been used in boilers. Their action is best described as mechanical, in that they exert an influence on the solid particles of scale-forming substances and prevent them from massing together. The explanation of this action seems to fall within the province of colloidal chemistry, the substances mentioned being said to act as "protective colloids." In order to comprehend the meaning of this term, it will be necessary to gain some conception of the colloidal state of matter in general. For this purpose the student is referred to page 361 of this text, where a brief discussion is given. Substances in the colloidal state are distinguished primarily from those that exist in the crystalloidal state, but the two states are not widely separated and most substances can be caused to pass from one to the other or to produce intermediate forms, possessing certain properties of each. Wherry and Chiles in their article on boiler compounds[1] offer an explanation of such transitions, together with the action of colloidal substances in preventing the formation of scale, and their discussion is in part quoted here.

"Most inorganic colloids—those derived from the mineral kingdom —are relatively unstable, and pass through a series of changes somewhat as follows: Starting from a condition of suspension in water, the colloid substance first assumes the form of a fine, soft, flocculent precipitate. By the action of 'cohesion' and 'adsorption' the particles gradually unite more and more firmly together and the originally soft mass becomes harder and harder. Finally, especially in the presence of a trace of some salt as sodium chloride, a tendency to revert to the crystalloidal form asserts itself, and the so far structureless material changes into a dense lamellar or fibrous crystalline mass."

In this manner boiler scale is formed from the sulfates of calcium and magnesium. In order to prevent the formation of a hard scale it is necessary only to prevent the loose particles from consolidating. It was shown in the discussion of sodium carbonate and phosphate that they accomplished this end by converting the scale-producing substances into forms that had but little tendency to consolidate. The substances now being discussed accomplish the same final result but in a different manner.

[1] *Eng. Mag.,* **45,** 518 (July, 1913).

They act as "protective colloids," possessing a function that may be briefly outlined by quoting further from Wherry and Chiles:

"It has been found that the various changes which a colloid may undergo—deposition from a state of suspension, coagulation and hardening in their various states, and the reversion to the crystalloidal condition—can be greatly retarded if not altogether prevented by the presence in the surrounding liquid of certain other colloids, such as glue, gelatine, and tannic acid, which need not in any way enter into chemical reaction with the first substance."

Since the colloid substance does not actually enter into combination with the scale-forming material, it is not very rapidly used up, and a very small quantity of it may exert a marked protective action for a relatively long period. Its presence in the water prevents the precipitated calcium salts from consolidating. Graphite, lamp-black, and soapstone, if powdered sufficiently fine, and the oils and fats also, seem to be able to form protective-colloid suspensions directly, without entering into solution.

References

A

MASON: Examination of Water, New York, 1901.
BOOTH: Water Softening and Treatment, London, 1906.
CHRISTIE: Boiler Waters, New York, 1906.
HAZEN: Clean Water and How to Get It, New York, 1909.
COLLET: Water Softening and Purification, London, 1911.
DON and CHISHOLM: Modern Methods of Water Purification, London, 1911.
CHRISTIE: Water; Its Purification and Use in the Industries, New York, 1912.
ROGERS and AUBERT: Industrial Chemistry, New York, 1915.

B

The Present Status of the Art of Water Softening, Cassier's Mag., 31, 416 (1907).
BOOTH: Boiler Water Purification, Jour. Ind. Eng. Chem., 3, 326 (1911).
JACOBSON: Analytical Control of Water Softening, Jour. Inst. Brewing, 8, 88 (1911).
FRANCIS: Locomotive Boiler Corrosion, Am. Ry. Assoc., 13, 420 (1912).
WHERRY and CHILES: New Light on the Chemistry of Boiler Scale Prevention, Eng. Mag., 45, 518 (1913).
BENSON and HOUGEN: Method for Testing the Efficiency of Boiler Scale Preventives, Jour. Ind. Eng. Chem., 8, 435 (1916).
BRUCKMILLER: Water Softening Calculations, Chem. Engr. Mfr., 24, 99 (1916).

CHAPTER II

FUELS

Under the head of fuels are included all of those substances, which when burned, are capable of producing heat energy that may be utilized for domestic or industrial purposes. Fuels owe their ability to burn to the presence of two elements, carbon and hydrogen, and in the burning of all fuels the heat obtained is derived essentially from these two constituents. In some fuels there is a small amount of sulfur which may burn and contribute to the heating value but the quantity thus derived is of little consequence. In the solid fuels, coal and coke, a great deal of the carbon exists in the free state, but in the gases, oils and wood it is combined. Except in certain gaseous fuels, hydrogen is always in combination with other elements, particularly with carbon and oxygen.

PRINCIPLES OF COMBUSTION

Combustion Products.—In the presence of the proper amount of air, *carbon* burns with the formation of carbon dioxide, but if the air supply is insufficient, carbon monoxide is formed. The dioxide is also reduced to the monoxide by carbon at a white heat, thus:

$$CO_2 + C \rightarrow 2CO$$

Above 1,000°C., carbon dioxide is dissociated into the monoxide and oxygen, which combine again when the temperature falls.

When *hydrogen* burns, water is formed; water being merely the oxide of hydrogen. When any fuel containing hydrogen, free or in compound, burns, water is always produced.

In the burning of *hydrocarbons*, if the combustion is complete, the products are water and carbon dioxide. If the combustion is incomplete, water may be formed, and some of the carbon may be set free, heated to redness by the heat of the burning

13

hydrogen together with the heat of its own partial burning. Or the carbon may be burned to CO. Or lower hydrocarbons may be formed; for example, ethylene may produce acetylene and water, thus:

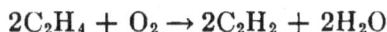

$$2C_2H_4 + O_2 \rightarrow 2C_2H_2 + 2H_2O$$

Flame.—When a gas burns it produces a flame. All flame results from the burning of gases or vapors. When a solid produces a flame it is first converted into either a vapor or a gas. For example, the wax of a candle or the oil of a lamp does not burn as such, combustion being impossible until sufficient heat has been applied to volatilize some of the wax or oil. Depending upon the character of the fuel and the conditions attending the burning, the flame may be luminous or non-luminous. In general, luminous flames are produced when the combustion of the carbon is retarded because of the lack of air. In such cases the carbon is liberated from the compound in a highly heated condition, as was explained in the discussion of the burning of hydrocarbons, and remains in this highly heated or incandescent state until sufficient oxygen has united with it to convert it into an invisible gas, either the monoxide or dioxide of carbon. This manner of burning is well illustrated by the candle flame which can be readily seen to consist of several parts. The colorless portion in the interior, nearest the wick, consists of gases and vapors set free by the heat but not yet having begun to burn. The luminous portion consists of particles of carbon heated to incandescence, but only partially burned. This is the part that deposits soot or produces smoke. An ordinary luminous flame is to a greater or lesser degree always a smoky flame, at least potentially. The colorless portion above the luminous section is made up of the products of completed combustion, essentially water vapor and carbon dioxide.

When sufficient air is present, as in a Bunsen flame, there is no luminosity. The carbon is burned at once, without being first separated as incandescent particles.

Conditions Necessary for Complete Combustion.—In order that combustion may be complete, two conditions must be fulfilled: there must be an excess of air, and the temperature of the fuel must be sufficiently high. Although an excess of air is desirable, this excess must not be too great, for air passing through the zone of combustion carries away heat and thus may retard the burning by cooling the fuel. It is due to this fact that a match or candle flame may be put out by blowing. The blast of air carries away so much heat that the temperature falls to below the kindling point of the fuel. The kindling point is defined as that temperature at which burning begins, or conversely, that temperature below which burning cannot take place. According to the same principle a candle flame may be put out by surrounding it with a coil of copper wire, or a gas flame may be put out by inserting in it a mass of cold silver or copper. These metals are good heat conductors and readily take up heat from the flame. But when any cold object is placed in a flame it absorbs heat to some extent and thus interferes with combustion in its immediate neighborhood. The flame cannot touch the cold object; a clear space intervenes. As the object grows hotter, this space lessens, and when the object is as hot as the flame, the flame touches it.

Ordinarily when water is boiled in an open vessel its temperature does not rise above 100°C. Since the temperature of an ordinary flame is generally not less than 1,000°C., it is evident that the flame cannot be in contact with the bottom of the vessel.

The hottest part of a flame is just at the point where combustion has been completed. In the Bunsen flame this is directly above the tip of the inner cone. Objects can be most advantageously heated by placing them at this point. If placed lower in the flame, they will likely cause incomplete combustion by cooling the flame.

Relative Volumes of Gases Produced by Combustion.—When carbon burns completely the volume of the products of combustion is the same as the volume of oxygen used, as:

$$2C + 2O_2 \rightarrow 2CO_2$$

When combustion is imperfect and carbon monoxide is formed, two volumes of gaseous product are made from one volume of oxygen, thus:

$$2C + O_2 \rightarrow 2CO$$

Two volumes of hydrogen with one volume of oxygen form two volumes of water vapor, as:

$$2H_2 + O_2 \rightarrow 2H_2O$$

There is then a diminution of one-third. But when the water vapor condenses the volume is enormously lessened. This is very noticeable in liquid fuels which contain much hydrogen.

Heat Radiated by Flames.—Non-luminous flames do not heat much by radiation, since gases are poor radiators. Because of this, when heating a room, for example, by means of a gas grate, material such as metal bars, fireclay forms or asbestos must be used. Although these substances become no hotter than the flame, they are better radiators.

Carbon is a good radiator; hence, a luminous flame radiates heat much more readily than a non-luminous one.

Moisture in a fuel is a source of loss of heat in burning. By the evaporation of the moisture, heat is lost as the latent heat of steam, 966 heat units being carried away by each pound of evaporated water. On the other hand, it is said that a certain amount of benefit is derived from moisture in the coal, and for boiler purposes, coal is sometimes intentionally wetted. The idea is based upon the fact that the resultant steam is decomposed, producing carbon monoxide and hydrogen, in accordance with the following equation:

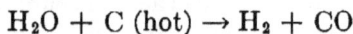

$$H_2O + C \text{ (hot)} \rightarrow H_2 + CO$$

As these gases afterward burn, the zone of intense heat is extended farther into the boiler tubes with the result that much more heat is absorbed.

Smoke.—There are in general two kinds of smoke. The first kind is of a whitish or greenish-yellow color, and consists mainly of hydrocarbons. It can be seen arising from the hot wick of a blown-out candle, or when coal is first placed on the fire, or when an attempt is made to ignite half-dried brush or weeds. The smoke from a cigar or other tobacco belongs to the same class. This hydrocarbon smoke can be quite easily burned. If a lighted match is applied to the upper part of the stream of smoke that arises from a blown-out candle, a flame will travel back through the ascending column and re-ignite the wick. Hydrocarbon smoke contains practically no free particles of carbon. Under microscopical examination, it proves to consist of small vesicles, or balloon-like structures, the skins of which consist of vapors and liquids distilled out of the burning substances. The vesicles are very light, being filled with gases, such as nitrogen, traces of oxygen, carbon dioxide, carbon monoxide, hydrogen and the hydrocarbons, methane, ethylene, etc. They float in the air until brought forcibly into contact with some surface, which causes them to burst and set the contents free. The skins of the vesicles produce a deposit generically known as "tar."

The second kind of smoke is the heavy black cloud. It consists of minute particles of carbon originating from the cooling of the red-hot particles of a luminous flame. This smoke arises usually from long-flame coals, but it is produced by any luminous flame that is sufficiently cooled by radiation, before it has met with enough air to complete the combustion.

The tendency toward the production of smoke in those cases where there is an *insufficient supply of air*, may be shown by the following experiment suggested by Kent:[1]

"Take a lighted, central-draft kerosene lamp and adjust the wick to such a point, that the lamp gives a rather short and clear, white light, without a trace of smoke. Now, without altering the adjustment of the wick, gradually obstruct the opening at the bottom of the central-draft tube and observe the result. The flame grows longer, and its whiteness changes to yellow and then to red. It begins to smoke; and finally when the flame has risen to nearly the top of the chimney, a dense column of black smoke and soot is given off. We learn from this experiment that with the same consumption of fuel, that is, the oil supplied by the wick, the flame may be short and intensely hot; or very long, of a low temperature, smoky and sooty. While the flame is lengthening, and before it becomes smoky, the combustion may be complete, but it is not effected in as short a space as with the original supply of air. For a given supply of fuel, a short flame means rapid and complete combustion; a longer flame, delayed combustion; and a very long flame, imperfect combustion."

In the long flames, the hot carbon particles are given a chance to radiate heat for a longer period and finally become cold and

[1] Steam Boiler Economy, p. 9.

2

black; or in other words, produce smoke. Even in flames of medium length, the hot particles may be cooled by contact with some cold object and be converted into smoke.

In steam generation, one of the common causes of smoke is *bad stoking*. A thick layer of cold coal is thrown upon the fire, and the tarry matters at once begin to distill in great abundance from the surface of the coal where it is in contact with the heat. It is very probable that these volatile products will not even be ignited, because after ascending through the cold layer of coal they are then separated by it from the heat, and consequently escape unburned, producing smoke, which in this case is mainly not carbon but tarry matter. But if these products do become ignited, because they are being produced in such great quantity, air cannot find entrance rapidly enough to burn them. Consequently their flame gradually cools by radiation and produces smoke, which now will be a heavy black cloud containing much carbon.

The *smokiness of bituminous coal* is due to the fact that it contains such a large proportion of volatile substances or gaseous constituents. These gases consist largely of hydrocarbons and pass off readily before they can be completely burned. The loss is avoided by the use of *mechanical stokers*, that feed the coal slowly and steadily, thus turning it into coke before it reaches the fire. This allows the gases that are thus produced only slowly to be burned as they are set free. Downdraft or underfeed furnaces are very efficient because the hydrocarbons must pass through the furnace bed of hot coals and are consequently burned. Ordinarily the loss of fuel by smoke is about 10 lbs. per ton.

An idea of the *composition of the soot-fall* resulting from general atmospheric smoke will be furnished by the following analysis of a deposit collected from the roof of a greenhouse after the falling of a snow: carbon, 39.00 per cent.; hydrocarbons, 12.30 per cent.; organic bases, 1.20 per cent.; sulfuric acid, 4.33 per cent.; ammonia, 1.37 per cent.; metallic iron and magnetic oxide, 2.63 per cent.; other mineral matter chiefly silica and ferric oxide, 31.24 per cent.; water, not determined.

The composition of the soot-fall varies greatly with the locality. In the Pittsburgh district, where blast furnaces are especially numerous, the soot-fall is very rich in iron oxide. An analysis[1] of a year's sample of soot-fall collected in the business district · of Pittsburgh, during the year beginning April, 1912, showed the following: tar, 2.19 per cent.; ash, 75.84 per cent.; fixed carbon, 21.97 per cent. The iron oxide, Fe_2O_3, in the ash was 32.08 per cent., which in the whole deposit amounts to 24.30 per cent.

Surface Combustion.—Surface combustion may be defined as the burning of an explosive mixture continuously and quietly; that is, without an explosion.[2] The method by which this is accomplished consists essentially of injecting through a suitable orifice, at a speed greater than the velocity of backfiring, an explosive mixture of gas (or vapor) and air, into a bed of incandescent, granular, refractory material, which is placed around or near the body to be heated.[3]

An idea of the velocity of backfiring as compared to the rate of gas flow may be obtained by the use of an ordinary Bunsen burner. If the burner is so adjusted that the mixture of gas and air passes through at a suitably high velocity, the flame may be carried entirely away from the burner, because the velocity of backfiring cannot equal the rate of gas flow. But if an iron ring of suitable size is fixed about 2 in. above the tip of the burner so that the stream of mixed air and gas will pass through it, the flow may be retarded sufficiently by the ring that the velocity of backfiring will be greater than the rate of gas flow above the ring. Now if the gas is lighted at some point above the ring, the flame can travel back against the flow as far as the ring, but cannot pass below it, because there the rate of flow exceeds the velocity of backfiring. Then the base of the flame will come to rest within the ring. Here the combustion will be quite intense. It is upon this principle that the method of surface combustion has been developed.

The principle is very conveniently made use of in the heating of diaphragms. If a diaphragm of porous material, sufficiently coarse that gas can flow through it, is placed over a container connected to the source of a combustible gaseous mixture, and if the gas is admitted and lighted after it has passed through,

[1] *Met. Chem. Eng.*, April, 1914, p. 247.
[2] *Machinery*, May, 1912, p. 677.
[3] *Eng. Rev.*, March 15, 1912, p. 367.

it will burn with ordinary flame for a time; but as the diaphragm becomes heated, the flame gradually disappears, and finally combustion takes place at the surface of the diaphragm.

Or if a crucible is placed in a furnace and a packing of granular, highly refractory material is filled in between it and the walls of the furnace, and then a suitable mixture of air and gas is admitted from below and ignited, combustion will proceed quietly, without flame, on the surface of the granules. When the gas is first lighted, it will issue from the top of the furnace and burn with an ordinary flame, but as the air supply is gradually increased the flame strikes back through the granular material until it is checked at the orifices at the base of the furnace where the gas enters.

An important point about surface combustion is that no oxygen or air in addition to that previously mixed with the gas is needed. When the diaphragm or granular material has become incandescent, if the proportions of air and gas in the mixing chamber at the back are correct, it will remain incandescent, even in an atmosphere of carbon dioxide.

The advantages claimed for the process are: that combustion is greatly accelerated and concentrated just where desired; that combustion is complete with a minimum excess of air; that the attainment of very high temperatures is possible; and because a large amount of radiant energy is developed, the transmission of heat to the object to be heated is very rapid.

Application of Surface Combustion to Boilers.—Surface combustion has been successfully applied to boilers in industrial plants. Such boilers have a mixing chamber in front of the tubes, and the tubes are filled with granular refractory material. A gaseous mixture is forced through the tubes at a high velocity, but combustion is complete after the gas has gone a very little distance. The remainder of the material in the tubes acts as a baffle to force the hot gases against the walls of the tubes. Thus a large proportion of the heat is given off to the water.

That surface combustion is very efficient is shown by the test conducted by Bradford and Corwin.[1] They found that the maximum temperature difference between the flue gas and steam was only 41°C., and the sensible heat lost in the waste gases

[1] *Power,* **42,** 787 and 855 (1915).

varied from 3.5 to 4 per cent. The main loss in flue gases occurs as the latent heat of the steam produced by the burning of the hydrogen in the fuel. Therefore, it is best to use a gas low in hydrogen, as for example, producer gas made from anthracite.

The idea involved in surface combustion is not new, but it has been only rather recently applied to practical purposes.

THE HEATING VALUE OF FUELS

Total Heating Power.—The quantity of heat set free by any substance in burning is measured in heat units. There are three of these in use:

A British thermal unit (B.t.u.) is the amount of heat necessary to raise the temperature of 1 lb. of water from 39° to 40°F.

A centigrade unit (c.u.) is the amount of heat required to raise the temperature of 1 lb. of water from 4° to 5°C.

A calorie (cal.) is the amount of heat required to raise the temperature of 1 kg. of water from 4° to 5°C.

One pound of *solid carbon* in burning to carbon dioxide produces 14,500 B.t.u. One pound of carbon in burning to carbon monoxide produces 4,450 B.t.u. If this CO should all escape,[1] as a portion of it may, 10,050 heat units are lost. However, if it could be retained and burned, the heat set free would equal, in these two stages, that of the carbon burning to CO_2.

When 1 lb. of *hydrogen* burns to form water, 62,000 B.t.u. are liberated.

When a *hydrocarbon* such as methane (CH_4) burns, carbon dioxide and water are formed. One pound of methane contains ¾ lb. of carbon and ¼ lb. of hydrogen. Three-fourths of a pound of carbon in burning to CO_2 produces 10,875, and ¼ lb. of hydrogen, 15,500 B.t.u., or a total of 26,375 B.t.u.

---

[1] It is perhaps well to state that precaution should always be taken to prevent carbon monoxide from escaping into the air from any source, regardless of the fact that it represents a loss in heat value. Less than 1 per cent. in air causes death when breathed for about 10 min. The gas acts as a poison by forming a compound with the hæmoglobin of the red corpuscles of the blood which prevents it from carrying oxygen to the tissues from the lungs. Even 0.05 per cent. will, if breathed for half an hour, produce dizziness upon exertion, and 0.1 per cent. inability to walk, while 0.2 per cent. within the same length of time will cause loss of consciousness.

Oxygen in a fuel is always bad. When found it is considered as being combined with hydrogen in the proportion to form water, 8 to 1. Thus each 8 parts of oxygen will render unavailable 1 part of hydrogen. Available hydrogen must be expressed as H minus $\frac{1}{8}$O.

The Evaporative Power.—Since 1 lb. of carbon in burning completely produces 14,500 B.t.u., and 966 B.t.u. are required to evaporate 1 lb. of water, the evaporative power of a pound of carbon is represented by $\frac{14,500}{966}$, or 15 lbs. This number of pounds of water are converted from water at 212°F. to dry steam at the same temperature.

In the case of hydrogen, since the combustion produces water and this must be evaporated, the weight of water produced must be subtracted from the theoretical evaporative power of hydrogen, which is $\frac{62,000}{966}$ or 64.1. Since 9 lbs. of water are produced when 1 lb. of hydrogen burns, 9 must be subtracted from 64.1 leaving 55.1, which only may be counted as the practical evaporative power of hydrogen. It is impossible to burn a pound of hydrogen without producing 9 lbs. of water.

Heat Carried out the Stack by Escaping Gases.—The amount of heat lost through the medium of the gases escaping from the stack is considerable. It may amount to from 15 to 40 per cent. of the heat value of the fuel. As is very apparent, the escaping gases consist of the products of combustion, the nitrogen left from the air actually required to burn the fuel, together with the excess air that is admitted. In practice a large excess of air is used; for gas furnaces 1.5 times as much as is theoretically required, for a good grate furnace twice as much, and for a poor grate three times as much. Of course, an excess of air is necessary but it is obvious that this excess should be no greater than is required to insure complete combustion. Whether or not the proper amount of air is being admitted is determined by an analysis of the escaping gases, commonly called a flue-gas analysis. The amount of oxygen present shows the amount of excess air passing through the fire. The presence of carbon monoxide or methane shows incomplete combustion and insufficient air.

The Available Heating Value of a Fuel.—As has been indicated in the preceding paragraphs, the total heating power of a fuel is not nearly all available. And there are many losses that occur in the ordinary use of a fuel that have not been mentioned. For example, the losses that occur when coal is employed under a steam boiler may be summed up as follows:

1. The latent heat absorbed in evaporating the water in the coal. This includes moisture, combined water and water formed during combustion. If the coal is very wet the loss of heat resulting from this cause will be quite great.

2. Heat lost through the escape of actual fuel in the form of smoke, since this contains a notable percentage of carbon and hydrocarbons.

3. Heat lost by incomplete combustion; that is, through the formation of carbon monoxide instead of carbon dioxide.

4. Heat carried out the stack as sensible heat by the escaping gases.

5. Heat represented by the unburned fuel removed with the ashes.

6. Heat lost through radiation.

These losses vary, of course, with the kind of coal, the type of boiler and stoker and other factors, but the total loss may at times amount to 50 per cent. of the total heating power of the fuel.

Ash.—The ash of any fuel is the mineral residue left after the combustion of the fuel is complete. In the case of coal the ash is made up of both the mineral matter that was present in the plant of which the coal was formed and the earthy matter mixed with the coal during its formation. Coal ash consists mainly of silica, alumina, lime, magnesia, and the oxide and sulfide of iron. Although the proportions of these constituents vary widely, silica usually predominates. Those coals are best the ashes of which are nearly pure white, and contain but little alkali and lime, as well as iron oxide, since these materials may cause the ash to fuse. It is interesting to note that coal ashes contain relatively very little potash, usually only a fractional part of 1 per cent., while the ashes of certain woods, especially certain hard woods, may contain as much as 15 per cent. or more.

Clinkering of the Ash.—Clinkers are formed by the fusion of the ashes. This fusibility depends in considerable measure upon the ratio that exists between the basic oxides and the silica present. Ashes that are high in either basic oxides or silica are not readily fusible. Those in which the constituents are present in the proper ratio to form the normal silicate, fuse the most readily as a rule.

The presence of iron pyrite, FeS_2, in the coal is closely connected with the formation of clinkers. During the burning of the coal a portion of the sulfur in this compound is oxidized and the pyrite is largely converted into ferrous sulfide, FeS. This sulfide is fusible at low temperatures, and as it melts, it entangles other substances and serves as a starting point for fusions that would not occur without this aid. It depends much on the state of division of the pyrite and its distribution through the coal, whether or not the ash will clinker. High sulfur as shown by an analysis does not necessarily mean clinkering tendencies, although this is the case generally. However, the sulfur may be present in the form of organic compounds and then it has no tendency to induce the formation of clinkers.

Sometimes the unskillful use of the poker may cause the formation of clinkers because channels may be formed in the bed of coals through which the draft ascends as a blast, causing very high temperatures to be produced at such points.

THE SOLID FUELS

Wood.—This is composed mainly of three things: water, the solids of the sap, and cellulose $(C_6H_{10}O_5)_x$. The only heat obtained is from the carbon, since sufficient oxygen is present to prevent the hydrogen from being available. Not more than 40 per cent. of the wood produces heat. A cord of seasoned hard wood is considered equal to a ton of coal. Soft wood has about half that value; the usual allowance in a boiler test being four-tenths that of an equal weight of coal.

The percentage composition of wood may be considered to be: water, 20 per cent.; carbon, 39 per cent.; hydrogen, 4.4 per cent.; oxygen, 35.6 per cent.; ash, 1 per cent. The ash of wood is partly potassium carbonate, K_2CO_3.

Peat has about the same heating power as wood, 1 lb. evaporating 4 to 5 lbs. of water.

COAL

Coal is of vegetable origin, being formed from wood possibly by this chemical reaction:

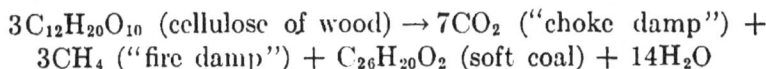

$3C_{12}H_{20}O_{10}$ (cellulose of wood) $\rightarrow 7CO_2$ ("choke damp") + $3CH_4$ ("fire damp") + $C_{26}H_{20}O_2$ (soft coal) + $14H_2O$

The gases shown are common in mines. All coal is the product of distillation of vegetable matter, brought about partly by the natural heat of the earth, and partly by the heat of its own decomposition. Intense pressure compacted the mass, and may have aided also in the carbonization.

Anthracite is the densest and hardest of all varieties. It is not an original variety of coal but is, perhaps, a modification of bituminous, caused by the application of more heat and pressure, and contains very little volatile matter. It kindles with difficulty, burns with but a small flame, emits scarcely any smoke, and makes an intensely hot fire. A complete analysis shows on an average: moisture, 2 per cent.; carbon, 86.6 per cent.; hydrogen, 2.8 per cent.; nitrogen, 0.8 per cent.; ash, 7.2 per cent.; sulfur, 0.6 per cent. Or moisture, 2 per cent.; volatile combustible, 4.3 per cent.; fixed carbon, 86.5 per cent.; ash, 7.2 per cent. The first statement of the analysis is known as an *ultimate* analysis, and the second as a *proximate* analysis.

Semi-anthracite and bituminous coals, as their names indicate, have properties midway between the two more important coals. The average of their analyses shows: moisture, 0.5 per cent.; carbon, 83.6 per cent.; hydrogen, 4.7 per cent.; oxygen, 4.2 per cent.; nitrogen, 1.3 per cent.; ash, 5.5 per cent.; sulfur, 0.8 per cent. Or moisture, 0.5 per cent.; volatile combustible, 16.7 per cent.; fixed carbon, 77.3 per cent.; ash, 5.5 per cent. The Blossburg (Tioga County, Pa.) coal is a semi-bituminous coal.

Bituminous coals are divided into three classes: (1) dry or non-coking coals, which burn freely but with little smoke, and do not fuse together when heated; (2) coking coals which in burning produce a little smoke and fuse together, Connellsville coal being an example; (3) fat or long-flaming coals, those that have a great deal of volatile matter, sometimes 50 per cent., therefore producing much smoke and either fuse or do not fuse in burning. Some Nova Scotia and Illinois coals are of the third class. The average of many analyses of bituminous coal shows: water, 0.9 per cent.; carbon, 77.1 per cent.; hdyrogen, 5.2 per cent.; oxygen, 6.7 per cent.; nitrogen, 1.6 per cent.; ash, 7.6 per cent.; sulfur, 1.0 per cent. Or water, 0.9 per cent.; volatile combustible, 27.4 per cent.; fixed carbon, 64.1 per cent.; ash, 7.6 per cent.

Lignite.—Although lignite is a fuel not as well known as the coals, there are nevertheless, according to the U. S. Geologic Survey, 150,000 square miles of land within the boundaries of the United States underlaid with it. The principal workable beds occupy the western part of North Dakota, the northwestern part of South Dakota and the eastern half of Montana. There are about 2,000 square miles in Texas and a great deal is found in other States.

Unlike coal, lignite is brown and not black. It is sometimes known as "brown coal." It has a wood-like appearance, and upon exposure to the air it "slacks" or crumbles. It, therefore, deteriorates greatly during storage and long transportation.

The following analysis shows the composition of a sample of raw lignite: moisture, 39.46 per cent.; volatile combustible, 29.50 per cent.; fixed carbon, 17.55 per cent.; ash, 13.49 per cent.; sulfur, 0.97 per cent.

A very notable and characteristic feature of its composition is its high moisture content, which so greatly reduces the heating value of freshly mined lignite. It is the partial evaporation of this moisture that causes the fuel to fall to pieces upon exposure. It has been found that by converting the lignite into the form of briquettes that it produces a very satisfactory and desirable fuel.[1]

There are two ideas, mainly, in the manufacture of *fuel briquettes;* they may be made with the idea of utilizing lignite, slack coal, culm, etc., which would be otherwise too fine for use; or they may be made with the idea of rendering the fuel less smoky. If for the former purpose, the binder may be any of the following, or of materials of similar nature: coal tar, coal-tar pitch, asphalt pitch, water-gas pitch or sulfite pitch. The last-

[1] See *Bull.* 14, Department of the Interior, Bureau of Mines.

named is a by-product of the treatment of wood pulp with an acid sulfite. It resembles a black, opaque resin but is quite soluble in water.

If the aim is to produce a smokeless fuel, the binder will be of a non-combustible, mineral nature, perhaps a clay or even a cheap cement, such a mixture being designed to hold all of the volatile constituents, so that they are completely burned before they pass off.

Summary.—The following summary shows the percentage composition and heat values of the solid fuels:

	Water	C	H	O	N	Ash	B.t.u.
Wood.........	20.00	39.00	4.40	35.60	0.00	1.00	7,000 to 9,000
Peat...........	20.00	41.00	4.40	24.60	0.00	10.00	7,000 to 9,000
Lignite........	39.46	.....	.....	.....	.....	13.49	6,000 to 9,000
Bituminous.....	0.90	77.10	5.20	6.70	1.60	7.60	11,000 to 14,000
Semi-bituminous	0.50	83.60	4.70	4.20	1.30	5.50	13,000 to 15,000
Anthracite......	2.00	83.90	2.80	2.70	0.80	7.20	12,000 to 14,500

Spontaneous Combustion in Stored Coal.—This is due to the chemical action between the carbon constituent of coal and the oxygen absorbed. All carbon has the peculiar power of condensing upon its surface and in its pores oxygen from the atmosphere. This power, of course, increases with the porosity of coal, some coals absorbing twice their volume of oxygen. When the coal is powdered it will absorb as much as 2 per cent. of its weight of oxygen and spontaneously ignite in a very short time. The heat of the slow oxidation is retained, which causes the temperature to rise and the rate of oxidation to increase.

The presence of wood in a pile of coal adds materially to the fire risk, since the heat developed will distill tarry matter from the wood, which will ignite very readily.

Outdoor exposure of coal results in loss of heating value varying from 2 to 10 per cent. Dry storage has no advantages over open storage except that the pyrites are not oxidized so much and hence the pores are not opened allowing the hydrocarbon gases to escape. During storage all losses cease practically at the end of 5 months.

Coal submerged in water does not lose appreciably in heat value.

COKE

Processes of Manufacture.—Coke is the residue, carbon chiefly, left after the dry distillation of certain coals. The purpose in converting the coal into coke is to get rid of the volatile, or smoke-producing content, and to lessen the amount of sulfur, as well as

FIG. 1.—Standard American beehive coke oven.

to harden the fuel so that it can support weight. The elimination of sulfur is very important. It has no heating value worthy of consideration and it has a highly undesirable effect upon the iron in the blast furnace and cupola where the coke is much used. The coking also reduces bulk and so saves some expense in transportation.

There are two processes for making coke, designated according to the character of the oven or retort in which the distillation is carried out.

First, there is the *beehive oven*, illustrated in Fig. 1, which is unfortunately the more common of the two. It is a dome-shaped oven of brick, about 10 ft. in diameter and 6 ft. high, with an opening in the top through which the coal is charged and a door in one side through which the coke is removed, and where also a small amount of air is admitted. The ovens are constructed in long rows, some of which are a quarter of a mile in length. Alternate ovens are charged with a quantity of coal to produce, when leveled off, a layer about 2 ft. deep. Meanwhile the intermediate ovens are burning at full heat and the heat from these passing through the party wall ignites the charge in the adjacent oven. Only enough air is admitted through the door in the side, that the amount of burning that takes place is sufficient to keep the temperature at the desired point. The heat furnished by this partial combustion causes the coal to melt or fuse and give up its volatile constituent. The time that the charge is allowed to remain in the oven is usually 24, 48 or 72 hrs. The longer the time, the harder the coke, and the less residual volatile matter it will contain. The 72-hr. coke is usually that which is allowed to remain in the oven over Sunday. Because of its hardness, it is this coke that is most desired by the foundryman.

The following is an analysis of a hard Connellsville coke and the coal from which it was made:

	Water	Vol. combustible	Carbon	Sulfur	Ash
Coal...............	1.26	30.00	59.62	0.780	8.23
Coke...............	0.03	1.29	89.15	0.084	9.53

In the beehive oven the yield of coke is about 60 per cent. of the coal charged, and averages about 5 to 5½ tons of coke per oven. About 70 per cent. of the coke produced in the United States is made in beehive ovens.

The other type of oven is the *retort or by-product oven*, shown in Figs. 2 and 3. It is a narrow rectangular chamber about 33 ft. long, 7½ ft. high and 20 in. wide. It is tightly closed so that no air is admitted, the heat for the coking being furnished by burning gas in flues contained in the side walls and bottoms of the ovens. About 60 per cent. of the gas distilled out from the coal in coking, is used in this way.

The average coking time is 24 hrs. The yield is about 70

per cent. of the coal charged, and is about 5 to 5½ tons per oven.

In the retort oven the by-products are saved, instead of being allowed to escape into the air to become a very great nuisance, as well as a very great waste, as is the case with the beehive oven. The gas amounts to 14 to 16 per cent.; *i.e.*, 8,500 to 10,000 cu. ft. per ton; the tar, from which is obtained benzol, creosote, aniline dyes, etc., 4 to 6 per cent.; and ammonia 0.25 to 0.30 per cent. (amounting to 16 to 24 lbs. of ammonium sulfate per ton).

There is probably no other industry in which there is a waste so enormous as in the manufacture of coke. As Moldenke has pointed out,[1] the value of the coal charged into beehive ovens

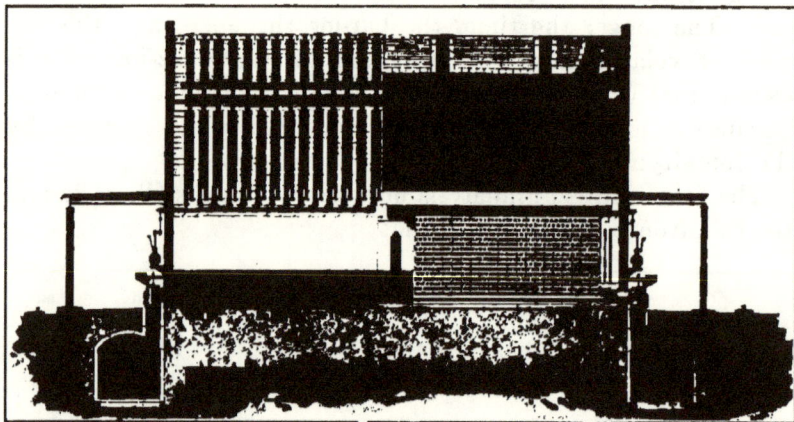

Fig. 2.—Koppers by-product coke oven. Longitudinal section through oven and heating flues.

during the year 1907 was $56,956,008; the value of the recoverable products wasted that year amounted to $44,000,000, or nearly 80 per cent. of the value of the coal that was used in the beehive ovens.

The beehive-oven coke has a clean, silvery-gray appearance, and the fracture induced by its quenching causes it to occur in long, slim pieces, commonly designated as "finger forms." The retort-oven coke is of a dull, black cast and occurs in short, thick pieces. The difference in the appearance is due almost entirely

[1] *Bull.* 3, Department of the Interior, Bureau of Mines.

to the manner of quenching. In the beehive oven, the coke is
quenched within the oven, as little water as possible being used.
In the retort oven, the red-hot coke is pushed out and deluged
with streams of water as it drops into the receptacle for the
coke. No matter how quickly the coke may be quenched after
being forced from the oven, contact of air with the highly heated

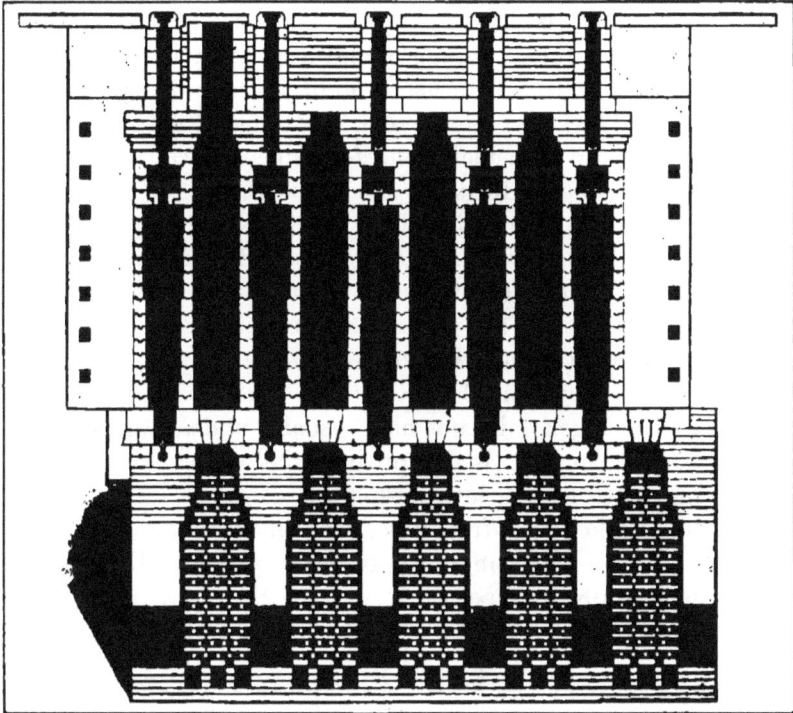

Fig. 3.—Koppers by-product coke ovens. Cross section through ovens.

coke will have occurred. This results in a certain amount of
surface oxidation which destroys the luster of the coke. As a
result of the manner of quenching, retort-oven coke contains more
water than that from the beehive oven, and the water is held
deep within the pores and is retained. Therefore, it is probably
well to specify a maximum limit for moisture in buying.

The size and shape of the lumps of retort-oven coke is an
advantage. The lumps being generally of a uniform size, the
space between them is greater and the blast can, therefore, pass

with greater case.   In general, then, if made from similar coals, the products of the two ovens do not differ in efficiency.

Coke as a Metallurgical Fuel.—The greatest amount of coke produced is used as a metallurgical fuel, chiefly for iron.   For this purpose, it is far more desirable than coal, because it contains less sulfur, and also is able to support the weight of the charge, which soft coal could not do.

However, coke is less desirable than charcoal because it contains more sulfur, phosphorus and ash.   But because of the greater cost of charcoal its use is limited, coke being far more widely used.   The fact that coke is stronger than charcoal is also an important consideration, for, since it can support a greater weight, higher furnaces can be built.

From the coke, practically all of the phosphorus and a large part of the sulfur is taken up by the iron.   Beside, the ash must be cared for as slag.   The sulfur content of coke is its most objectionable feature.   This varies from 0.6 per cent. in the best cokes to 2 per cent. in inferior cokes.   In the ordinary good coke the sulfur will vary from 0.75 to 1.5 per cent.

Phosphorus is generally not high, usually under 0.05 per cent. However, 0.03 per cent. is the maximum allowable for use with irons required to have a low phosphorus content.

The ash of coke varies from 6 per cent. in extremely good cokes to 18 or 20 per cent. in those of poor quality.   Usually the ash ranges from 8 to 12 per cent.

LIQUID FUELS

Burning of Liquid Fuels.—Liquid fuels burn much more rapidly than solids but not so rapidly as vapors and gases. Although the burning of liquid fuels is rapid under proper conditions it cannot be made to proceed rapidly upon an unbroken surface.   Oils are most conveniently and efficiently burned when pulverized or "atomized."

The pulverizing may be accomplished by directing a jet of compressed air or steam across a jet of oil at certain angles.   It is more advantageous to use compressed air because this helps to introduce the required amount of oxygen needed in burning the oil.   However, it is more convenient to use steam, since with it no special compressing apparatus is required as is needed in the case of air. An objection to the use of steam is that it is dissociated and much heat used up, so that the flame is cooled just at the point where combustion should be most rapid. The idea is sometimes gotten that after the steam has pulverized the oil, and is itself dissociated, that the hydrogen thus set free, on burning, adds much to the heat, but this idea of added heat is all wrong. There is no possible use of steam that can cause a gain of heat. Steam is merely a carrier of heat. It must be borne in mind that when the elements of any compound are separated, as much heat is absorbed as can be set free when they unite. In this case 62,000 heat units are required to break up sufficient water to set free a pound of hydrogen. This is also the amount of heat set free when a pound of hydrogen burns to form water. Steam cannot add to the calorific value of a fuel.

Advantages of Liquid Fuels.—Some of the chief advantages pertaining to the use of liquid fuels may be summed up as follows:

The fuel can be more easily stored and moved. It can be fired mechanically, thus eliminating the cost of labor in stoking. The fire may be regulated to develop a most intense heat in a very short time, and then can be cut off in a moment. There is an absence of dust, ashes and clinkers. There is less heat lost by way of the stack owing to the clean condition in which the boiler tubes may be kept.

Another advantage of liquid fuel is its high calorific value per pound. The best oil has, weight for weight, double the value of an inferior coal, and 30 per cent. greater value than the best coal.

It is very necessary that the fuel oil be free from water, the heating value of the oil being reduced 13.14 B.t.u. for each 1 per cent. of water. Water in a fuel oil is difficult to remove, since the densities of the two are not far apart.

Materials Used as Liquid Fuels.—The liquid fuels are chiefly coal tars, the residuum or other portions of petroleum oil separated in refining which seem to be unsuitable for any better purpose, and the heavy black petroleum oils found in Texas, California, Oklahoma and other States. By far the greater portion of the liquid fuel of the United States consists of the last-mentioned heavy petroleum oils. They contain practically no naphtha or illuminating portions, such as found in the oils of the

3

Appalachian field, and hence there is less inducement to refine them.

Since the naphthas and illuminating oils are fuels even though used in a different manner, the crude oils producing these and the methods of refining them will be included here.

The Production of the Petroleum Products

Origin of Petroleum.—The results of all of the more recent investigations seem to indicate that petroleum has originated from the decomposition of organic material, of both animal and vegetable source. Orton[1] says that there is reason to believe that the oil found in the Appalachian field, which includes Pennsylvania, New York, southern Ohio, West Virginia and Kentucky, is of vegetable origin. Peckham[2] states that the California and Texas product is of animal source. It is sometimes found in rocks filled with animal remains. Engler, in working on this problem, distilled $\frac{1}{2}$ ton of menhaden fish oil under a pressure of ten atmospheres, and obtained a distillate of hydrocarbon oils, accompanied by gas, that had practically all the properties of natural petroleum. This result lends support to the theory that petroleum was formed by the action of heat and pressure on organic remains in the earth.

Composition.—The various crude petroleums are quite complex in structure. They are not definite compounds, but are mixtures of several series of compounds known as hydrocarbons. The petroleums of the United States are of three general types: (1) the paraffine-base oils, of which the Pennsylvania oil is an example; (2) the asphaltum-base oils, of which the California oil is typical; and (3) mixed-base oils, to which class the Texas oil belongs. The Pennsylvania oil consists almost entirely of hydrocarbons belonging to the paraffine series. The general formula for any member of this series is C_nH_{2n+2}. The series begins with methane gas, CH_4, and each member increases by CH_2, as: C_2H_6, C_3H_8, C_4H_{10}, C_5H_{12}, C_6H_{14}, C_7H_{16}, etc., the last member being a solid paraffine, probably $C_{35}H_{72}$. This paraffine

[1] Report on the Occurrence of Petroleum, Natural Gas and Asphalt in Western Kentucky (1891).
[2] *Am. Jour. Sci.*, **48** (1894).

series is called also a *saturated* series, each molecule being considered to possess a chain-like structure, with single "linkings" between the carbon atoms, and with all other "bonds" occupied by hydrogen, as is shown by the following structural formula for the fourth member of the series, thus:

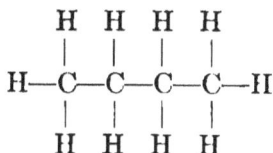

$$
\begin{array}{cccc}
\text{H} & \text{H} & \text{H} & \text{H} \\
| & | & | & | \\
\text{H}-\text{C}-\text{C}-\text{C}-\text{C}-\text{H} \\
| & | & | & | \\
\text{H} & \text{H} & \text{H} & \text{H}
\end{array}
$$

California oil is rich in hydrocarbons of an *unsaturated* series, C_nH_{2n}, of which the Pennsylvania oil contains only about 3 per cent. The term unsaturated here signifies that between certain of the carbon atoms there exist "double bondings," which lessen their capacity for holding hydrogen, thus:

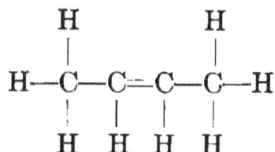

$$
\begin{array}{cccc}
\text{H} & & \text{H} & \\
| & & | & \\
\text{H}-\text{C}-\text{C}=\text{C}-\text{C}-\text{H} \\
| & | & | & | \\
\text{H} & \text{H} & \text{H} & \text{H}
\end{array}
$$

Such compounds have a greater tendency to deposit carbon in burning than the saturated compounds.

Chemical Considerations in Refining Oil.—The petroleum oil of the Appalachian field, commonly known as Pennsylvania oil, is the source of most of the refined petroleum products. The process of refining it varies somewhat with different refiners, but it consists essentially of fractional distillation. The products that are generally obtained, named according to their density, are: (1) petroleum naphtha (which is divided into several fractions), (2) lamp oils, (3) gas-making oils, (4) fuel oils, (5) light, medium and heavy lubricating oils, (6) paraffine, (7) asphaltum pitch, and (8) coke. There are two general methods whereby these products may be separated, the products being essentially the same in both cases, but differing in amounts. The methods are: (1) simple fractional or steam distillation; and (2) "dry" or destructive distillation.

In simple fractional or *steam distillation*, open steam is introduced into the interior of the retort to mechanically sweep out the oil vapors before they become unduly heated and decomposed. The various portions of the crude oil are thus separated according to boiling point and specific gravity with practically no chemical alteration. This method is employed if it is desired to have a larger yield of lubricating oil of especially high quality.

If, on the other hand, a larger yield of petroleum naphtha and lamp oils is wanted the heavier portions of the crude oil are *distilled destructively*. In this case the steam is not used, and the heavier vapors condense and fall back upon the retort bottom, where they are decomposed by the high degree of heat existing there. In this way much of the natural lubricating oil and other heavy portions are decomposed into naphthas and lamp oils. This decomposition is commonly known as *"cracking."* The reactions that occur during cracking are probably complex and are not definitely known. However, according to Thorp,[1] reactions such as the following may occur. The hydrocarbon, $C_{18}H_{38}$, boiling at 317°C. might decompose as follows:

$$C_{18}H_{38} \rightarrow C_{10}H_{22} + C_7H_{16} + C$$

The $C_{10}H_{22}$ boils at 173°C. and belongs with the lamp oils, while the C_7H_{16} boils at 98°C. and belongs with the naphthas. Or the reaction may proceed in the following manner:

$$C_{18}H_{38} \rightarrow C_8H_{18} + C_{10}H_{20}$$

In this case the first product belongs to the saturated paraffine series and boils at 125°C., while the second is an unsaturated compound of the olefine series, boiling at 172°C. It should be noted that there are several series of unsaturated hydrocarbons, the name olefine being applied to only one of such that may be produced by destructive distillation of the paraffine series. The unsaturated compounds are in general less stable than paraffines of the same number of carbon atoms, and are more likely to deposit carbon when burned, as for example, in the cylinder of an engine, than are paraffine hydrocarbons of the same molecular weight. However, it is said[2] that satisfactory utilization of a

---

[1] Outlines of Industrial Chemistry, p. 308.

[2] RITTMAN, DUTTON, and DEAN: Petroleum Technology, U. S. Bureau of Mines, *Bull.* 114 (1916).

"cracked" gasoline can be obtained in internal-combustion engines by suitably adjusting the admixture of air. Further, the investigations of Hall[1] tend to show that a properly "cracked" gasoline or motor spirit gives a greater mileage than a "straight" gasoline distillate. This he explains on the ground that the former burns more slowly than the latter and thus imparts its power to the piston for a longer period. But because of the tendency toward the deposition of carbon, it is generally considered that "cracked" distillates rich in olefines, can be most advantageously used by blending with "straight" distillates.

Recent work has called attention to the fact that the percentage of olefines or other unsaturated hydrocarbons formed in the production of gasoline by the cracking process can be quite accurately controlled by regulating the temperature and pressure. When the process is carried out at atmospheric pressure, the resulting gasoline may contain as much as 45 per cent. of unsaturated compounds, but with suitable increase of temperature and pressure, the production of unsaturated compounds can be largely prevented.

The Rittman Gasoline Process.—Rittman has succeeded in controlling the saturation by cracking the heavy oils in the vapor state under pressure. The following synopsis of the process is quoted from *Chemical Abstracts:*[2]

"Vaporized petroleum is passed into a hot tube at 450°C., the pressure being 90 to 500 lbs. per square inch. The vapors are condensed under pressure and the condensate distilled for its gasoline content. The residue may be revaporized and passed through the tube. By using the oil in the form of vapor, any combination of temperature and pressure may be applied, and in the case of fire only a small amount of oil is exposed. Any oil from kerosene up may be used and it is claimed that by reason of the reaction being gaseous instead of liquid, the amount of carbon deposited is so small as to be negligible."

General Methods of Refining.—In refining by either the steam or destructive distillation process, the earlier steps are practically the same. After removing the mud and water by gravity in storage tanks, the oil is charged into cylindrical steel stills heated

---

[1] Address delivered to the Institution of Petroleum Technologists, London, Feb. 18, 1915.

[2] **9**, 2147 (1915).

by direct fires and the more readily volatile portion is distilled off. It is during the latter part of the process, when the distillation of the heavier members must be accomplished, that the modifications for steam or destructive distillation are made. The divisions of the distillate are determined largely by density measurements, usually with a Baumé hydrometer (see Chapter XX).

The Volatile Portion.—This generally distills from the crude oil between 50 and 150°C. It is composed of that portion of the series ranging from C_4H_{10} to C_9H_{20} inclusive.

Pennsylvania oil yields about 34 per cent. of this light product. After being distilled from the crude oil, it is agitated with concentrated sulfuric acid to remove the tarry, sulfur-bearing compounds, which impart a dark color and disagreeable odor. After the acid treatment, it is agitated with a solution of caustic soda and then with water after which the water is allowed to settle out. The finished product is known as "deodorized naphtha."

Divisions of the Volatile Portion.—The light product obtained from crude petroleum is further subdivided into more limited divisions. Concerning the names of these divisions there is much confusion, and unfortunately no uniformity of usage. Robinson[1] says that it is best to make use of but two general names—gasoline and naphtha. In this scheme the name "gasoline" is confined to the mixtures of light hydrocarbons designed for such use that they are required to be vaporized, as in a gasoline torch, stove or automobile carburetter. In similar manner the name "naphtha" is confined to those hydrocarbons whose use depends upon their solvent action, as when used in dry cleaning, oil extraction or varnish making. This excludes the use of the term "benzine" entirely. And in view of the fact that this term is very similar to "benzene" which is used to apply to a coal-tar product of entirely different nature, it is clearly evident that there would be a distinct advantage in thus eliminating the name "benzine."

Although the terms applying to the various divisions of the volatile portion obtained from petroleum are by both refiners and users of these liquids, often indiscriminately applied without

---

[1] *Met. Chem. Eng.*, July, 1913, p. 389.

much consideration of density or other properties, the following classification shows the ranges covered as they are most commonly used. It will be noted that there is considerable overlapping.

1. Cymogene.................... about 108° Bé.
2. Rhigolene.................... about 94° Bé. to 92° Bé.
3. Petroleum ether.............. about 92° Bé. to 80° Bé.
4. Gasoline.................... about 86° Bé. to 60° Bé.
5. Naphtha.................... about 68° Bé. to 58° Bé.
6. Benzine.................... about 66° Bé. to 60° Bé.
7. Motor fluid................. about 62° Bé. to 58° Bé.
8. Turpentine substitute......... about 62° Bé. to 52° Bé.

Sometimes the term petroleum ether is used to cover the first three products in the preceding list. Naphtha is also sometimes called *ligroine*, and the turpentine substitute may also be called *"painter's spirits."* The term *motor fluid* is of recent origin, and is used to indicate the heavy type of fluid that the more modern automobile engines are capable of using.

Since, as stated before, there is no classification strictly adhered to among refiners and users, and names, therefore, carry no definite significance, the confusion will be lessened, if the characteristic properties and uses of the liquids are associated with their densities and boiling points, especially the latter.

"Straight" and "Blended" Gasolines.—Since gasolines have acquired such great economic importance, and the supply has been scarcely equal to the demand, the practice has grown up of mixing or "blending" the very light, volatile grades with heavier fractions that by themselves would be of little value, thus producing a product of a mean density, or Baumé test, that has been known to the trade as a desirable gasoline. But with the appearance on the market of such "blends," the Baumé or gravity test at once ceased to be a reliable guide to the quality of the fluid. With gasolines that are "straight" distillates, the various grades represent fractions of the crude-oil series that were separated in order of their boiling points. Thus, the light hydrocarbons distil off at the lower temperatures and possess the lighter densities as are shown by their higher Baumé readings. And the medium gasoline hydrocarbons distil off at medium temperatures and may be recognized by their medium densities. In brief, the specific gravity normally increases with the boiling points of these hydrocarbons, so that the Baumé readings serve to indicate to some degree, the volatility of the liquid. But in a "blend" this does not follow, since a medium density may be secured by mixing two constituents, one boiling at a lower and the other at a higher temperature than a "straight" distillate of medium density, while hydrocarbons such as the density would seem to indicate may be almost entirely lacking. Hence, there is now a growing tendency to disregard density in the choice of a gasoline, the trend being toward a selection on the basis of boiling point or volatility. For this purpose a sample is distilled under special conditions, and note is taken of the temperature at which distillation begins and where it ends, particularly of the latter temperature which is known as the "end point." Also much information about the uniformity of the sample can be obtained by noting between what temperatures the bulk of the liquid distils. "Blends" will likely begin to distil at relatively low temperatures, have a long distillation range, and a high "end point." On the other hand, "straight" distillates of equivalent density will generally have higher initial boiling points and lower end points, that is, the gasoline will consist of hydrocarbons of less variable volatility.

"Blends" have been designed especially for use as motor fuels, and although they possess certain objectionable features, it seems that they are not entirely lacking in merit. In fact, it is rather difficult to determine which is the more advantageous to the general consumer, since both possess certain desirable features, and there are contradictory claims for each.

As is pointed out in the discussion of the proposed gasoline specifications by the United States Bureau of Mines,[1] the basic property that determines the grade and usefulness of a gasoline is volatility, and since there is an advantage in having a motor fuel vaporize easily, it would be reasonable to conclude that a gasoline having a fairly high initial boiling point and a low end point would be most desirable. But a gasoline having a closely limited boiling range is expensive, and further, such products are not wholly free from objections. A high initial boiling point

[1] *Met. Chem. Eng.*, Nov. 15, 1916.

causes the engine to be more difficult to start in cold weather. Also, in the opinion of some, the low end point is not altogether advantageous. Certain producers of the "blended" products claim that the presence of the heavier hydrocarbons causes the explosion in the engine cylinder to take place more slowly, producing nevertheless the same amount of power which has the added advantage in that it is applied with less explosive violence.

The following quotation is taken from the reference just given:

"In summing up the desirability of various degrees of volatility, it appears that it is desirable to have a certain percentage of fairly low-boiling constituents so that engines may start readily, but undesirable that such constituents should be present in too large proportion, on account of the resulting increased possibility of loss through evaporation and of accidental ignition and explosion. A reasonably low-end point is desirable in order to insure complete vaporization, but it makes the gasoline expensive. In addition, if it is possible to utilize high-end-point gasolines efficiently, it seems conceivably possible to develop maximum power with less strain on the engine. For these reasons, the end point should be as high as is possible, without exceeding a limit set by the ability of the engine to atomize and vaporize the gasoline properly."

"Casing-head" Gasoline.—The gas that issues from certain natural gas wells is known as "wet gas;" that is, it contains in considerable proportion, the vapors of a very volatile gasoline. By compression and cooling, this gasoline can be extracted. From 2 or 6 gals. of "natural gas" gasoline may be separated from 1,000 cu. ft. of gas measured under a pressure of 250 lbs. In another method the gasoline is extracted by passing the "wet gas" through oil, such as a heavy lamp oil, which absorbs and retains the gasoline. The absorbed gasoline is then separated from the lamp oil by distillation. Casing-head gasoline is too volatile to be used by itself for ordinary purposes, and is commonly "blended" with heavier products.

Lamp Oils.—After the naphtha distillate has been removed in the general refining process, the distillation of the remainder of the crude oil is continued. The distillate which passes off between 150° and about 300°C. is known as lamp oil, burning oil, illuminating oil or kerosene. It has an average boiling point of about 230°C.

The grades of illuminating oil are designated according to their "fire test." The *fire test* is that temperature at which the oil will give off enough vapor to support a steady flame. It is a temperature lower than its boiling point.

Boverton Redwood gives the following as the classification of illuminating oil according to fire test: (1) 110°F. fire-test oil; (2) 120°F. fire-test oil; (3) 150°F. fire-test oil; (4) 300°F. fire-test oil.

The first two are exported largely to England and China. The fourth is known as "mineral sperm," "mineral seal," "mineral colza" or lantern oil. It is much used in railway lanterns and switch lamps.

Any one of the preceding four grades may again be divided into three grades according to color. The first or best grade is known as "water-white" or "head-light" oil. Head-light oil is usually 150°F. fire test, but not necessarily so.

The second grade is called "prime-white." This is a faint yellow.

The third grade is known as "standard-white," and is a pronounced yellow.

The illuminating oils may range from 45° to 49°Bé.

The tarry matters and decomposition or oxidation products to which the color is due, and which cause the flame to smoke and the wick to become clogged when the oil is used, can be removed, as is the case in the "water-white" oil, by treatment with sulfuric acid.

The oil is agitated by means of compressed air with 1 to 2 per cent. of concentrated sulfuric acid for about 1 hr. The acid now containing the colored substances is allowed to settle and is then drawn off. The oil is then agitated with water, allowed to settle, then agitated with about a 1 per cent. solution of caustic soda to correct any remaining trace of acidity, and finally it is again washed with water. After this it is drawn into shallow tanks exposed to light and air and allowed to settle until bright.

Because of its greater volatility and the consequent ease with which it can be converted into a "gas" for use in a gas engine, the gasoline commands a higher price than the heavier kerosene.

But when compared on the basis of their heat units, the gasoline is the less efficient of the two. A gallon of 85°Bé. gasoline, when burned completely produces 102,980 B.t.u. and costs 25 cts. or more (1916). A gallon of 150° fire-test kerosene produces 124,450 heat units and costs about 12 cts.

Gas Oil.—Although other oils have been used, Robinson states that the most economical oil for use in making gas consists of a mixture of heavy hydrocarbons, with an average boiling point of 600° to 650°F. It must be practically free from gasoline and illuminating oil on the one hand and the heavy lubricating oils and asphalt on the other.

Fuel Oil.—This oil is of too great density for use as an illuminating oil, and has practically no value as a lubricating oil, and hence is used as a fuel oil. A fuel oil must not contain gasoline and must have a viscosity sufficiently low that it may be pumped through pipes and burners. According to Robinson except for these restrictions one oil is as good as another. The light oils have a slightly higher heat of combustion per pound, but the heavier ones have a slightly higher heat of combustion per gallon. For some purposes, such as in oil engines, special oils are required; but in general, fuel oils are those that cannot be used for any better purpose.

The next portion separated out from the crude oil is the *lubricating oil*, for which see *"The Mineral Lubricating Oils,"* page 340.

GASEOUS FUELS

Advantages and Disadvantages.—As stated previously, fuels burn more readily when converted into either a vapor or a gas. When a wax candle burns the wax is first liquefied, which liquid then ascends the wick, and is gasified and burned. If a fuel can be changed into a gaseous state, as an operation entirely separate from its burning, it is a long way advanced toward effective combustion. This is the idea in using a gas producer.

It must be understood that no more heat is made available per pound of fuel by gasification. Quite otherwise. Heat is lost by the cooling of the hot gases coming from the producer, and in other ways. The use of manufactured gas under a boiler would be an economic loss.

However, because the gas can be better applied there is a gain in intensity. Other advantages are:

1. The temperature can be easily controlled.

2. There is but little smoke, indicating almost complete combustion, even though but little more than the theoretical amount of air is required.

3. The gas can be used for the direct production of energy; i.e., in gas engines.

Beside these, there are many other advantages associated with transportation, stoking, etc., not discussed here.

Producer gas is an artificial gas resultant from the incomplete combustion of a solid fuel in a partly closed generator of special construction known as a gas producer. There are many forms of producers, each embodying usually some special feature.

The essential principle in their operation is as follows: air is admitted to the producer only in limited quantity, under which condition part of the fuel is burned to carbon dioxide. This gas, together with the nitrogen admitted as part of the air, absorbs the heat produced by the burning and carries it through the fuel above, heating it to incandescence. From this incandescent fuel each molecule of the carbon dioxide now takes up another atom of carbon, thus forming carbon monoxide. The first reaction liberates heat and the second takes up a *part* of it. The reactions are shown by the following equations:

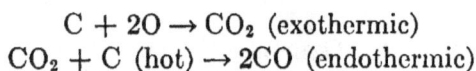

$$C + 2O \rightarrow CO_2 \text{ (exothermic)}$$
$$CO_2 + C \text{ (hot)} \rightarrow 2CO \text{ (endothermic)}$$

Not all of the CO_2 is thus reduced, some of it passes through unchanged. Its presence in the final product is undesirable. In addition to having no fuel value it represents a loss of heat, since to produce it, complete combustion must have occurred in the generating chamber, so that all the heat was liberated there.

The final product or "producer gas" contains also a certain amount of free hydrogen. This comes partly from hydrogen contained in the solid fuel used, and partly from water contained as moisture in the air, which reacts with the heated carbon according to this equation:

$$H_2O + C \text{ (hot)} \rightarrow H_2 + CO$$

Frequently producer gas contains also hydrocarbons which have distilled from the fuel as such.

Heat Lost in the Producer.—As has been previously stated, the reduction of carbon dioxide to carbon monoxide by means of the incandescent carbon stores up only a part of the heat liberated when the carbon dioxide was formed. The heat not so taken up is equivalent to about one-third of the total heat of combustion of the solid carbon, and is manifested as sensible heat in the producers. However, this heat need not all be lost. The gases may be used direct from the producer, while much of the heat is still contained in them, and higher temperatures may be thus obtained. Or the hot gases may be used to preheat air and steam in a *modified process*, in which steam is forced into the producer with the air. This steam reacts with the heated carbon as shown above:

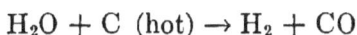

$$H_2O + C \text{ (hot)} \rightarrow H_2 + CO$$

The reaction is highly endothermic and helps to store up much of the otherwise wasted heat. Another gain in using the steam is that it yields the same amount of carbon monoxide in burning the carbon as when the carbon is burned with air. Then the carbon monoxide instead of being mixed with twice its volume of nitrogen, a useless gas, is mixed with once its volume of hydrogen which is highly combustible and heat-producing.

When air alone is used, the product is known as *"air gas."* When steam is blown in with the air the product is known as *"semi-water gas."* Because of the heat conserved and the improved quality of the gas the process with steam is the one most commonly used.

The following analyses represent the composition of "air gas" made by the use of bituminous and anthracite coal respectively:

CH_4	C_2H_4	H_2	CO	CO_2	N_2	O_2
2.5	0.4	12.0	27.0	2.5	55.3	0.3
1.2	0.0	12.0	27.0	2.5	57.0	0.3

In the first case the percentage combustible is 41.9 and in the second 40.2.

The following is representative of a "semi-water gas:"

CH_4	H_2	CO	CO_2	N_2	O_2
2.0	34.0	27.0	8.0	29.0	0.0

The percentage combustible is 63.

Frazer[1] says producer gas has the lowest calorific value of any gaseous fuel and the temperature of its flame is the lowest of any (excepting blast-furnace gas) and still it is the cheapest artificial gas per unit of heat.

Water Gas.—The principle involved in the manufacture of water gas is essentially the same as in making semi-water gas, but in the former, the air and steam are blown in at alternate intervals. Air is first forced through the coals in the producer until a high degree of heat is reached, which process itself makes a low-grade gas containing some carbon monoxide.

When the bed of coals is sufficiently hot, the air is cut off and steam is blown in. The steam reacts with the incandescent carbon with the formation of hydrogen, and either carbon monoxide or dioxide depending upon conditions. If the temperature of the carbon is above 1,000°C. the reaction proceeds as follows:

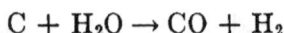

$$C + H_2O \rightarrow CO + H_2$$

But if the temperature falls below 1,000°C. the following reaction occurs:

$$C + 2H_2O \rightarrow CO_2 + 2H_2$$

Since both carbon monoxide and hydrogen burn with a non-luminous or blue flame, water gas is sometimes known as *"blue gas."*

Water gas is used either alone as a fuel or it may be used as a constituent of illuminating gas. If used for illumination, the "blue gas" must be enriched with hydrocarbons, in which case it is said to be *carburetted.*

The process of making water gas is then modified somewhat, depending upon the purpose for which the gas is to be used.

If it is used for a fuel gas the products of combustion resultant from the blowing in of air may serve to preheat the incoming air or steam, or may be allowed to escape.

If used for illuminating gas the products resultant from the air blast are burned to heat a checkerwork of brick, which is used to gasify the vapors of petroleum oil used in carburetting. The "blue gas" is first mixed with the oil vapors and this mixture

[1]ROGERS and AUBERT: Industrial Chemistry, p. 80.

is then conducted through the hot checkerwork where the hydrocarbon vapors are decomposed and fixed in permanent gaseous form.

The following represents the composition of water gas ("blue gas"):

CH_4	H_2	CO	CO_2	N_2
0.1	51.5	41.0	4.0	3.4

The percentage combustible is 92.6.

Since water gas contains such a high percentage of carbon monoxide, which is practically without odor and extremely poisonous, much danger attends its use.

Blast-furnace gas is the waste gas issuing from the top of the the blast furnace. It is recovered and used either in the Siemens regenerator to preheat the blast for the furnace, or in gas engines. Also it may be used beneath the boilers for the production of steam for the blowing engines. It is not a rich gas and shows the following average composition:

CH_4	H_2	CO	CO_2	N_2	O_2
0.20	2.74	28.61	11.39	57.06	0.00

Only 31.37 per cent. is combustible.

Coal Gas.—This is made by the dry distillation of coal in retorts. It is a somewhat expensive gas because only the volatile portion distills off leaving much of the coal in the form of a soft, non-coherent coke, which has but a low fuel value. Coal gas is rich in illuminants and is used largely as an illuminating gas. It is also used as a fuel in domestic stoves and sometimes in gas engines. The following analysis shows an average composition in per cent.:

CH_4	C_2H_4	H_2	CO	CO_2	N_2	O_2
40.5	4.0	47.0	6.0	0.5	1.5	0.5

The percentage combustible is 97.5.

Natural Gas.—The origin of natural gas is closely associated with the origin of petroleum. It belongs to the same series of hydrocarbons, as those which make up petroleum. It consists largely of methane which is the lowest member of its series.

The composition of natural gas varies somewhat according to

the locality from which it comes. The following shows the average of 20 analyses[1] of Pittsburgh natural gas:

CH_4	C_2H_6	CO_2	O_2	N_2
80.6	18.2	0.0	0.0	0.8

The percentage combustible is 98.8.

Natural gas has the highest kindling temperature and the highest calorific value of all the gaseous fuels.

Summary.—In the following table the analyses of the various gases are collected for the sake of comparison:

	CH_4	Illumi-nants, C_2H_4, etc.	H_2	CO_2	CO	O_2	N_2	B.t.u. per cu. ft. (about)
Natural gas...	80.60	18.20	0.00	0.00	0.00	0.00	0.80	1,000
Coal gas......	40.50	4.00	47.00	0.50	6.00	0.50	1.50	660
Producer gas..	2.00	0.00	34.00	8.00	27.00	0.00	29.00	145
Water gas.....	0.10	0.00	51.50	4.00	41.00	0.00	3.40	330
Blast-furnace gas......	0.20	0.00	2.74	11.39	28.61	0.00	57.06	98
Coke-oven gas	35.00	3.50	48.00	2.50	6.00	0.30	4.70	600
Oil gas.......	27.00	13.00	38.00	4.00	16.00	0.30	1.70	650

References

A

Byrom and Christopher: Modern Coking Practice, London, 1910.

Booth: Liquid Fuel and Its Apparatus, London, 1911.

Thorp: Outlines of Industrial Chemistry, New York, 1912.

Brislee: Introduction to the Study of Fuel, London, 1912.

Somermeier: Coal; Its Composition, Analysis, Utilization, and Valuation, New York, 1912.

Lucke: Design of Surface Combustion Appliances, Easton, Pa., 1913.

Brame: Fuel, Solid, Liquid, and Gaseous, London, 1914.

Hays: How to Build up Furnace Efficiency, Chicago, 1914.

Maujer and Bromley: Fuel Economy and CO_2 Recorders, New York, 1914.

Coste and Andrews: Examination and Thermal Value of Fuel, London, 1914.

Butler: Oil Fuel, London, 1914.

Strohm: Oil for Steam Boilers, New York, 1914.

Kent: Steam Boiler Economy, New York, 1915.

Rogers and Aubert: Industrial Chemistry, New York, 1915.

[1] Department of Interior, Bureau of Mines, *Tech. Paper*, 109, p. 13.

B

CHRISTIE: Economy in Fuel Combustion, *Eng. Mag.*, **21**, 81 (1901).
ATWATER: Smokeless Fuel for Cities, *Cassier's Mag.*, **30**, 313 (1905).
KERSHAW: Industrial Smoke and Its Prevention, *Cassier's Mag.*, **29**, 109 (1905).
MOSS: The Chemistry of Combustion, *Power*, **26**, 619 (1906).
REINHARDT: Purification of Gas for Gas Engines, *Eng. Min. Jour.*, **82**, 776 (1906).
BENJAMIN: Smoke Prevention in the Power House, *Cassier's Mag.*, **31**, 339 (1907).
BRECKENRIDGE: Smokeless Combustion of Coal, *Eng. Mag.*, **34**, 1055 (1908).
PORTER and OVITZ: The Volatile Matter in Coal, U. S. Bureau of Mines, *Bull.* 1 (1910).
MOLDENKE: The Coke Industry of the U. S. as Related to the Foundry, U. S. Bureau of Mines, *Bull.* 3 (1910).
DANNE: Natural Gas and Its Production, *Met. Chem. Eng.*, **8**, 412 (1910).
MOORHEAD: Determination of Calorific Power and Operating Conditions from Analyses of Flue Gases, *Am. Gas Light Jour.*, **93**, 929 (1910).
ALLEN: Specifications for the Purchase of Fuel Oil, *Jour. Ind. Eng. Chem.*, **3**, 730 (1911).
KERSHAW: Dust, Soot, and Smoke, *Cassier's Mag.*, **39**, 251 (1911).
BUCK: Use of Blast Furnace Gas, *Iron Age*, **87**, 902 (1911).
STORY: Comparison of Gasoline and Alcohol Tests on Internal Combustion Engines, U. S. Bureau of Mines, *Bull.* 32 (1911).
HARTMAN: By-product Coke Ovens, *Proc. Eng. Soc. West. Penn.*, **28**, 311 (1912).
RANDALL: The Smoke Problem at Boiler Plants, U. S. Bureau of Mines, *Bull.* 39 (1912).
BELDEN: Beehive Coke Oven Industry of The United States, *Jour. Ind. Eng. Chem.*, **5**, 71 (1913).
BELDEN: Metallurgical Coke, U. S. Bureau of Mines, *Tech. Paper* 50 (1913).
WALKER and BOHNSTENGEL: Kansas Fuels, *Bull. Univ. Kansas*, 14, No. 5 (1913).
ALLEN, JACOBS, CROSSFIELD and MATTHEWS: Physical and Chemical Properties of the Petroleums of California, U. S. Bureau of Mines, *Tech. Paper* 74 (1914).
FIELDNER: Notes on Sampling and Analysis of Coal, U. S. Bureau of Mines, *Tech. Paper* 76 (1914).
CLEMENT, FRAZER and AUGUSTINE: Factors Governing the Combustion of Coal in Boiler Furnaces, U. S. Bureau of Mines, *Tech. Paper* 63 (1914).
BRADFORD and CORWIN: Surface Combustion, *Power*, **42**, 787 and 855 (1915).
KREISINGER and OVITZ: Sampling and Analysis of Flue Gases; U. S. Bureau of Mines, *Bull.* 97 (1915).
CONE: Liquid Fuel for Foundry Cupolas, *Iron Age*, **95**, 1058 (1915).

SCHRAM and CAIN: A Test of a Surface Combustion Furnace, *Jour. Ind. Eng. Chem.*, **8**, 361 (1916).

RITTMAN, DUTTON and DEAN: Petroleum Technology, U. S. Bureau of Mines, *Bull.* 114 (1916).

SPRINGER: The Prevention of Spontaneous Combustion of Coal, *Ry. Mech. Engr.*, April, 1916.

RITTMAN, JACOBS and DEAN: Physical and Chemical Properties of Gasolines Sold throughout the United States during 1915, U. S. Bureau of Mines, *Tech. Paper* 163 (1916).

CHAPTER III

REFRACTORY MATERIALS FOR FURNACES

General.—The term *refractory*, as used in this connection, signifies primarily the property of being able to resist fusion by heat, although good refractory material must also possess other properties of scarcely less importance. For example, it must be able to withstand abrasion while hot and be chemically indifferent to the fused material that may come into contact with it during use.

A refractory material must be chosen with respect to the *temperature* obtained in the furnace in which it is employed. Thus in the openhearth steel furnace, a temperature of 1,600°C. is often developed; in a blast furnace just in front of the tuyères, it may be as high as 1,960°C.; while in the electric arc furnace, the temperature may reach 3,500 to 3,600°C. For comparison, the melting points[1] of the common refractory materials are shown:

Material	Incipient melting point in degrees Centigrade
Aluminous fireclay brick	1,400 to 1,750
Silica brick	1,600 to 1,800
Magnesia brick	1,850 to 2,000
Bauxite brick	1,850 to 2,000
Chromite brick	2,000 and above
Plumbago-graphite	above 3,500

But, even though the material is able to resist fusion, it may not be able to withstand the *abrading action* of the furnace charge. Thus, magnesia is noticeably softened at 1,500°, although still far from its melting point. The density of the brick has much to do with its resistance to mechanical abrasion. Silica bricks are relatively porous, and are not firm and tough, so yield more readily to abrasion on this account.

If the material of the lining is *chemically active* with the material

[1] FULTON: Principles of Metallurgy, Chapter X.

of the charge, the lining will be removed because of the formation of fusible compounds. Thus, the acid anhydride, silica, contained in a lining of silica brick, will react with lime in a basic slag producing calcium silicate, which being more readily fusible than silica, will melt and become a part of the slag. Other basic substances, as metallic oxides of the charge, may act in the same way as the lime. Conversely, basic anhydrides, as magnesia from a lining of this material, will react with silica carried in a highly siliceous slag. Some materials, such as chromite, graphite, etc., are indifferent to the action of both basic and acid anhydrides and are known, therefore, as "neutral" substances.

The nature of the lining is thus determined by the character of the slag. If the slag is basic, the lining must be basic or neutral; if the slag is acid, the lining must be acid or neutral. The character of the slag is, in turn, determined by the chemical reactions it is desired to bring about in the furnace. For example, in the openhearth steel furnace, if it is desired to remove phosphorus from the metal of the charge, the slag must be basic, lime being used which forms calcium phosphate with this element as it is oxidized. A siliceous slag is incapable of such action.

In addition to the properties previously mentioned, it is important that refractory materials have as low *heat conductivity* as possible, and an ability to withstand sudden changes of temperature.

Although the heat conductivity of all refractory materials is relatively low, there is, nevertheless, a considerable difference in respect to this property. The following values, based on graphite as 100, have been abstracted from a table given by Fulton:[1]

Material	Relative conductivity, per cent.
Graphite brick	100.0
Carborundum brick	92.4
Magnesia brick	28.4
Chromite brick	22.8
Fireclay brick	16.7
Building brick	14.0
Bauxite brick	13.2
Silica brick	7.8
Kieselguhr (diatomaceous earth)	7.1

[1] *Loc. cit.*, p. 345.

Furnace linings are often built up of several materials to combine different desirable features not possessed by one alone. Thus magnesia, which is highly refractory but a poor heat insulator, may be backed up by silica brick, which is practically a non-conductor of heat.

Refractory materials should be able to withstand sudden *changes of temperature*, such as are brought about by currents of inrushing air when a furnace is emptied. Other factors being equal, there is less cracking and flaking off (spalling) under these conditions, if the material is porous; although it must be remembered that high porosity lessens the ability of the material to withstand abrasion. Magnesia brick do not possess very high resistance to rapid temperature changes, while chromite brick are very resistant.

Classification of Refractories.—According to the reactions that may occur between linings and slags, discussed in the preceding paragraphs, refractories are divided into three classes, namely:

1. Acid Refractories.—This class includes the materials that consist essentially of silica, or at least contain this substance in excess of the bases present; and do not react with siliceous slags, but do react with slags rich in metallic oxides. The most important are Dinas and silica brick, gannister and quartz. Also fireclay, which is sometimes classed as a neutral refractory, contains an excess of free silica, and to a certain extent may be considered acid.

2. Basic Refractories.—The materials included here are those that contain little or no free silica, and are not much affected by slags rich in metallic oxides, but are attacked by highly siliceous slags. The most common representatives are magnesite or magnesia, dolomite, lime and bauxite.

3. Neutral Refractories.—Those substances that are not affected by slags rich either in silica or basic oxides are called neutral. The most important are graphite and chromite.

Preparation.—Refractories are used in various ways, as crushed and ground material, standard bricks, shapes, and special forms. The bricks and shapes are usually formed from a stiff paste which is afterward pressed and then burned at a high temperature. Many of the properties of such refractories, as abrasion resistance, heat conductivity, etc., are not inherent in the material, but are largely dependent upon the method of manufacture. In the following paragraphs, the more important properties of the various refractories are briefly discussed.

ACID REFRACTORIES

Dinas Brick, Silica Brick.—The original Dinas brick is made from a highly siliceous sandstone occurring in South Wales. Fulton[1] gives as the composition of the best stone, the following:

	Per cent.
Silica	98.31
Alumina	0.72
Ferric oxide	0.18
Lime	0.22
Alkalies	0.18
Moisture	0.35

The bricks are made by mixing with the coarsely ground rock about 2 per cent. of lime with a little water. The freshly moulded bricks are difficult to handle because they lack coherence. When dry they are fired at a temperature sufficiently high to cause a calcium silicate to form on the surface of the grains, which acts as a cement and binds them together. Clay is sometimes used in place of lime for this purpose. The finished brick contains about 96 per cent. of silica.

Dinas brick are imported into the United States, but a large part of the silica brick used here is of domestic manufacture, being produced chiefly in Pennsylvania. The domestic brick is manufactured in a manner similar to that described for Dinas brick, except that the binding material is generally clay.

Although silica brick are highly resistant to heat, they are not suitable for all parts of the furnace. They are rather friable and do not stand abrasion well; but they are very suitable for those locations, where the heat is intense and the abrasive action low, as in the roofs of openhearth furnaces. They are much subject to spalling under sudden changes of temperature and so should not be used to line charging doors, or in furnaces that are allowed to cool down frequently, as furnaces for brass melting. And, of course, they must not be used where they will come into contact with basic slags.

[1] *Loc. cit.*, p. 334.

Gannister.—This term was first applied to a close-grained, clayey sandstone found near Sheffield, England; but it is now used somewhat as a generic term to indicate any similar material wherever found, and is even applied to artificial mixtures prepared to conform to the requirements. Fulton gives the following as an approximate composition of gannister: quartz, 83 per cent.; clay, 13 per cent.; impurities and moisture, 4 per cent. The amount of clay is sufficient to form a good bond, but the amount of impurities is not sufficiently high to lower the refractoriness to any great extent. It is generally not moulded into bricks, but is used to form furnace bottoms, and line converters by tamping or moulding it in place.

Fireclays.—As has been pointed out before, these are often classed as neutral refractories; but, because of the free silica they contain, they are decidedly less neutral than the true neutral refractories, as graphite and chromite, and on this account are here included among the acid materials. Like all other clays, they consist essentially of kaolin, a hydrated silicate of alumina, $Al_2O_3 \cdot 2SiO_2 \cdot 2H_2O$, which is produced in the pure state by the weathering of pure feldspars, as described on page 256. But, because feldspar is usually associated with other minerals, its decomposition product is generally contaminated with these. Thus, quartz, mica, iron oxide, lime, magnesia and other substances are almost always contained in the clays. Also the alkalies are very likely to be present to some extent, these being due to incompletely weathered feldspar.

The following analyses given by Ries[1] show the relative composition of fireclays and brick clays. It will be noticed that the

Kind of clay	SiO_2, per cent.	Al_2O_3, per cent.	Fe_2O_3, per cent.	CaO, per cent.	MgO, per cent.	K_2O and Na_2O, per cent.	TiO_2, per cent.	H_2O, per cent.
No. 1 Fireclay	50.60	34.35	0.78	tr.	tr.	tr.	1.62	12.90
No. 1 Fireclay	51.56	33.13	0.78		0.12		1.91	12.50
Fire-mortar clay	67.26	23.36	1.63	0.25	......	0.65	Incl. with Al_2O_3	6.94
Common brick clay	66.67	18.27	3.11	1.18	1.09	4.22	0.85	4.03
Pressed brick clay	68.28	18.83	2.60	0.70	0.13	2.29	0.27	7.23

[1] Building Stones and Clay Products, and Economic Geology of the United States.

fluxes—iron, lime, magnesia, and the alkalies—are generally lower
in the fireclays than in the brick clays.

Gowland[1] says that in the fireclays the silica generally ranges
from 50 to 60 per cent. or more, and the alumina from 22 to 35
per cent.[2] In almost all instances these combined constituents
will comprise from 92 to 98 per cent. of the clay. The other
constituents are classed as fluxes, since they very noticeably lower
the refractoriness. It is generally stated that the total fluxes
should not exceed from 3 to 4 per cent., but this is dependent upon
conditions. Thus, a very coarse-textured clay, high in alumina,
may carry as much as 10 per cent. of these impurities and still
be a fireclay, although not of high quality. Refractoriness is
also largely dependent upon the relative quantities of the fluxes;
that is, upon which of these predominate. The alkalies are very
objectionable, as also is iron oxide, particularly if present in
the ferrous condition, or if present in the ferric state accompanied
by some reducing substance as carbonaceous matter. The
alkalies and iron oxide become especially objectionable if present
in conjunction.

From the standpoint of the amount used, fireclay is the most
important refractory material. It is employed chiefly as bricks
and other moulded forms.

BASIC REFRACTORIES

Magnesia.—This is the most valuable basic material. It is
produced by calcining magnesite at about 1,700°C. The greatest
amount of magnesite used in the United States is imported, although some is obtained from California. Gowland[3] says that
when calcined at the temperature mentioned it is practically
infusible, and unlike lime, has no tendency to absorb water and
slake. Further, even though basic, it does not readily react
with silica when used in furnace linings. Magnesia bricks are

---

[1] The Metallurgy of the Non-ferrous Metals, p. 4.

[2] Pure anhydrous kaolinite consists of silica 54 per cent., and alumina 46
per cent. Additional non-combined silica, present as quartz sand, acts as a
flux to the kaolinite, and lowers the melting point until a eutectic of 90 per
cent. silica is reached, which melts at 1,600°C. This should be remembered
in making mixtures of sand and fireclay.

[3] *Loc. cit.*, p. 10.

very hard and dense, and are very refractory. However, they are much poorer heat insulators than fireclay or silica brick, and cannot well withstand sudden changes of temperature. They are much more expensive than fireclay or silica brick, and this factor causes their use to be more limited than would be the case otherwise.

Lime.—This material has a very high melting point—about 1,900°C.—and is also cheap, but it possesses one property that causes it to be practically useless as a refractory material. It cannot stand exposure to the atmosphere, since it slakes and disintegrates under these conditions.

Dolomite is a magnesian limestone, containing calcium carbonate and magnesium carbonate in molecular weight proportions. Calcined dolomite does not slake in the air as does lime, but it is not entirely resistant to this action. It is far cheaper than magnesia, but it is less resistant to abrasion, and cannot be used in contact with acid material as can magnesia.

Bauxite is a hydrated aluminum oxide, $Al_2O_3 \cdot 3H_2O$. It is rarely pure, containing, generally, iron oxide, silica, and titanium oxide. Bauxite is known as red or white, depending upon the amount of ferric oxide present, this amounting sometimes to 25 per cent. in the red variety. It is only the white, in which the ferric oxide is practically absent, that is used for making the most refractory bricks, and in this the silica is desired not to exceed 6 or 8 per cent.[1]

Good bauxite bricks have high resistance to fusion, abrasion, and to reaction with metallic oxides and are much cheaper than magnesia bricks.

For certain special work, bauxite is purified and fused in the electric furnace, being thus formed into a product known as *alundum*. This material is very refractory, having a melting point of about 2,050°C., but it is too expensive for ordinary furnace use.

Neutral Refractories

Chromite is a compound of iron and chromium, having the composition shown by the formula, $FeO \cdot Cr_2O_3$. Frequently, part of the iron oxide is replaced by magnesia, and part of the

[1] GOWLAND: *Loc. cit.*, p. 10.

chromium oxide by alumina, while silica may be present as an impurity.

Chromite is practically infusible at furnace temperatures, and resists the chemical action of both acid and basic slags. Also, it withstands abrasion and sudden changes of temperature better than any other refractory. However, it cannot be used extensively because of its high cost.

Graphite is an allotropic form of carbon. Two kinds are recognized—"amorphous" and "crystalline." It is only the latter that is of importance as a refractory material. The former is used chiefly for paints, lead pencils, stove polishes, etc. The finest grade of crystalline graphite comes from the island of Ceylon, although very fine grades are obtained from Siberia, Canada, and certain portions of the United States. Selected specimens of Ceylon graphite will contain from 99.7 to 99.8 per cent. of carbon but generally the amount ranges from 70 to 90 per cent. The following analysis may be considered as typical:

	Per cent.
Carbon	86.66
Insoluble silica	4.22
Soluble silica	4.05
Alumina and ferric oxide	4.66
Lime and magnesia	1.01

The soluble silica and iron oxide determine its quality, since these substances lower its refractoriness.

Graphite can be prepared artificially by the Acheson process, in which anthracite coal, coke, or other forms of amorphous carbon are heated in the electric furnace in a reducing atmosphere, and then allowed to cool slowly. That used as a refractory is generally prepared from anthracite.

Graphite is used largely for making crucibles. For this purpose, it is necessary to mix with it a certain amount of good fireclay, since graphite is not plastic. The amount of clay used varies from 50 to 75 per cent., depending upon the grade of the crucible.

Graphite is the most refractory material known. It has not been fused, although it softens somewhat and vaporizes at the temperature of the electric arc—about 3,700°C.[1] As a refrac-

[1] FULTON: *Loc. cit.*, p. 342.

tory material, it can be used only under reducing conditions; it burns when under the influence of oxidizing flames and oxidizing fluxes.

References

A

SEXTON: Fuel and Refractory Materials, New York, 1910.
FULTON: Principles of Metallurgy, New York, 1910.
HAVARD: Refractories and Furnaces, New York, 1912.
GOWLAND: The Metallurgy of the Non-ferrous Metals, London, 1914.

B

SEAVER: Refractory Materials, *Proc. Eng. Soc. West. Penn.*, April, 1910.
SAUNDERS: Alundum as a Refractory, *Met. Chem. Eng.*, **9**, 257 (1911).
FITZGERALD: Refractories, *Met. Chem. Eng.*, **10**, 129 (1912).
CRAFT: Use and Development of Refractories in the Iron and Steel Industry, *Chem. Eng.*, **18**, 206 (1913).
KELLEY: Selecting Refractories for the Foundry, *Iron Age*, **94**, 942 (1914).
HEISEL: Bricks for High Temperatures, *Brit. Clayworker*, **24**, 183 (1916).
NESBITT and BELL: Testing Fire Brick, *Iron Age*, **98**, 85 (1916).

CHAPTER IV

IRON AND STEEL

Occurrence of Iron.—Iron exists in almost every portion of the earth's surface but only very rarely is it found in the free state, and nowhere is it obtained in commercial quantities in this condition. This is due to the marked instability of the metallic element in the presence of air and moisture. The ores of iron include oxides, hydroxides, carbonates and sulfides. Of these, one of the most common and valuable is merely impure iron rust, called red hæmatite, Fe_2O_3, which if entirely pure would contain 70 per cent. of iron. A hydrated hæmatite, called limonite, is also found in large quantities. Another valuable ore is magnetite, Fe_3O_4. This is of the same chemical composition as mill or forge scale, and if absolutely pure would contain 72.42 per cent. iron. Because of moisture and impurities, especially of an earthy nature, the amount of iron in the natural ores usually does not exceed 55 or 58 per cent. Unless the amount of iron present is at least 35 to 40 per cent., it is usually not profitable to smelt the ore.

Production of Iron.—Iron is obtained commercially by the reduction of its oxides by carbon. Coke or charcoal have been found best for this purpose. The process of reduction is carried out in a blast furnace—a steel, brick-lined structure ranging from 80 to 100 ft. high, and from about 18 to 22 ft. in diameter at its widest part. Into the top of this is charged a mixture of *iron ore*, *coke* and *limestone*, while a blast of highly-heated air is blown in at the bottom. Coke is the fuel that is generally used, both because it is relatively cheap, and because it best of all supports the weight of the charge. Part of it burns at the expense of the oxygen of the blast producing carbon monoxide, which gas, together with another portion of the coke, brings about the reduction of the ore. Charcoal is in some respects even more desirable as a blast-furnace fuel than coke because it contains much

60

less sulfur and phosphorus, which, as will be later explained, are very objectionable impurities. Iron produced by the use of charcoal is known as charcoal iron and is very desirable because of its greater purity. However, the use of charcoal for this purpose is largely prohibited because of its greater cost.

The gases that are ultimately conducted away from the top of the furnace contain about 28 per cent. of carbon monoxide,

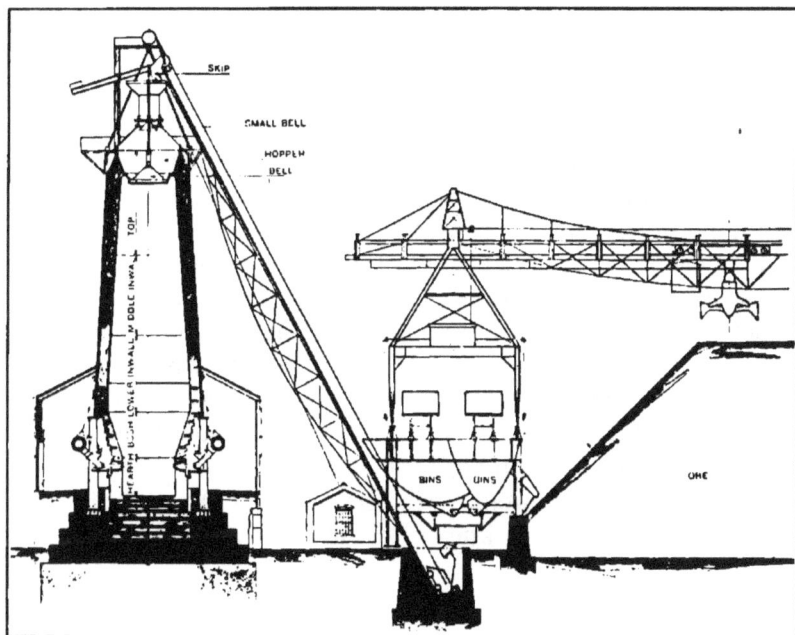

Fig. 4.—Section of blast furnace showing filling arrangement, bins and ore bridge.

together with 2 or 3 per cent. of hydrogen, the remainder being carbon dioxide and nitrogen. This blast-furnace gas has a heat value of about 98 B.t.u. per cubic foot and the greater part of it is used to preheat the blast. To do this, the gas is first burned within stoves, which are tall, steel structures lined with a checker-work of brick. When the brickwork has been heated to an intense heat the air blast is sent through the stoves and thus preheated. Another portion of the furnace gas is burned beneath the boilers of the blowing engines, while a part is also often used to develop electric power.

CAST IRON

Pig Casting.—As the iron is reduced, it collects in the molten state in the bottom, or hearth of the furnace, and periodically— usually at intervals of from 4 to 6 hrs.—it is drawn off and cast in moulds of suitable size. These castings of crude iron are known as pigs, or *pig iron* because formerly it was the custom to cast the iron in sand beds at the base of the furnace, and then the arrangement of channels and open moulds was said to resemble a sow with suckling pigs. At the present time, the pigs are usually formed in a mechanical pig-moulding machine. This consists of an endless chain carrying open metallic moulds into which the iron is poured from a ladle as the moulds travel slowly past. Pigs made in this way are not coated on the surface with sand, as are the sand-cast pigs.

Composition of Cast Iron.—As the iron is reduced in the blast furnace, it absorbs a certain amount of carbon from the fuel, which, depending somewhat upon the temperature of the furnace, amounts to from 3.25 to 4.50 per cent. of the (pig) cast iron. Also because of certain impurities that are found to exist always in the ore and fuel, the iron contains varying amounts of silicon, sulfur, phosphorus and manganese.

The following percentages will serve to illustrate the composition of cast iron that is known to the trade as medium iron:

	Per cent.
Carbon	4.00
Silicon	2.00
Sulfur	0.05
Phosphorus	0.75
Manganese	0.70
Iron (by difference)	92.50

Source of Impurities.—The *silicon* finds its way into the furnace as silica, SiO_2, or sand in the ore or ash of the coke. The *sulfur* gains entrance both as ferrous sulfide, FeS, and iron pyrite, FeS_2, as well as sulfates, which exist as impurities in the ore and coke. It is the office of the limestone, serving as a flux, to take up as much of the silica and sulfur as possible. And the greater portion of the silica is cared for in this way, being converted into calcium silicate, which becomes a part of the slag.

But some of it is reduced, as:

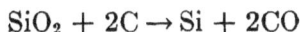

$$SiO_2 + 2C \rightarrow Si + 2CO$$

the reduced silicon then being taken up by the iron. Also most of the sulfur is converted into calcium sulfide, as:

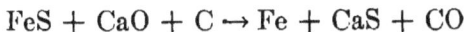

$$FeS + CaO + C \rightarrow Fe + CaS + CO$$

That which is thus converted into calcium sulfide is then taken up by the slag, but that which remains as iron sulfide is taken up by the iron. The quantity of both silicon and sulfur that the cast iron finally contains, depends upon the condition of the furnace and can, therefore, be controlled by the furnace operator. If the temperature of the furnace is above normal, the amount of silicon in the iron will be high and sulfur will be low; while if the temperature is lower than normal, the iron will be high in sulfur and low in silicon.

The *phosphorus* that is found in the cast iron gains entrance into the furnace as phosphates in the ore and coke. As the charge descends, these phosphates are reduced and the phosphorus is taken up by the iron, forming iron phosphide, Fe_3P. Unlike the silicon and sulfur, the quantity of phosphorus in the iron can be controlled only by the composition of the materials that are charged into the furnace. The operator can control it in no other way, for practically all that enters the furnace, in whatever constituent of the charge, is ultimately absorbed by the iron.

Apparently all iron ores contain small quantities of *manganese* ore, chiefly the dioxide, MnO_2. This is reduced to the metallic state by the carbon and absorbed by the iron.

Although silicon, sulfur, phosphorus and manganese are spoken of as impurities in pig and cast iron, they are quite important in many ways. It is by means of these elements that many of the most useful properties of cast iron are regulated.

Cast Iron Distinguished from Other Commerical Forms.— There is no essential difference between cast iron and pig iron. The term pig iron is used to designate the iron in the molten state as it is obtained direct from the furnace as well as when it has been cast into the form of pigs. The term cast iron applies more properly when the iron has been cast in some form other than pigs.

From a scientific standpoint, cast iron may be considered as an alloy of iron and carbon. It differs from steel, which is also an alloy of iron and carbon, in that it lacks malleability, as well as in the fact that it contains less iron and more carbon and impurities than steel; the lack of malleability being due, in fact, to the large amount of carbon. Cast iron differs from wrought iron in much the same way as it does from steel, although wrought iron contains considerable entrained slag, whereas cast iron contains but little.

THE SOLUTION THEORY OF CAST IRON

Descriptive Terms.—Before proceeding with this discussion, it is necessary to explain certain terms that are used to describe the physical structure of the iron.

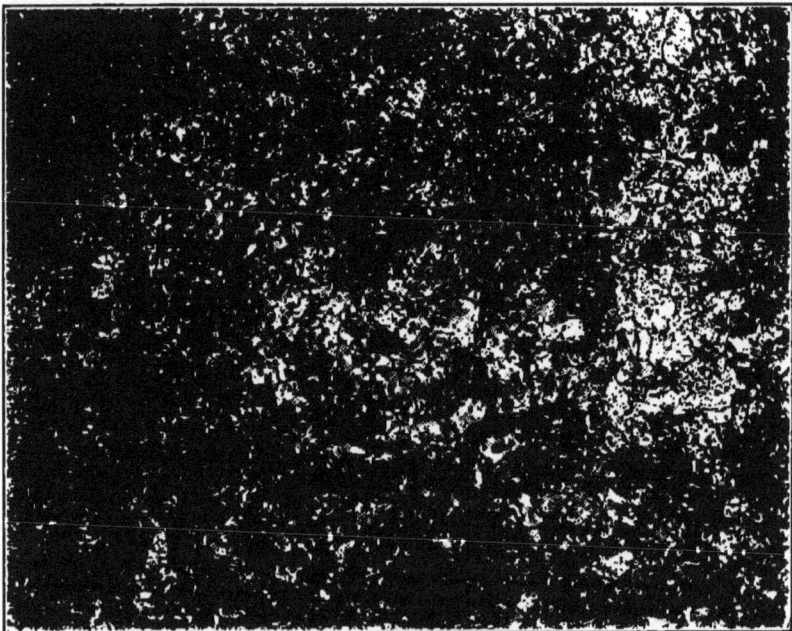

Fig. 5.—Grain structure of carbonless iron, practically 100 per cent. ferrite. Magnified 100 diameters.

Ferrite is the term used to designate pure, or more properly, carbonless iron, that exists as a microscopical constituent of the various forms of iron and steel. It is very soft, but very tough, ductile and malleable, somewhat resembling copper in respect to these properties. It has very high magnetic permeability and fair electrical conductivity. Wrought iron serves best to illustrate the properties of ferrite, since aside from the seams of slag which wrought iron carries, it approaches somewhat closely to the composition of ferrite. The grain structure of ferrite is shown in Fig. 5.

Cementite is the term used to designate the carbide of iron, Fe_3C. It is an extremely hard substance—the hardest of all

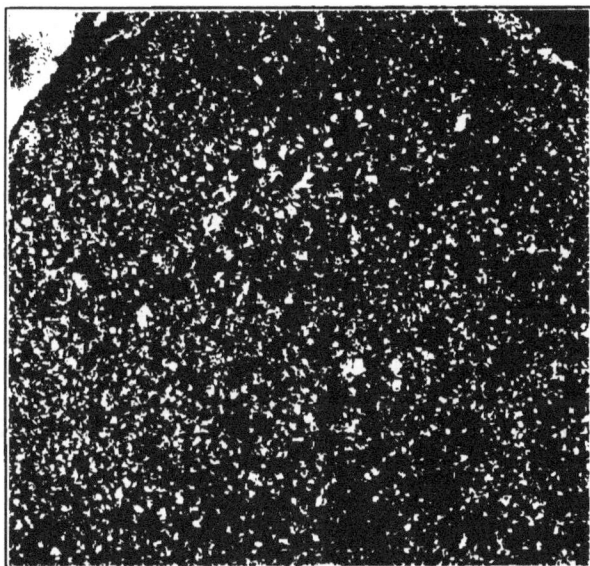

Fig. 6.—Graphite flakes in gray pig iron. Natural size.

the constituents that occur in iron and steel. It is said to be harder than glass and nearly as brittle. It has a very low resistance to shocks or any form of vibratory stress. Cementite is roughly one-fifteenth carbon by weight.

Graphite is one of the forms of carbon. It occurs in cast iron in the form of soft, thin flakes, which may vary in size from particles just visible with the microscope, to others an eighth of a square inch in area. Each flake is composed of other flakes, which have but little coherence between them. Graphite has

5

a relatively low specific gravity, being only 2.25 as compared with pure iron which has a specific gravity of 7.86. On this account, an iron that contains 3 per cent. by weight, contains by volume about 12 per cent. of graphite.

Solid Solutions and Eutectics.—For a discussion of solid solutions and the formation of eutectics, an understanding of which is necessary at this point, see "Non-ferrous Alloys," page 165.[1]

Pure Iron-carbon Alloys.—Any iron-carbon alloy in the molten state may be considered to be a solution of liquid carbon in liquid iron or ferrite. When this solution cools and solidifies, it depends upon the amount of carbon present, whether the alloy will solidify as a solid solution or form a eutectic, after the manner of the lead-tin alloys. If the carbon does not exceed 2.2 per cent., the alloy will solidify as a solid solution. A solid solution of iron and carbon is known as *austenite,* and this name applies to all such solutions regardless of the amount of carbon, which may range from almost none at all up to the saturation point at 2.2 per cent. The saturated solution is sometimes known as "2.2 per cent. austenite." If the alloy contains more than 2.2 per cent. of carbon, it solidifies with the production of a eutectic, which solidifies or freezes at 1,130°C., and contains 4.3 per cent. of carbon, as is indicated at B in Fig. 7.

In other words, everything to the left of the line *XY* in the figure, freezes as a solid solution, and everything to the right of it freezes selectively.

The 2.2 per cent. of carbon is recognized as the division between iron and steel. Any iron-carbon alloy containing less than this amount being defined as steel and any that contains more being defined as cast iron.

Structure of Cast Iron.—By definition, cast iron does not contain less than 2.2 per cent. carbon. And, if it does not contain more than 4.3 per cent., which it usually does not, upon reaching its solidifying range, crystals of iron will begin to

---

[1] Although, for the development of this subject, there might have been an advantage in placing the chapter on *Non-ferrous Alloys* before the present chapter, other considerations have caused the arrangement here used to be determined upon. Moreover, the author believes that the ordinary objection to the use of excerpts from succeeding chapters does not hold in a text-book of this character.

separate out. But the iron does not separate out in the pure state, just as the lead and tin did not, in the formation of their eutectic. The separated iron always carries with it 2.2 per cent. of carbon in solid solution. But still, more iron than carbon is thus separated out, and the amount of carbon in solution in the molten remainder is proportionally increased as the temperature falls, as is shown by the line *AB,* so that finally the eutectic containing 4.3 per cent. of carbon is formed at *B.* As has been said, this is at the temperature of 1,130°C. Then as the eutectic further cools below the point *B* it solidifies, but simultaneously

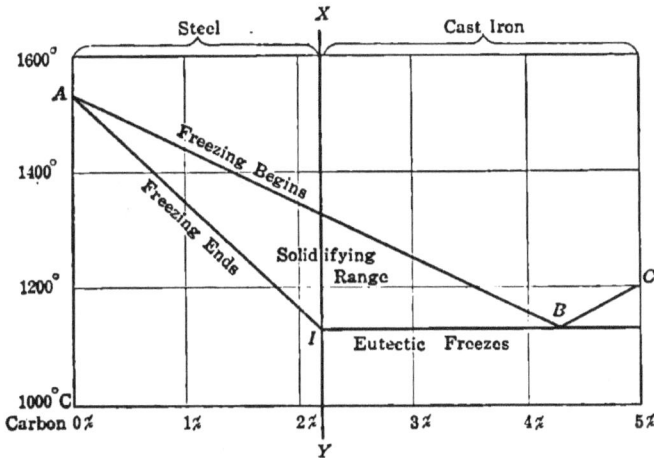

FIG. 7.—Freezing-point curves of the iron-carbon alloys.

with its solidification, it decomposes into 2.2 per cent. austenite and cementite.

If, on the other hand, the molten alloy should contain more than 4.3 per cent. of carbon, it will cool without change in composition until it meets the line *CB.* As it cools further, carbon in the graphite form will be precipitated out of solution, the composition following the line *CB,* until finally the eutectic containing 4.3 per cent. of carbon is produced at *B.* As this eutectic cools and solidifies, it will decompose into cementite and the 2.2 per cent. austenite as stated before.[1]

[1] For a more complete discussion of the changes that occur during the solidification of the iron-carbon alloys see SAUVEUR: The Metallography and Heat Treatment of Iron and Steel, Chapter XXII; STOUGHTON: The Metallurgy of Iron and Steel, p. 286; HATFIELD: Cast Iron in the Light of Recent Research, p. 7; or HOWE: The Metallography of Steel and Cast Iron, p. 61.

In reality, the carbon in the solid solution or austenite is in the combined form, that is, as cementite. In other words austenite is, more strictly speaking, a solid solution of iron and iron carbide. Also, this austenite is stable only at high temperatures, so that as the now solidified mass slowly cools, the *austenite decomposes into cementite and ferrite.*

Commercial Iron-carbon Alloys.—To avoid the introduction of as many complications as possible, the preceding discussion has dealt with only the pure alloys of iron and carbon, slowly cooled. · But the commercial alloys are not pure. There are always present in them certain other elements, that exert a modifying influence on the changes during cooling. Also the rate of cooling is an important factor.

Effect of Other Elements.—As has been stated before, the impurities are silicon, sulfur, phosphorus and manganese. The effect of silicon is most important in its effect upon the structural changes. It has a tendency to cause the cementite that separates out, to immediately decompose into ferrite and graphite in accordance with the following equation:

$$Fe_3C \rightarrow 3Fe + C$$

It exerts this influence to a degree depending upon the amount present. Briefly, phosphorus may be said to help silicon in this action, as a rule, while sulfur and manganese tend to oppose it, but this will be explained more fully later.

Effect of the Cooling Rate.—As in all chemical reactions, a certain amount of time is necessary for the decomposition reaction of cementite to run to completion. Hence, if the cast iron is quickly cooled, the amount of decomposition will be lessened. In fact, if the silicon is very low, the sulfur high, and the cooling rate very rapid, it is possible that practically no cementite will be decomposed at all. But, on the other hand, if the silicon is high, the sulfur low and the cooling rate slow, most of the cementite will decompose, and the iron will contain its carbon largely in the graphite condition.

Kinds of Cast Iron.—Depending upon the condition in which the carbon exists, pig iron and cast iron have been divided into two major classes:[1] Pig iron and cast iron in the fracture

---

[1] STOUGHTON: The Metallurgy of Iron and Steel, p. 7.

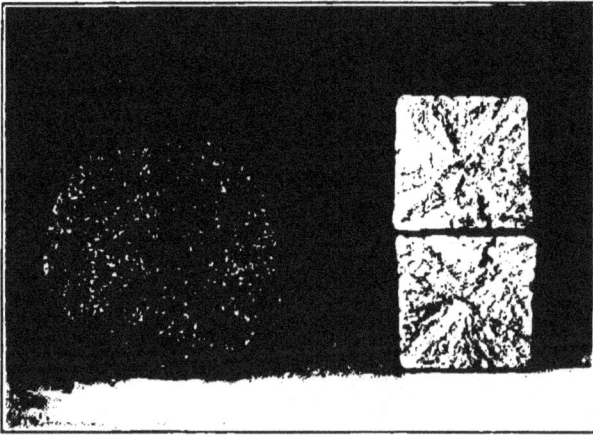

Fig. 8.—Fractures of gray and white cast iron.

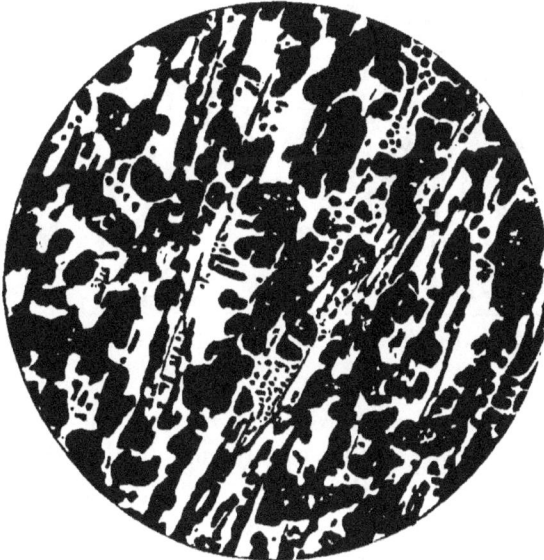

Fig. 9.—White cast iron containing about 3 per cent. combined carbon. The light areas are cementite. Magnified 56 diameters.[1]

[1] Reproduced by permission from Sauveur's Metallography of Iron and Steel.

of which the iron itself is nearly concealed by the graphite, so that the fracture has the color of the graphite, is called 'gray' *pig iron or cast iron.*

Pig and cast iron, in the fracture of which little or no graphite is visible, so that the fracture is silvery-white, is called '*white iron.*' In this the carbon is nearly all combined.

FIG. 10.—Gray cast iron free from combined carbon. Magnified 100 diameters. Not etched (Langenberg).[1]

In the commercial cast irons, the carbon, which may amount to 3 or 4 per cent., is neither wholly combined in the white iron, nor wholly graphitic in the gray. The white iron will contain some graphite, although the flakes will be small and not readily visible on the face of the fracture. Gray-iron castings will usually contain from 0.25 to 1.50 per cent. of combined carbon and the remainder in the graphitic state.

[1] Reproduced by permission from Sauveur's Metallography of Iron and Steel.

Physical Properties of Cast Iron.—If we assume that a *white* cast iron contains 3 per cent. of combined carbon, then it will contain approximately 40 per cent. of iron carbide, Fe_3C. It therefore partakes of the nature of this compound. Thus, white cast iron is very hard and brittle, and can be readily shattered by hammer blows. It is machined only with the greatest difficulty, but on the other hand will withstand abrasion well while in use. Its uses are few as compared to gray iron.

In *gray cast iron*, graphitic carbon predominates. This implies a relatively small amount of combined carbon and consequently the existence of much ferrite. Both graphite and ferrite are soft, the result being that gray iron is also relatively soft. It is chiefly for this reason that this is the iron that is used for general foundry purposes, since most castings require some machining. Also, in the gray casting the formation of the graphite crystals prevents the excessive shrinkage which is so characteristic of the white iron. Other properties that make the gray iron suitable for foundry purposes are its low melting point, its fluidity and the ease with which its properties may be varied.

PROPER DISTRIBUTION OF CARBON BETWEEN THE STATES OF GRAPHITE AND CEMENTITE, TO GIVE VARIOUS PROPERTIES

Properties sought	Example of use	Distribution of carbon		Color of iron.
		Cementite	Graphite	
Resistance to shock	Machinery	Little		Open gray.
Softness, easy machining	Machinery	Little		Open gray.
Sharpness of detail, expansion in solidifying	Ornaments		Much	Open gray.
Fluidity (phosphorus)	Pipes	Moderate		Close gray (No. 3).
Strength	Car-wheel centers	Moderate		Close gray.
Imperviousness	Hydraulic cylinders, radiators.		Very little	Very close gray
Hardness	Trend of car wheels.	Much		Nearly white.

Its disadvantages are its lack of strength, ductility and malleability. The reason for the lack of these properties in gray iron is quite apparent, for, although the ferrite of which it is largely composed is very ductile and malleable, the flake-like graphite crystals very effectually break up the continuity of the metallic mass, as is shown in Fig. 10, so that it becomes weak and brittle.

In order to briefly summarize the various properties of cast iron that are dependent upon the condition of the carbon, the table shown on page 71 is quoted from Howe.[1]

EFFECT OF IMPURITIES IN CAST IRON

The effect of the impurities in cast iron is quite important. The condition of the carbon in the iron; that is, whether it will be in the free or in the combined state, depends very largely upon the amount of these elements present, and upon the condition of the carbon, the properties of the iron almost wholly depend.

Silicon in Cast Iron.—The amount of silicon in cast iron may vary from 0.50 to 3.00 or 3.50 per cent. It is probable that it all exists in combination with the iron in the form of an iron silicide, FeSi, which in turn dissolves in the remainder of the iron. The amount of this iron silicide is equal to exactly three times the weight of the silicon. Consequently, if the iron contains, for example, 2.5 per cent. of silicon, it will contain 7.5 per cent. of iron silicide. Silicon is a very important element in cast iron because of its marked tendency to cause cementite to decompose into ferrite and graphite. Thus it softens and renders easy to machine that which would without it be a very hard and brittle iron.

It has been noted before that slow cooling works in conjunction with silicon, but for the conversion of cementite into ferrite and graphite, the use of an abundance of silicon, unaided by slow cooling, is to be preferred to an equivalent production of graphite brought about by the employment of less silicon and a prolonged cooling period, because under the latter conditions the graphite flakes will be broad and introduce more extended planes of weakness into the casting than will the smaller flakes produced by silicon without the help of slow cooling.

When the iron is remelted, the silicon burns, thus increasing the temperature, and consequently decreasing the amount of sulfur the iron can take up, for it has been observed that a high

[1] The Metallurgy of Steel and Cast Iron, p. 109.

temperature in a furnace or cupola causes the iron to take up less sulfur.

In remelting silicon acts as a deoxidizer also. It helps to eliminate blow-holes by lengthening the time of fluidity of the iron. This action is aided by the fact that it increases the solubility of the gases in the solidifying iron, thus lessening the separation of gas bubbles. Its action in lessening the sulfur content also works toward the elimination of blow-holes.

Silicon lessens the tendency of the iron to "chill," and also tends to offset the hardening effect of manganese.

High-silicon Iron for Acid-proof Castings.—Although in ordinary cast-iron, the amount of silicon does not, as a rule, exceed 3 per cent., castings of special composition are made that contain from 14 to 15 per cent. of silicon. With these percentages, the product must be considered as an iron-silicon alloy, instead of an iron-carbon alloy, for in the silicon alloy the carbon is low, generally under 1 per cent. Castings of the high-silicon alloy are not made for strength or general utility, but for purposes where resistance to acid is required; and they seem to possess this property in a high degree. Acid-resisting castings of the iron-silicon alloy are sold under the trade names of "Duriron" and "Tantiron."[1]

Sulfur in Cast Iron.—Cast iron of good quality does not contain sulfur in an amount greater than 0.10 per cent., although for special purposes, as for the making of chilled castings, as much as 0.20 per cent. may be allowed. It occurs in the iron both in the form of manganese sulfide, MnS, and ferrous sulfide, FeS, but usually in the form of the manganese compound. These metallic sulfides are in turn dissolved in the iron. Each part of sulfur produces approximately $2\frac{1}{2}$ parts of either of the two sulfides. As long as there is sufficient excess of manganese present, practically all of the sulfur will be in compound with that metal, since manganese is able to transpose the ferrous sulfide into the form of the manganous compound. Should there not be a sufficient excess of manganese present to cause all of the sulfur to combine with it, the remaining sulfur would then exist in the form of the ferrous compound.

The effect of the sulfur upon the condition of the carbon in

[1] *Jour. Ind. Eng. Chem.*, **8**, 879 (1916).

cast iron is directly opposite to that of silicon. It has a tendency to keep the carbon in the combined form. It is usually considered that 0.01 per cent. of sulfur will neutralize the effect of fifteen times (0.15 per cent.) as much silicon in this effect upon carbon. However, sulfur in the form of the manganese compound is not so effective in this respect as when it is in combination with iron. Consequently if sufficient manganese is present to combine with all the sulfur, its effect is not so pronounced. Since it keeps carbon combined, white iron has in it more sulfur, usually, than the gray. High sulfur and high silicon are not to be expected to be present in the iron at the same time, since a furnace "working hot" decreases the amount of sulfur and increases the amount of silicon in the iron, and *vice versa.*

Additional effects of sulfur are that it makes the iron somewhat more easily fusible, but makes it more viscous or sluggish when it has been melted. This increased viscosity causes the production of unsound castings because the gas bubbles cannot separate so readily from the more sluggish metal. It increases shrinkage because it prevents the formation of crystals of graphite, which would if they formed, increase the bulk of the iron and so help to offset the contraction during cooling of the casting. It increases the hardness of the iron. It increases the depth of "chill;" that is, the layer of surface metal of a casting free from graphitic carbon, produced by casting the metal against a metallic wall. It causes the iron to "check;" *i.e.*, crack at a red heat, and if put under strain at this time it exhibits the phenomenon of "red shortness."

Phosphorus in Cast Iron.—In ordinary foundry iron the amount of phosphorus does not usually exceed 1.00 per cent. It exists as definite phosphides, Fe_3P and Fe_3P_2. The effect of phosphorus on the condition of the carbon depends somewhat upon the amount of silicon present. Its chemical effect is to keep the carbon in the combined condition, and this it will do, especially if the amount of silicon is small, for example under 0.75 per cent., and the phosphorus high, for example above 1.25 per cent. But if the amount of silicon is high enough that its effect is quite pronounced, for example between 1.50 and 2.50 per cent., then the final effect of phosphorus is to cause carbon to take the graphite form. It does this under these conditions, because it has a very marked effect upon the time of fluidity of the iron. It seems to cause the iron to remain fluid, or at least in a pasty state until much lower temperatures are reached than would be the case if phosphorus were absent. This allows the silicon a much greater period, perhaps several minutes longer, through which its influence is effective. Consequently more graphite is separated out, and in larger crystals as well. Other effects of phosphorus are that it causes the iron to be especially easily broken by shock or vibration. This is described as "cold shortness." When working in conjunction with silicon, phosphorus lessens the amount of shrinkage.

Because of the fact that phosphorus allows the iron to remain fluid at lower temperatures, iron high in phosphorus is selected for making thin castings which naturally cool quickly, and if phosphorus were absent, might solidify before the iron had perfectly filled the mould. Being high in phosphorus, articles that are cast in thin section are, therefore, more brittle than they would be if they were cast in larger size of ordinary foundry iron, less rich in phosphorus, and then machined to an equally thin section. This will be further discussed under "Composition and Size of Castings."

Manganese in Cast Iron.—Manganese is a metal resembling iron in many of its properties. It alloys with iron in all proportions, but the amount in cast iron should not exceed 1.00 per cent. as a rule. There is seldom more than 2.00 per cent., and it may be as low as 0.10 per cent.

Manganese increases the ability of iron to absorb carbon and thus increases the total amount in the iron. All of the manganese in ordinary cast iron exists in combination with either sulfur or carbon. Primarily it combines with sulfur, and if there is sufficient present so that there is an excess above that needed for the sulfur, this excess combines with the carbon and exists in the form of manganese carbide, which becomes associated with the iron carbide to form manganiferous cementite.

As a result of the fact that manganese exists in two compound forms in iron, it has a contradictory effect upon the condition of the carbon. Any manganese that exists as the carbide serves to keep carbon in the combined form, but that which is in the form of the sulfide is not active in this manner. In fact its influence is in the opposite direction. For were it not for the presence of the manganese, the sulfur would exist in the form of ferrous sulfide, which is extremely active in keeping carbon in the cementite form. But manganese transposes sulfur from the ferrous compound into the manganous compound and thus renders it inactive. Beside, the manganous sulfide is to a greater extent removed by the slag. Hence, it depends upon the amount of sulfur and manganese in the iron, whether the manganese aids finally in producing graphite or cementite.

Manganese has other noteworthy effects. It removes oxygen, that is, it transposes ferrous oxide into the form of manganous oxide which is taken up by the slag. It makes iron less fusible, but when molten, makes it more fluid. This it is able to do because of its decomposition of ferrous sulfide, which makes iron sluggish when molten. If present in large amount, it hardens and toughens the iron and may make the castings noticeably more difficult to machine. By its removal of sulfur it tends to remove "red shortness." By its removal of oxygen it tends to prevent "boiling" during cooling and so aids in preventing blowholes.

Grading of Foundry Irons.—The physical character of pig iron is dependent mainly upon the condition of the carbon, and the condition of the carbon is chiefly determined by the amount of silicon present. Still, as has been explained, other factors are at work. Pigs that are cast in iron moulds will not hold all of the carbon in the same condition as those cast in sand moulds, even though the iron in both cases may be exactly of the same composition. Also the condition of the carbon in the pigs cast in that portion of the sand bed farthest from the furnace will not be the same as in those cast nearest to the furnace, even though the composition of the iron in both cases might be the same. This, of course, would be due to the fact that the iron in the farthest portion of the bed would be most rapidly cooled because of having traversed more cold sand. The condition of the carbon can be determined by examining the fracture of the iron. But since silicon is not the only controlling factor it frequently happens that it does not exert its normal influence on the carbon, whereupon the fracture ceases to be reliable guide to the quantity of silicon and consequently to the quality of the casting that will result. It is therefore far better to grade iron by the quantity of silicon as determined by analysis than by the appearance of the fracture.

The usually specified limits of the chief grades of pig iron are as follows:

Grade of iron	Per cent. Si	S	P	Mn
No. 1 foundry	2.50–3.00	Under 0.035	0.50–1.00	Under 1.00
No. 2 foundry	2.00–2.50	Under 0.045	0.50–1.00	Under 1.00
No. 3 foundry	1.50–2.00	Under 0.055	0.50–1.00	Under 1.00
Malleable	0.75–1.50	Under 0.050	Under 0.20	Under 1.00
Bessemer	1.00–2.00	Under 0.050	Under 0.10	Under 1.00
Low phosporus	Under 2.00	Under 0.030	Under 0.03	Under 1.00
Basic	Under 1.00	Under 0.050	Under 1.00	Under 1.00

Composition Affected by Remelting.—When the iron is remelted in the cupola, its composition is altered slightly, especially in reference to sulfur and silicon. As the drops of molten iron trickle down over the coke, sulfur is absorbed from the coke by the iron. The amount of this absorption depends upon the percentage of sulfur in the coke as well as on the proportions of coke and iron used in the charge. The sulfur absorbed will vary from 0.02 to 0.035 per cent. of the iron; that is, if the sulfur in the pig iron charged was 0.08 per cent., there will be from 0.1 to 0.115 per cent. in the castings.[1] Also, the sulfur in the first iron melted will be higher than that in the middle of the run, because of the extra amount of coke burned before any iron is tapped.

As the molten metal passes from the melting zone down through the blast from the tuyères, some silicon is oxidized and carried into the slag. The silicon lost in this way varies from 0.25 0.40 per cent., but according to Stoughton,[1] with good practice it should be no more than 0.30 per cent. To illustrate, if the loss is just 0.30 per cent. then, when the pig iron charged into the cupola contains 2.00 per cent. silicon, the amount in the castings will be only 1.70 per cent.

Also, during the melting, a small amount of manganese will be lost through oxidation, and a small amount of phosphorus

[1] STOUGHTON: Metallurgy of Iron and Steel, p. 258.

may be gained from the coke but these alterations are of comparatively minor importance.

Composition and Size of Castings.—As the size of the casting
largely controls its rate of cooling, it follows that the composition of the iron should vary with the size of the casting.   Since

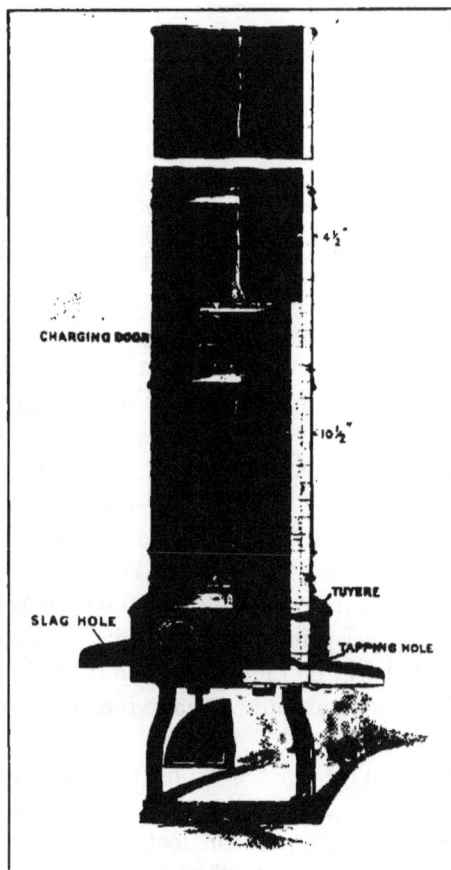

Fig. 11.—Section of cupola.

a thin casting cools rapidly, it should be made of iron that is
very fluid, and that has no pronounced tendency to retain its
carbon in the combined form.   On the other hand, a large mass
of iron, which would cool slowly, should not have a composition
that would allow it to form large crystals of graphite in its interior.   In general, it may be said that small castings should be

high in silicon and phosphorus and large ones should be high in manganese and sulfur. The following limits of composition, based on size of castings, have been suggested:[1]

Thickness of section	Si	P	Mn	S
Under ¼ in. thick.....	3.25	1.00	0.40	0.020
¼ to ½ in. thick......	2.75	0.80	0.40	0.040
½ to ¾ in. thick......	2.50	0.75	0.50	0.050
¾ to 1 in. thick.......	2.00	0.70	0.60	0.060
1 to 1½ in. thick......	1.75	0.65	0.70	0.070
1½ to 2 in. thick......	1.50	0.60	0.80	0.080
2 to 2½ in. thick......	1.25	0.55	0.90	0.090
2½ to 3 in. thick......	1.00	0.50	1.00	0.100

Composition and Use.—The composition of the iron that is suitable for certain uses may be illustrated by the composition of the samples used by the Committee on Standardizing the Testing of Cast Iron.[2]

	Si	P	S
Novelties......................	4.19	1.236	0.080
Stove plate....................	3.19	1.160	0.084
Cylinders......................	2.49	0.839	0.084
Light machinery................	2.04	0.578	0.044
Heavy machinery...............	1.96	0.522	0.081
Dynamo frames.................	1.95	0.405	0.042
Ingot moulds...................	1.67	0.095	0.032
Car wheels.....................	0.97	0.301	0.060
Chilled rolls...................	0.98	0.482	0.070
Sand rolls.....................	0.72	0.454	0.070

Composition and Hardness.—In order to illustrate the kind of iron that is suited for the making of castings of different degrees of hardness, the specifications of the Case Threshing Machine Co. are given:[3]

[1] FORSYTHE: The Blast Furnace and Manufacture of Pig Iron, p. 312.
[2] *Jour. Am. Fdy. Assoc.*, **10**, part 2, p. 25.
[3] *Iron Age*, Sept. 29, 1898, pp. 4 and 5.

	Soft castings, pulleys, or small castings	Medium castings cylinders, gears, pinions, etc.	Hard castings, valves, and high-pressure cylinders
Si, per cent	2.20–2.80	1.40–2.00	1.20–1.60
S	Below 0.085	Below 0.085	Below 0.095
P	Below 0.700	Below 0.700	Below 0.700
Mn	Below 0.700	Below 0.700	Below 0.700
Tensile strength per square inch	18,000 lb.	20,000 lb.	22,000 lb.
Transverse strength per square inch	2,000 lb.	2,200 lb.	2,400 lb.
Deflection not less than	0.10 in.	0.09 in.	0.08 in.
Shrinkage, per foot, not over	0.127 in.	0.136 in.	0.148 in.
Chill, not over	0.05 in.	0.15 in.	0.25 in.

Shrinkage of Cast Iron.—White cast iron begins to shrink almost immediately after solidifying, and shrinks rather uniformly and continuously. Gray cast iron does not shrink at all immediately after solidifying, but undergoes a noticeable *temporary expansion*, this expansion being, of course, due to the separation of the graphite. When the iron is high in both silicon and phosphorus, the expansion is very long continued; and the metal has cooled almost to a black heat before the contraction of the metal due to cooling, overcomes this expansion, and the piece has shrunk again to the size it was when just cast. The phosphorus prolonging the fluidity, allows a more perfect separation of the graphite. This, too is an explanation of why the phosphorus-bearing irons fill the mould so perfectly. By the long-continued expansion, the pasty metal is forced into all the crevices of the moulds.

As the shrinkage generally increases with the hardness of the cast iron, or in other words, with the decrease of graphitic carbon, it becomes a valuable indication to the quality of the iron,[1] since it may be measured with comparatively little trouble. Shrinkage is usually considered to be about ⅛ in. to the foot, but it is variable and depends upon the silicon content and the area of the cross-section, as is indicated in the following table,

[1] FORSYTHE: *Loc. cit.*, p. 311.

in which the amount of shrinkage is stated in the decimals of an inch.

The perpendicular readings show decrease of shrinkage due to increase of silicon, and the horizontal readings show decrease of shrinkage due to increase of size.[1]

Per cent. of silicon	½-in. square	1 in. square	2 in. by 1 in.	2 in. square	3 in. square	4 in. square
1.00	0.183	0.158	0.146	0.130	0.113	0.102
1.50	0.171	0.145	0.133	0.117	0.098	0.087
2.00	0.159	0.133	0.121	0.104	0.085	0.074
2.50	0.147	0.121	0.108	0.092	0.073	0.060
3.00	0.135	0.108	0.095	0.077	0.059	0.045
3.50	0.123	0.095	0.820	0.065	0.046	0.032

FIG. 12.—Shrinkage flaws in white and gray cast iron.

Although silicon in conjunction with phosphorus has perhaps the most important effect in reference to shrinkage, the effect of sulfur also is noteworthy. Because it is very effective in keeping the carbon in the combined form, it helps to increase shrinkage. Manganese, by increasing the total carbon, tends to increase the amount of graphite and therefore helps to lessen shrinkage. Also it does this because it combines with sulfur and causes it to become inactive.

[1] STOUGHTON: p. 328.

6

Since the last metal to solidify is the interior of the larger portion of the casting, it not frequently happens that this interior has a spongy texture, or contains open flaws with rough and crystalline surfaces, as illustrated in Fig. 12. As the exterior cooled and solidified, the shrinkage of the underlying portion was compensated for by metal from the interior, until when solidification was complete, there was a lack of metal in this portion of the casting. These open or spongy spots are known as *shrinkage flaws* or cracks. They can be readily distinguished from the unsound spots due to gas bubbles by the character of the surfaces of the openings, as will be apparent from a comparison of Figs. 12 and 13. Shrinkage flaws can be prevented by the judicious use of risers.

Cracking or "Checking" of Castings.—Occasionally parts of castings, especially thin sections, that are so located that they contract against cores or other portions of the mould, crack during cooling. This usually occurs at the time the casting is just above a black heat, for at this time the metal is in a very weak and loosely coherent state. Sulfur has a very important bearing upon this tendency to "check," because both the sulfide of manganese and of iron have relatively low melting points and at the temperature mentioned, are in a pasty condition so that they can offer very little resistance to breaking. Stoughton says that the sulfide of iron is much the worse because it is spread out in thin plates or in network form, thus offering more extended planes of weakness than the sulfide of manganese, which is more apt to exist in globular state.

Phosphorus helps to increase the liability to checking in that it has a tendency to cause the production of large-sized crystals. Manganese tends to decrease crystal size and so in a measure, at least, offsets the effect of phosphorus.

Segregation of Impurities.—When iron is molten, carbon and the impurities are dissolved in it. But the impurities are not so soluble in the solid metal and consequently tend to separate out as the iron freezes. It is quite apparent, therefore, that each layer of metal as it solidifies, beginning at the outside, rejects some of its impurities to be taken up by that which remains molten in the interior. This action is aided by the fact that the compounds in which the impurities exist have much lower melting points than the ferrite matrix. Then that portion of the casting which solidifies last will be richest in carbon and the impurities, especially sulfur and phosphorus. The result of this action is known as *segregation*, and may be defined as the collection together of the impurities in spots. Of course, the material last to solidify is usually in the larger sections of the casting. Also, it is in these larger portions that the shrinkage flaws or cracks occur. Then in these cracks, or spongy areas, collects the metal, that because it is rich in impurities, has a low melting point and so remains fluid while the other portions solidify. These areas, when very bad, are sometimes known as "hot spots," being so named probably because of their deferred solidification. If rich in ferrous sulfide, they may be so extremely hard that no tool can cut them.

This segregation may be to a large extent obviated by the use of large risers or headers, which, because they are large, solidify last and prevent the formation of shrinkage spots by feeding the rest of the metal well under pressure. Also, because they solidify last, the segregation occurs in them and is thus temporarily removed. This method is not advisable, however, as continuous practice, because the risers find their way back into the cupola as scrap and thus increase the quantity of impurities in subsequent castings.

Blow-holes in Castings.—Blow-holes are unsound spots in castings, produced by gas bubbles. The gas is absorbed by the liquid metal, but cannot be retained in the same proportion in the solid metal, hence it separates as the metal solidifies. In cast iron, and in cast steel as well, these gases are found to be chiefly hydrogen, with smaller quantities of nitrogen and the oxides of carbon. The hydrogen is produced from moisture ecomposed by the heat, probably in the case of cast iron, from moisture in the cupola blast. The carbon monoxide and dioxide are produced by the reaction between the carbon of the iron and some oxidized constituent of the metal. There is no difficulty in distinguishing between the irregular sponginess of shrinkage flaws and the smooth surfaces of blow-holes. In iron castings, an oxidized surface indicates trapped steam, or air, while a bright surface is produced by gases set free during solidification, since these latter are almost' without exception reducing in their effect, or else are entirely inactive.

Manganese and silicon are quite effective in preventing blow-holes. They seem to act partly by combining with the oxygen in the metal so that the gaseous oxides of carbon are reduced or else not formed, and partly by increasing the dissolving power of the solid metal for gases so that less gas separates out. Aluminum and titanium are also sometimes added for the same purpose, although the latter element has an additional effect in that it combines with nitrogen.

General Rules.—In a general way it may be stated that for the average foundry irons these rules hold good:[1]

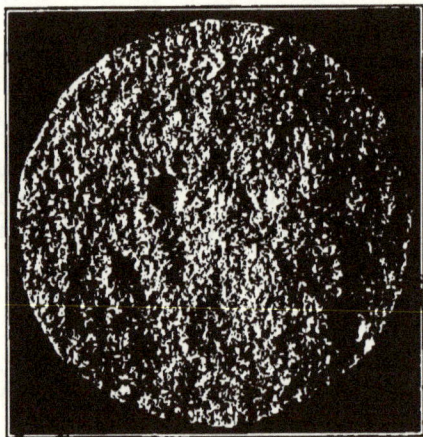

FIG. 13.—Gas flaws (blow holes) in cast iron.

To increase the strength of castings. Decrease phosphorus and lessen graphite by decreasing silicon, thereby allowing more combined carbon. The manganese may also be increased and the casting cooled more rapidly.

To decrease the hardness and shrinkage of the casting. Decrease the quantity of combined carbon and increase the graphite by decreasing sulfur or by increasing silicon and cooling more slowly.

To prevent "chilling" of the casting. Decrease sulfur, increase silicon and cool more slowly.

[1] FORSYTHE: *Loc. cit.*, p. 310.

To prevent blow-holes in the casting. Decrease sulfur and increase manganese and silicon.

To prevent kish, or accumulations of floating graphite. Decrease the percentage of carbon by adding scrap to the cupola. The addition of small quantities of ground ferromanganese or 50 per cent. ferrosilicon gives generally beneficial results. It deoxidizes the metal, thereby softening and strengthening it. It aids in lessening shrinkage and helps to make clean castings without materially altering the composition of the metal.

MALLEABLE CAST IRON

Malleable cast iron is a cast iron in which the carbon has been slowly set free in an amorphous condition without melting the iron. It is malleable to a slight degree because the natural malleability of the ferrite existing in it is to some extent retained and not destroyed, as it is in the ordinary gray foundry iron, by the flake formation of the carbon. It is a product obtained by two operations. The casting is first made of white cast iron of special composition. This hard, brittle casting is then cleaned and made malleable by subsequent annealing or heat treatment. Malleable castings cannot be made from gray cast iron. During the annealing process the cemetite of the white cast iron is decomposed and the combined carbon is set free, not, however, in the form of flakes. It is liberated in a very finely divided amorphous condition, which unlike crystalline graphite is readily oxidizable. To it the name of "temper carbon" has been given. Malleable cast iron consists then, almost entirely of ferrite and temper carbon.

The composition of the iron for the original white castings must be kept within rather well-defined limits. Since it is absolutely necessary that the castings be white it might be supposed that the total absence of silicon would be advantageous. However, this is not so because the temper carbon will not separate out in the annealing process unless a certain amount of silicon is present. Still, too much silicon is fatal because it will prevent the original casting from being white. In brief, then, the composition of the metal poured must be such that the carbon is just on the line between being separated out as graphite and being held as cementite.

The composition of the pig iron ordinarily used for making the casting is about as follows: Si (according to the weight of the casting) 0.75 to 1.25 per cent.; Mn not over 0.6 per cent.; P not over 0.225 per cent.; S not over 0.05 per cent. In normal American practice this will result in a white casting of about the following composition: Si 0.45 to 1.00 per cent., Mn 0.3 per cent.; P up to 0.225 per cent.; S up to 0.07 per cent. However, the

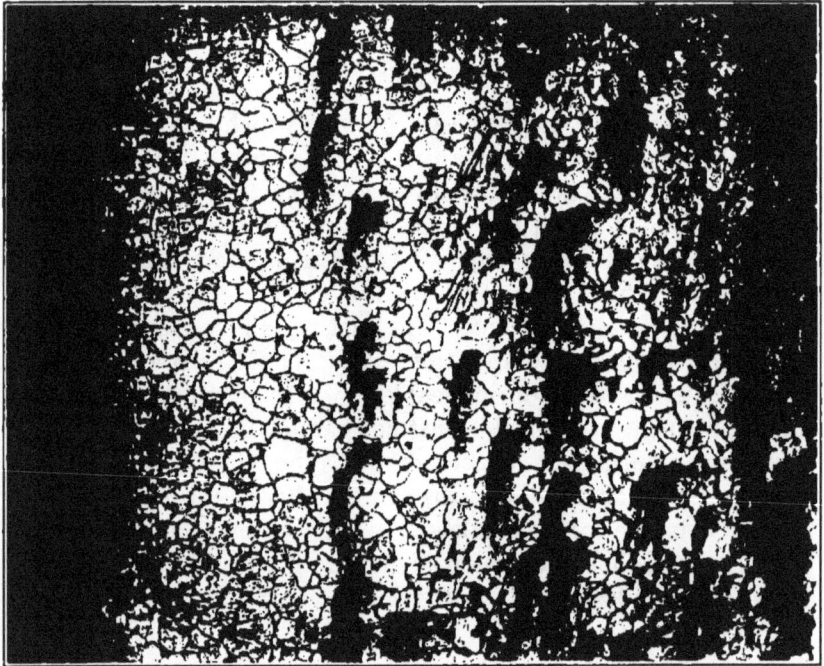

FIG. 14.—Typical malleable iron, showing structure of ferrite, and temper carbon.   Magnified 100 diameters.   Etched.

charge is not usually made up of pig iron alone. It more commonly consists of part steel scrap, "return scrap" from the malleable foundry itself, and pig iron.

Melting.—The iron for the original white castings may be melted in the cupola, openhearth furnace, or air furnace. Each has its advantages and disadvantages, but it seems that in view of all considerations the air furnace is generally recognized as being most desirable. It consists essentially of a basin-like hearth over which a large volume of flame is blown. Injector pipes delivering air above the bath aid in securing the proper temperature and oxidizing conditions. The oxidizing conditions are necessary in order that the amount of silicon and carbon may be reduced sufficiently to cause the residual carbon to remain in the combined form in the castings.

Contraction and Shrinkage in Casting.—The hard castings contract in the mould to a degree almost equal to that of steel. This is because of the lack of graphite formation. A fault that occurs frequently in malleable castings is due to this same contraction. The exterior of the casting solidifies to a continuous shell while the interior lessens in bulk, and thus becomes full of cracks [and planes of separation known as "shrinkage flaws" (see page 81).

Annealing.—After cooling, the castings are cleaned and packed in annealing boxes in some material, as iron ore, mill scale, lime or sand[1]. Then they are placed in the annealing ovens and heated to about 730°C. which is about 450° below their melting point. After having reached this temperature they are kept at this point for about 60 hrs. During the heat treatment the separation of the carbon, which was prevented by rapid cooling and other influences when the casting was formed, occurs.

The "Black-Heart" Malleable.—When the annealing is conducted with the iron oxide as the packing material, the annealed product will have a fracture such as is shown in Fig. 15. Due to the oxidizing effect of a packing of this sort the carbon is entirely removed as CO from the outer layer of the casting, leaving practically iron only. Because of the resultant white skin and dark interior this product is called "black-heart" malleable. It is malleable castings of this kind that are usually produced in American practice. However, some castings are annealed in lime or sand so that the fracture is totally black with no white exterior. The presence of the white outer layer has been quite generally considered to add much to the strength of the casting

---

[1] The object of this packing material is to support the castings equally at all points, and thus prevent them from warping while they are hot. However, in some cases, by carefully arranging the castings in the oven, the packing material and boxes may be omitted altogether.

but Touceda[1] seems to have shown that this belief is not supported by the facts.

Properties of Malleable Cast Iron.—The specifications of the American Society for Testing Materials require that malleable castings have a tensile strength of not less than 40,000 lbs. Malleable cast iron can be twisted and bent to a certain degree but its greatest advantage is its resistance to shock. Steel castings cannot stand constant battering because their structure is made up of distinct crystals of ferrite closely packed between other crystals of iron carbide, and between these, planes of cleavage are

Fig. 15.—Typical fracture of "black heart" malleable cast iron.

developed. The malleable casting is made up of iron crystals between which are scattered particles of free, amorphous carbon, which have a cushioning effect and so increase the shock resistance.

Uses.—Malleable cast iron is especially useful for all purposes where the casting is required to withstand shock or any form of vibratory stress. For example, car parts are made of it. Also, it is especially suitable for the making of small castings. Small castings that are made of ordinary cast iron are brittle and cannot withstand shock. It is difficult to make steel castings of small size, owing to the difficulty in obtaining a temperature

[1] Remarks on the Strength and Ductility of Malleable Castings after the Skin Has Been Removed, *Trans Am. Fdy. Assn.*, 23, 440 (1914).

sufficiently high to cast them properly. Consequently, small castings that are required to possess considerable strength are made of malleable cast iron. Window fastenings and harness fittings might be considered to be typical of small castings of this sort.

Also, sometimes tools, such as carpenter's chisels, hammers, etc., are made of malleable cast iron, and it happens occasionally that these are improperly sold as cast steel. There is a zone or layer just beneath the white exterior of the "black-heart" malleable casting that partakes of the nature of tool steel, and use is made of this to form the cutting edge of the tool. The existence of this steel-like layer is no doubt the result of its position. It lies midway between two layers, one rich in carbon and the other practically carbon-free, and has a composition midway between the two. Consequently, under heat treatment it acts as such, being capable of hardening and tempering. Also, the malleable casting may be given a steel-like fracture throughout, if the annealing is carried out at a temperature that is higher than normal, or if it is continued for a longer period than usual, and then cooled rapidly.[1] Sometimes the mere heating of a properly annealed casting and then quenching it will cause the carbon to assume the cementite form.

WROUGHT IRON

Wrought iron is defined as slag-bearing, malleable iron[2] that does not harden materially when suddenly cooled. It consists essentially of ferrite; therefore, it is quite soft, but very tough, malleable and ductile.

Manufacture.—This iron is made by melting pig iron and exposing it to the flame of a reverberatory furnace, the hearth of which is lined with iron oxide. (A reverberatory furnace consists essentially of an enclosed fireplace where the fuel is burnt, and from which the products of combustion and flame are conducted across the adjacent hearth. The hearth is covered with

---

[1] STOUGHTON: *Loc. cit.*, p. 355.

[2] The term "malleable iron," as used here, must not be construed to refer in any way to the malleable cast iron just discussed. Wrought iron is actually malleable; that is, can be rolled and forged, whereas malleable cast iron cannot.

an arched roof so that the heat is caused to reverberate or to be reflected upon the hearth.) The carbon, silicon and manganese are largely oxidized, the oxides either entering the slag or escaping by way of the stack. Much phosphorus and sulfur are also removed, and thus the iron is very largely purified.

During the oxidizing or purification process the iron is stirred or "puddled," in order that air and the slag which is rich in iron oxide may gain access to all parts of the charge. As is the case with other metals the melting point of the iron rises as it approaches purity. Hence, even though the furnace may have increased much in temperature during the process, the charge eventually becomes thickly fluid, and at last assumes a pasty state.

Fig. 16.—Bar of wrought iron hammered cold, showing its fibrous character.

The furnace operator or "puddler" now breaks up the charge with a bar and roughly forms it into balls which may weigh 125 to 180 lbs. These balls consist of innumerable, loosely coherent globules of iron, which are almost pure in themselves, but between which is a great quantity of slag. This mass of iron and slag is taken from furnace and compressed in a squeezing apparatus or under the hammer to force out the slag and weld the globules of iron together. However, the slag is never completely removed by this treatment. After rolling, the product is known as muck bar. To further eliminate slag and also to interweave the fiber of the iron, the muck bar is cut and piled, often at right angles, wired together, then heated to a welding heat and rolled again. This makes the merchant iron of commerce.

Effect of the Contained Slag.—The residual slag in the merchant iron will be as much as 1 or 2 per cent. The fibers of iron are surrounded by envelopes of slag which form planes of separation and so weaken the material by destroying cohesion between the particles of iron. Still, its effect is not altogether bad. It helps to produce a fibrous structure in the iron, as is shown in Fig. 16, and also is said to serve in retarding its crystallization.

The Properties of Wrought Iron as Compared to Low-carbon Steel.—Wrought iron costs from 10 to 20 per cent. more than the

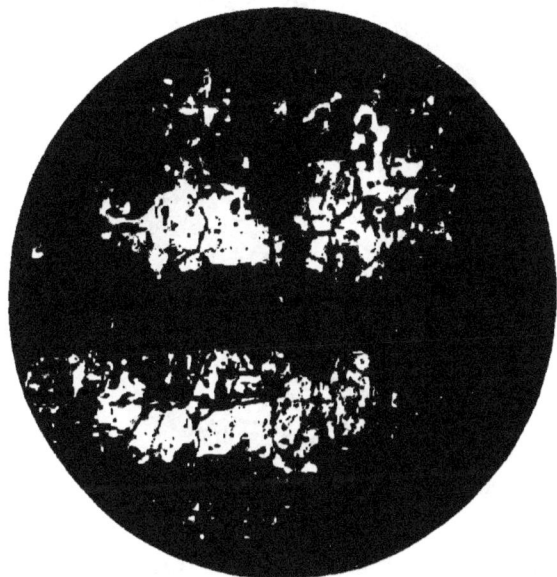

Fig. 17.—Wrought iron. Longitudinal section showing slag lines. Magnified 100 diameters (Boynton.)[1]

cheapest steel; hence low-carbon steel, in the form of pipe, bars, etc., is not infrequently sold under the name of wrought iron. But wrought iron can be distinguished from low-carbon steel by its content of slag. Stoughton states[2] that there is usually more than 1 per cent. of slag in wrought iron, and less than 0.5 per cent. in steel. Aside from this, wrought iron and steel may have identical chemical composition. If the temperature of the puddling furnace were sufficiently high so that its product would

[1] Reproduced by permission from Sauveur's Metallography of Iron and Steel.

[2] Metallurgy of Iron and Steel, p. 154.

be delivered in a molten state the slag would separate out, and the product would be called steel. Still normal wrought iron is as a rule practically free from manganese, while normal Bessemer and openhearth steel will contain 0.4 per cent. or more. Wrought iron generally contains more than 0.1 per cent. of phosphorus while steel, as a rule, does not.

By the users of these products, wrought iron is usually more desired than steel. The claim for the superiority of wrought iron

FIG. 18.—Wrought iron. Transverse section. Magnification not stated. (Guillet.)[1]

over steel equally low in carbon is based upon its greater purity and the presence in it of slag. The slag causes the iron to have a fibrous structure, and is said to increase its resistance to breaking when bent or subjected to sudden shock. Another very important claim that is made for wrought iron is that it will resist corrosion better than steel, although this is a point upon which there is much difference of opinion. It is very probable that there is less segregation of the impurities in wrought iron than in steel, which would cause less difference in electrical potential between the various portions of the iron, and this in turn would lessen its tendency to corrode (see page 132).

Wrought iron has a high electrical conductivity and can be

[1] Reproduced by permission from Sauveur's Metallography of Iron and Steel.

readily magnetized, but it does not long remain in the magnetized state. Its tensile strength is about 50,000 lbs. per square inch.

Wrought Iron of Inferior Grade.—All brands of wrought iron are not of equal quality. Stoughton says[1] that 50 per cent. of the American product is made by piling wrought-iron scrap, wiring together, then heating to a welding heat and rerolling. Since the scrap is usually indiscriminately collected and of non-uniform composition, and beside often contains intermingled steel scrap, the resultant product is not of high quality. One reason for its inferiority is the difference of potential between its various parts which accelerates corrosion, as has been said.

Ingot Iron.—The term "ingot iron" is sometimes used to designate very mild, soft or low-carbon steel. It differs from wrought iron in that it is obtained molten from the furnace and is therefore cast, while wrought iron is obtained in a pasty condition from the furnace and contains slag. Aside from the slag, ingot iron and. wrought iron may have identically the same composition.

Welding.—Because of the ease with which it can be welded, wrought iron is largely demanded by the forgeman. The weldability of iron products seems to be inversely proportional to their carbon content. Iron can be more readily welded by sprinkling over the parts to be joined, after heating, some borax or sand. By reacting with the iron oxide on the surface, these substances produce a readily fusible slag consisting of a borate or silicate of iron. The slag protects the iron from oxidation. As the parts are hammered together the slag is forced out and clean iron surfaces are brought into contact.

CARBON STEEL

Steel Distinguished from Other Commercial Forms of Iron.— The ordinary, so-called carbon steels, like cast iron, are alloys of iron and carbon, and chemically may be considered to be purified cast iron. They contain more iron and less carbon, silicon, sulfur and phosphorus, while the manganese may or may not be less than in cast iron. The following percentages may serve to furnish an idea of the chemical composition of the

[1] *Loc. cit.*, p. 154.

carbon steels: carbon from 0.05 to 1.75 per cent., depending upon the degree of hardness of the steel; silicon between 0.05 and 0.30 per cent. usually, but somtimes, as in cast steel, it may be as high as 0.60 per cent.; sulfur generally between 0.01 and 0.05 per cent.; phosphorus should not exceed 0.10 per cent.; manganese varies greatly, but is usually between 0.20 and 1.00 per cent.

The main distinguishing features of steel are indicated by the statement that it is malleable when cast. Having been cast, that is, taken from the furnace in the molten state, it is relatively free from slag and this distinguishes it from wrought iron.[1] Being malleable, distinguishes it from cast iron. And being malleable when cast, distinguishes it from malleable cast iron.

Manufacture.—There are a variety of processes by which steel may be made, but they are all based on the purification of the metal, usually by the oxidation of the elements it is desired to remove. The processes differ from each other only in the method by which this purification is accomplished. An exception may be considered in the making of cementation steel, in which case the purification is not considered a part of the steel-making process but is secured in a prior operation.

The processes may be indicated as (*a*) the Bessemer process, in which the converter may have either an acid or a basic lining, although the basic Bessemer is not used in America; (*b*) the openhearth process, both acid and basic; (*c*) the crucible process; (*d*) the cementation process; and (*e*) electric refining.

The description of the various processes for the purposes of this text will not need to be detailed. Only that which is considered necessary for a discussion of the characteristics of the product will be given.

The Bessemer Process.—This consists of blowing cold air through liquid pig iron, which may be either used molten, directly from the blast furnace, or may be melted in a cupola. The Bessemer converter is an egg-shaped steel receptacle lined for the acid process with refractory material composed chiefly of silica, and a little fireclay. The term acid is used because of the relative absence of lime, and magnesia—basic materials—

---

[1] Exception to this statement must be made in the case of cementation or blister steel which may be produced from wrought iron without melting.

in the lining, which fact demands an acid slag and so determines the chemistry of the process. The passage of the air through the molten iron in this acid process burns out the manganese, silicon and carbon, and the heat produced thereby is often so great that the addition of pig iron or scrap, or even steam, is necessary as a cooling agent. The heat is produced mainly by the burning of the silicon. In this acid Bessemer process, no phosphorus or sulphur is removed, and as no steel is allowed to contain over 0.1 per cent. of either, the pig iron used here must be of an exceptionally good grade; that is, it must contain less than 0.1 per cent. of each of these two elements. Because of the growing scarcity of iron ore capable of producing pig iron of this quality, the production of Bessemer steel has been greatly lessened within recent years. The openhearth product is taking its place.

Although it is possible to stop the blowing when the steel contains the required amount of carbon, this method is more difficult because tests are required in order to determine whether the amount of residual carbon is right, hence the blowing is usually continued until the carbon is practically all burned out, and then the required amount is added.

The carbon is introduced into the molten mass at the end of the process as a carbide contained in a manganese-iron alloy. The introduction of the manganese is necessary also in order that the steel may be sufficiently tough, when hot, for rolling. If an extremely soft steel, as for wire, material to be welded as in making pipes, tubes, etc., is desired, then ferromanganese, containing 80 per cent. manganese and about 6 or 7 per cent. of carbon is added. The required amount of manganese can thus be introduced without admitting too much carbon. If higher-carbon steel, as for rails, is wanted, melted speigel-eisen is added. This contains about 20 per cent. manganese and as much as 6 per cent. of carbon. Thus the required amount of carbon without too much manganese can be introduced. Manganese in carbon steel ranges usually from 0.2 to 1 per cent. depending upon the use to which the steel is to be put. A further very important work accomplished by the added manganese is the removal of the oxygen which always remains at the end of the blow in the molten metal, either as dissolved oxygen or as iron oxide. Unless

this is removed, defects occur in the finished steel. Further purification is secured by adding titanium in the form of a ferro-
titanium alloy to the molten steel.

The Bessemer process is the cheapest way for converting
iron into steel, and hence a great deal of the ordinary steel,
as for rails, wire, etc., was formerly made in this manner. In
1906 more than half of all the steel produced in the United States
was made in Bessemer converters, but in 1914 the Bessemer
product amounted to only about one-quarter of the total.

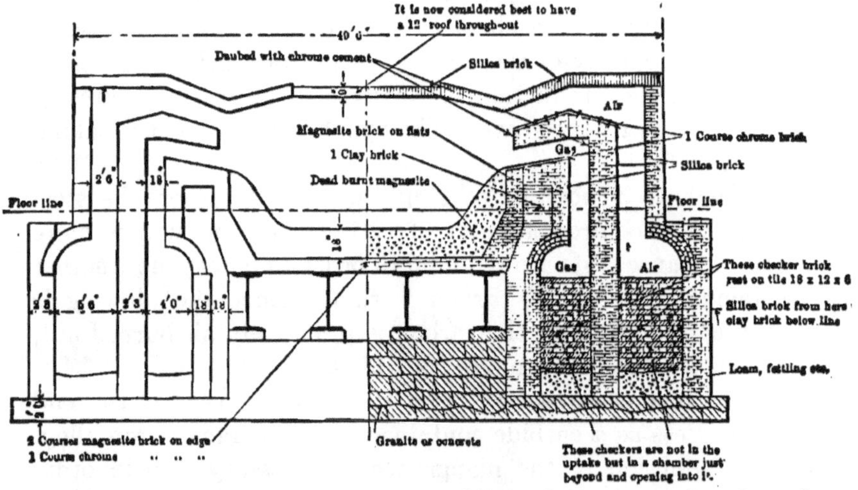

Fig. 19.—Materials of construction and lining for basic openhearth steel
furnace. (Stoughton.)

The Openhearth Furnace.—An openhearth furnace is one
with a hearth exposed to the flame, so that whatever lies upon
it is subjected to the direct action of the burning gases. The
openhearth must also be regenerative, since the required temperature cannot well be obtained in any other way. Regenera-
tion signifies that the air blast and fuel gas have been preheated
by passing through a checkerwork of heated brick. This
brickwork is previously heated by passing through it the products
of combustion on their way to the stack. Thus the brickwork
alternately stores up heat and in turn delivers it to the incoming
gases.

The Acid Process.—An acid openhearth furnace is one having
a lining of sand. If the lining is acid, the slag carried must be

acid, since a basic slag would destroy the lining. The character of the slag determines the chemistry of the process.

The charge in the acid openhearth furnace consists of pig iron and scrap. Nothing is added to form a slag, since the oxidation of the silicon, manganese and some iron, with sand from the bottom is sufficient. When the charge is melted it is fed with iron ore, which serves as an oxidizing agent to burn out the manganese, silicon and carbon. As in the acid Bessemer, no sulfur or phosphorus is removed in this process, and hence the charge must contain quantities of these elements no greater than is allowable in the finished steel.

When the required composition is obtained, which is judged by the melter and confirmed by analysis, the metal is tapped. Just prior to the tapping there is added either speigel-eisen or ferromanganese. This serves to remove or neutralize gases and oxides and also to introduce manganese and carbon. If especially high carbon is desired, it is added in the form of crushed anthracite or coke. Also, if special steels are being produced, the necessary elements, as silicon, chromium, etc., in the form of an iron compound or alloy are added at this time.

The Basic Process.—In the basic process the general principle is the same as in the acid process; *i.e.*, the oxidation and removal of the impurities through the slag or as gases. The chemistry of the basic process differs from that of the acid, in that by the basic process, phosphorus and to some extent sulfur can be removed. In the presence of lime or other basic mineral material contained in the basic slag, the oxides of sulfur and phosphorus act as acid anhydrides and form compounds with the basic material which are removed in the slag. But the use of lime for this purpose in an acid-lined furnace would destroy the lining, hence the necessity for a basic-lined furnace. It has been found that burned magnesite is most suitable for this basic lining, if cost is disregarded.

The ferromanganese, silicon or whatever deoxidizing material is being used cannot be added in the presence of a basic slag, but must be added as the molten steel is flowing from the furnace into the ladle. The reason is that the manganese, silicon and carbon will reduce the phosphorus from the phosphates contained in the slag and cause it to pass again into the molten steel. This

7

method allows less time for the deoxidizer and recarburizer to thoroughly mix with the molten mass, and so there is a greater chance for a lack of uniformity in the basic steel.

As shown in "Mineral Resources," the total amount of steel produced in the United States during the year 1914 was 23,513,030 tons, of which 17,174,684 tons were made in openhearths. Of the total openhearth product for this year, all but about 5.25 per cent. was made by the basic process. The reason is, of course, that so much of the material is not suitable for conversion by the acid process.

Cement or Blister Steel.—The cementation process is the oldest method for making steel. By it, wrought iron, or at times low-carbon steel is converted into high-carbon steel. The bars to be converted are packed in charcoal, sealed from the air, and heated to a temperature of from 650 to 700°C. In Sheffield, England, where this process is much used, 2 days are required for the furnace to reach the full heat, and the heat is then maintained from 7 to 11 days longer, depending upon the product desired. The carbon penetrates, or is absorbed by the iron at the rate of about ⅛ in. every 24 hrs.—a slow process. The saturation point is about 1.50 per cent. of carbon, but the steel when finished usually contains about 1 per cent.

The product is called cement steel because ferrite of the wrought iron is converted into iron carbide or cementite by the addition of the carbon. Also, it is called "blister" steel because as the carbon migrates into the wrought iron it comes into contact with the slag which is rich in iron oxide and a reaction occurs, producing carbon monoxide which causes blisters to form as the gas attempts to escape.

This steel is much used for tools bearing a cutting edge, but because the slag of the wrought iron still remains in the steel, the chance is great that the cutting edge will be imperfect. Hence, the steel is melted in crucibles to eliminate the slag. This double process greatly increases the cost of the product. Therefore, more commonly the carbon is introduced and the slag eliminated in a single operation known as the crucible process.

Crucible Steel.—In the crucible process the wrought iron (or sometimes low-carbon steel) is cut up into small pieces and melted in crucibles capable of holding about 100 lbs. In America the crucible is made commonly of a mixture consisting of about half graphite and half fireclay.

The charge for the crucible consists of the pieces of wrought iron, the required amount of carbon, usually in the form of charcoal, some ferromanganese, or speigel-eisen, and sometimes a small amount of glass or similar material to produce a neutral slag. If a special alloy steel is being made, then the special alloying element, as nickel, chromium, tungsten, molybdenum, etc., are also added. A cover is placed on the crucible and sealed with fireclay. These crucibles are then placed in a suitable furnace and heated from 3½ to 4 hrs.

The reactions in the crucible are slight as compared with other steel-making processes. The chief actions that occur are the elimination of the slag from the wrought iron, and the absorption of carbon and manganese, as well as some silicon from the walls of the crucible. When the melt lies quiet in the crucible, it is poured into ingots.

Crucible steel contains usually less than 0.40 per cent. Mn., more than 0.20 per cent. Si, less than 0.025 per cent. P, less than 0.03 per cent. S, and from 0.6 to 1.5 per cent carbon, depending upon the use to which the steel is to be put. It is used for cutting tools and drills for iron, brass and other alloys, rock drills, cutlery, wood-working tools, dies, files, springs, hammers, shafts, axes, and a great variety of other articles.

The amount of crucible steel produced is small in comparison to that produced by the openhearth and Bessemer processes. In 1914 the crucible steel manufactured in the United States was a little less than 0.4 per cent. of the total of all kinds.

The Electric Refining of Steel.—Bessemer and openhearth steel may be further treated in the electric furnace, and it has been found that the tensile strength of the Bessemer and basic openhearth product may be increased 10,000 lbs. per square inch by this treatment.

It is, of course, possible to produce steel direct from pig iron in the electric furnace, but the elimination of carbon, silicon and manganese can be far more cheaply secured by the other methods. Also, the removal of phosphorus is just as cheaply accomplished in the puddling and basic openhearth processes. Hence because of the higher cost, the electric process at present is largely limited to the super-refining of Bessemer and openhearth steel. However, it costs less than the ordinary crucible process, and it is believed by many that electrically refined steel is superior to crucible steel.

Because of the extremely high temperature that can be secured, and because of the neutral or reducing atmosphere that can be maintained in the electric furnace, the removal of sulfur, various oxides, and entangled particles of slag can be made almost complete.

Following are some of the points given by Stoughton wherein electric refining differs from other methods.

Sulfur may be reduced to a lower content than may ordinarily be achieved in the older processes. Likewise, dissolved gases, occluded slag, etc., may be eliminated from the liquid steel by means of heat instead of by means of manganese, silicon, titanium, etc. The impurities carried in by the usual fuels are avoided. Oxygen is not introduced into the furnace in which the operation is carried out.

The quantity of electrically refined steel produced in the United States has been gradually increasing within recent years. "Mineral Resources" shows that 13,762 tons were produced in 1909, and 69,412 tons in 1915.

Comparison of Steels[1]

The several kinds of steel usually exhibit certain differences in their properties that are dependent upon the process by which the steel was made. These peculiarities are often noticeable, regardless of the fact that the steels compared may be of the same composition; that is, in respect to the amounts of the elements usually shown by a chemical analysis of the steel.

Crucible and Openhearth Steels.—Crucible steel is the most expensive of the steels, and costs about three times as much as acid openhearth steel. Also aside from steel refined in the electric furnace, it is the best quality steel. It is believed that the reason for its superiority is that it is manufactured in a covered container which excludes the air and furnace gases, and so con-

[1] Stoughton: *Loc. cit.*, pp. 148 to 156.

tains less oxygen, hydrogen and nitrogen. Also, because its deoxidizer is added in the beginning of the process, the deoxidation is more complete and the mass more uniform. It is scarcely possible to get a gas-free "dead melt" of this sort in the openhearth furnace, hence the openhearth product will be likely to contain more blow-holes in the ingot and resultant seams in the steel when rolled. Also the ingots of steel poured in the openhearth plant are very large compared with those of the crucible works, and the chances for segregation are greater in the larger ingot.

Acid and Basic Openhearth Steels.—Since no effort is made to take out the somewhat difficultly removable sulfur and phosphorus in the acid process, the operations are shorter here than in the basic process. The charge is therefore for a less time exposed to the oxidizing conditions of the furnace, the slag is less highly oxidized, and the molten steel is more free from occluded gases and oxides at the end of the process. Hence there is less trouble from blow-holes in the ingot. These blow-holes roll out into longitudinal flaws and are considered as being responsible for the starting of cracks. It is for this same relative freedom from blow-holes, that acid, and not basic steel, is used for the making of steel castings.

Another point in favor of the acid steel is that it is likely to be more uniform, since there is a better chance for the proper mixing of the deoxidizer and recarburizer with the steel. In the acid process the deoxidizer is added while the steel is in the furnace, and in the basic process after it has been poured into the ladle. The ingot is cast by pouring from this ladle, so that thorough mixing is much less probable.

The phosphorus and sulfur are lower in the basic than in the acid openhearth steel, but this seems to be more than offset by the defects caused by the occluded gases and oxides, and the nonuniformity produced by the method of deoxidizing and recarburizing.

Because of the more expensive material for lining and slag, the operation of the basic openhearth is more expensive than the acid. Also, because of the longer period, more fuel is consumed. But because of the cheaper stock used in the charge, basic steel is cheaper than that made by the acid process.

Basic Openhearth and Bessemer Steels.—It is believed that the basic openhearth is superior to Bessemer steel because the Bessemer will contain more oxygen, nitrogen and other gases, on account of the intimate mixture of air in its making. This is especially true if the process is continued a few seconds too long.

In the Bessemer process all the carbon is removed and then the total amount desired is added. In the openhearth, the carbon is removed to a point only a little below the desired amount, and enough is then added to bring the total up to the desired point. The greater amount of recarburizer used in the Bessemer, tends to cause the product to be less uniform, because the larger addition may not distribute itself uniformly.

Then, in general, it may be said that the following steels stand in this order in respect to cost and quality: (1) crucible, (2) acid openhearth, (3) basic openhearth, and (4) Bessemer.

CONSTITUTION OF STEEL

As has been stated previously, *ferrite* is the term used to designate theoretically pure iron. It makes up certain percentages of all forms of iron and steel, but occurs in especially large proportions in wrought iron and low-carbon steels. Like certain other elements, sulfur for example, which undergoes allotropic variations, existing as rhombic crystals below 96.5°C., as monoclinic crystals above 96.5°, and as plastic or amorphous sulfur when cooled rapidly from just below its boiling point, so ferrite exists in certain allotropic modifications. There are three of these allotropic forms, designated as alpha, beta and gamma iron.

The alpha form of iron is highly magnetic and is the one that exists at atmospheric temperatures. If alpha iron is heated to 750°C. it undergoes a sudden change. The change is indicated by an absorption of heat in a manner similar to the absorption of heat unattended with a rise in temperature, which occurs when ice is being converted into water. The iron is now said to be in the beta form. It is *non-magnetic,* and also has a changed electrical conductivity and specific heat. If now the beta iron is further heated to 875°C. it changes into the gamma form. Gamma iron differs from the beta modification, especially in electrical conductivity and crystalline form. These changes are reversible, for upon cooling from above 875°, the transformations occur in the opposite order. But upon cooling, there seems to be a little lag, so that the successive changes occur at temperatures about 30° lower than those given for the transition points with a rising temperature.

Gamma Iron in Austenite.—Since the carbon steels melt between 1,400° and 1,500°C., it is obvious that the ferrite of the molten steel is in the gamma form. Also molten steel consists of molten ferrite and molten iron carbide dissolved in each other. It was shown in Fig. 7, to the left of the line XY, that the iron carbon alloys that are known as steels, contain less than 2.2 per cent. of carbon and solidify as solid solutions. The solidification range, as shown in this figure, is included between the lines AB and AI. Now, the temperature of solidification of the steel is far above the point where beta iron turns into gamma iron, and consequently when the solid austenite is produced, its ferrite is in the gamma form. And even though the austenite is cooled below the transition point of gamma iron, it seems that as long as the solid solution is not decomposed, the ferrite remains in the gamma form.

Decomposition of the Solid Solution.—When the solid solution decomposes, it separates into iron carbide and ferrite. The manner in which it decomposes is indicated in Fig. 20, where the solid solution is shown to exist within the range above the line $GOSI$. Now, as was the case with liquid solutions, the temperature at which the decomposition begins, depends upon the amount of carbon present. If, for example, we start with a steel containing 0.3 per cent. of carbon, which is equivalent to about 4.5 per cent. of iron carbide, it will cool without change until it meets the line GO, which will be at the temperature of about 810°C. But it has been found by experiment that this is the lowest temperature at which the solid solution will contain as much ferrite as this, if it is slowly cooled. As the temperature falls further, ferrite separates out, thus proportionately increasing the amount of iron carbide in the solution. This precipitation of ferrite continues until the remainder of the solid solution contains sufficient iron carbide to equal about 0.9 per cent. of carbon, as is indicated by the point S. And in like manner ferrite would separate out from any other steel containing less than 0.9 per cent. of carbon, if it were slowly cooled from above the line GOS.

If, on the other hand, the steel should contain more than 0.9 per cent. of carbon, when it is slowly cooled from above the line IS, iron carbide will separate out. This separation so increases the proportion of ferrite in the remainder of the solid solution,

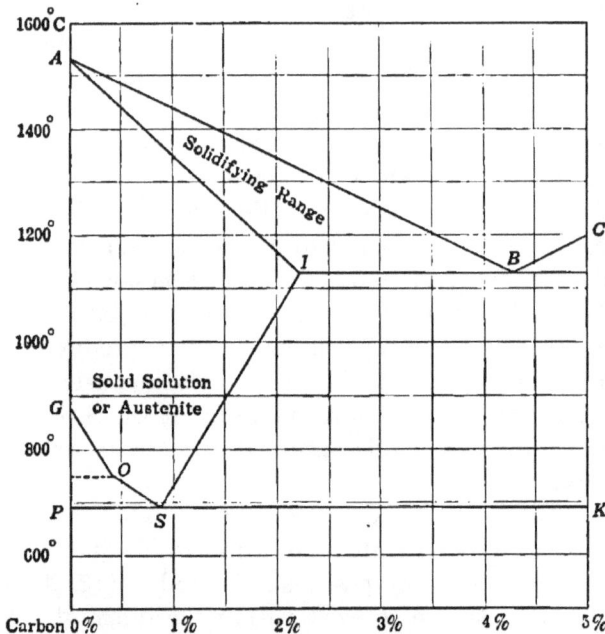

FIG. 20.—The freezing-point and solid-solution-decomposition curves of the iron-carbon alloys.

as is indicated by the line IS, that when by continued separation the point S is reached, this remainder contains 0.9 per cent. of carbon. It is evident from this, that the solid solution containing 0.9 per cent. of carbon has the lowest decomposition point, and since in this respect, it resembles the eutectic of a liquid solution, or alloy, it has been named the *eutectoid* alloy of iron and carbon. Also, steel containing this amount of carbon is known as a eutectoid steel. As a result of the selective separation, all steels, when slowly cooled from above the line GOSI, to about 690°C. (the point S), will contain a certain amount of the eutectoid alloy. This will be mixed with either crystals of ferrite or cementite, depending upon the amount of carbon in the steel.

It has been seen that the solid solution cannot cool below the line *GOS* without precipitating ferrite, and it cannot cool below the line *IS* without precipitating cementite. But as it further cools and passes the point *S* it crosses both lines. Therefore, the remainder of both the ferrite and cementite that now exist in the eutectoid solid solution are separated at this point. These two constituents separate from each other as fine crystals and may arrange themselves in the form of a banded structure, as shown in Fig. 21, to which the name *pearlite* has been given, be-

Fig. 21.—Eutectoid steel. Showing laminated pearlite. Magnified 750 diameters. (Goerens.)[1]

cause under certain conditions, it has an appearance resembling mother-of-pearl.

In Fig. 22 is shown the structure of a steel made up of grains of ferrite and pearlite. Since the carbon is only half that needed for a completely eutectoid steel, the excess ferrite has crystallized out in the form of separate grains. Such steels as this, that contain less than the eutectoid amount of carbon, are called hypoeutectoid (*hypo*, less than) steels. Fig. 23 shows a hypo-eutectoid steel of similar structure, but insufficient magnification to show the lamellar nature of the pearlite. In Fig. 24 the structure

[1] Reproduced by permission from Sauveur's Metallography of Iron and Steel.

of a steel containing more than the eutectoid amount of carbon
is shown. Here the grains of pearlite are surrounded by free
cementite. Such steels, containing above the eutectoid amount
of carbon, are known as hyper-eutectoid (*hyper,* over) steels.

The Constituents of Hardened and Tempered Steels.—It
has been determined that when the austenite of the eutectoid

Fig. 22.—Steel containing 0.45 per cent. carbon. Ferrite and pearlite.
Magnified 1000 diameters. (Osmond.)[1]

alloy is converted into pearlite, the change does not take place
spasmodically, as might seem to be indicated by the point *S,*
but in stages. In fact, the changes that take place, in the region
of which the point *S* may be considered the center, extend over
a range of at least 30°C., beginning approximately 15° above the
point *S,* and ending a like number of degrees below it.

In discussing Fig. 20, *GOSI* was considered as a line, and any
point upon it corresponding to a given percentage of carbon, was
looked upon as the *critical point or temperature* for a steel con-

[1] Reproduced by permission from Sauveur's Metallography of Iron and
Steel.

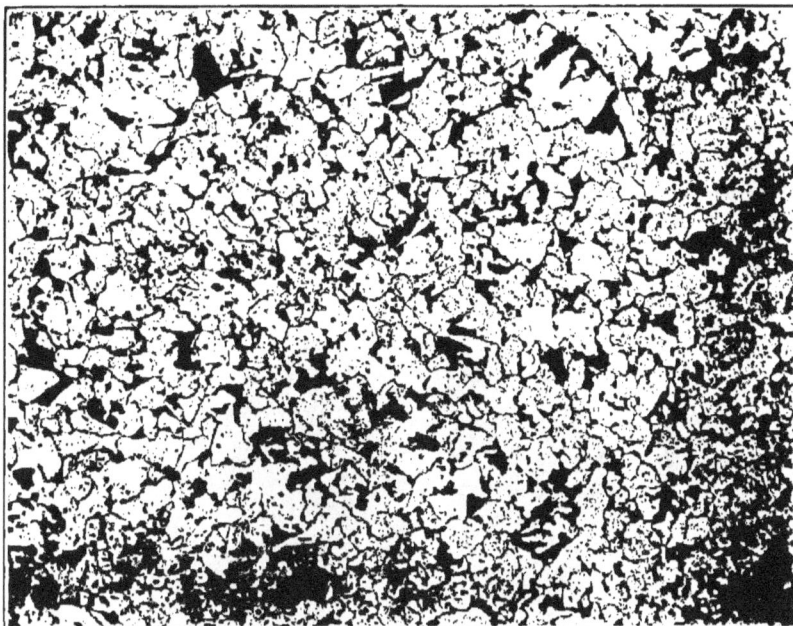

Fig. 23.—Steel containing 0.20 per cent. carbon. Light areas ferrite, dark areas pearlite. Magnified 100 diameters. Etched.

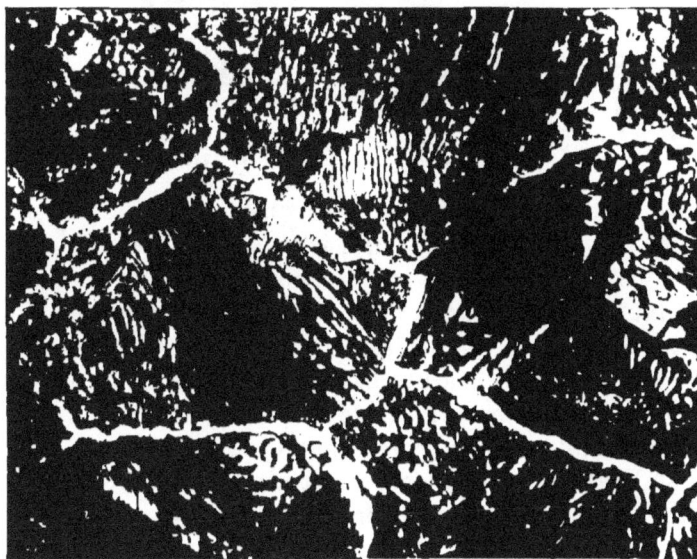

Fig. 24.—Steel containing 1.43 per cent. carbon. Pearlite and cementite. Magnified 500 diameters. (Boynton.)[1]

[1] Reproduced by permission from Sauveur's Metallography of Iron and Steel.

taining that percentage. But since, as has just been explained, the structural changes do not in reality take place sharply, the critical temperature is more properly spoken of as the *critical range*. Hence the structural change or transition from the solid solution, austenite, to the mechanical mixture, cementite and ferrite, called pearlite, is comparatively gradual, and can be caused to stop at certain points by rapid cooling. At least three incomplete stages of transition are commonly recognized, and the constituents at these stages are known as transition substances.

Fig. 25.—Steel containing 0.45 per cent. carbon. Martensite and troostite. Magnified 1000 diameters. (Osmond.)[1]

Named according to the degree to which the austenitic decomposition has proceeded, the series is as follows: austenite, martensite, troostite, sorbite, and pearlite, all of which can be recognized in suitably prepared specimens under the microscope. Photomicrographs are shown in Figs. 25 to 27. The names of all these constituents, except pearlite have been derived from the names of noted investigators who have worked upon this subject.

Although all of the substances, except pearlite, exist naturally in a stable form only at comparatively high temperatures, it is possible, by rapid cooling under certain conditions, to cause them to be preserved until atmospheric temperatures are reached.

[1] Reproduced by permission from Sauveur's Metallography of Iron and Steel.

Austenite will then be found in steels most rapidly cooled. However, to preserve austenite it is necessary that the carbon be above 1.1 per cent. Martensite will be found in those cooled with the next degree of rapidity, troostite in the next, sorbite in the next and pearlite in those that are allowed to cool slowly.

Characteristics of the Constituents of Steels.—Austenite is very hard but *martensite is harder*. A polished surface of aus-

Fig. 26.—Hot-worked steel. Carbon 0.50 per cent. Magnified 650 diameters. The ill-defined constituent is sorbite.[1]

tenite can be plainly scratched with a steel needle, while a polished surface of martensite will be unaffected. Austenite does not occur in tempered carbon steels, but martensite will be found to a great extent in the harder varieties. Troostite is also softer than martensite and occurs in great amount in tempered steels, especially in those quenched in boiling water, oil and similiar less active tempering media, as well as in steels that

[1] Reproduced by permission from Sauveur's Metallography and Heat Treatment of Iron and Steel, Ed. 2.

have been quenched and then tempered by reheating. Sorbite is still softer than troostite, being very close to pearlite in hardness. Pearlite is the softest of all. It occurs in slowly cooled or annealed steels. It is made up of alternate flake-like layers of alpha ferrite and cementite.

Hardening of Steel.—In heating carbon steel, pearlite persists or is stable, at any point below 690°C., *i.e.*, the point S. Hence

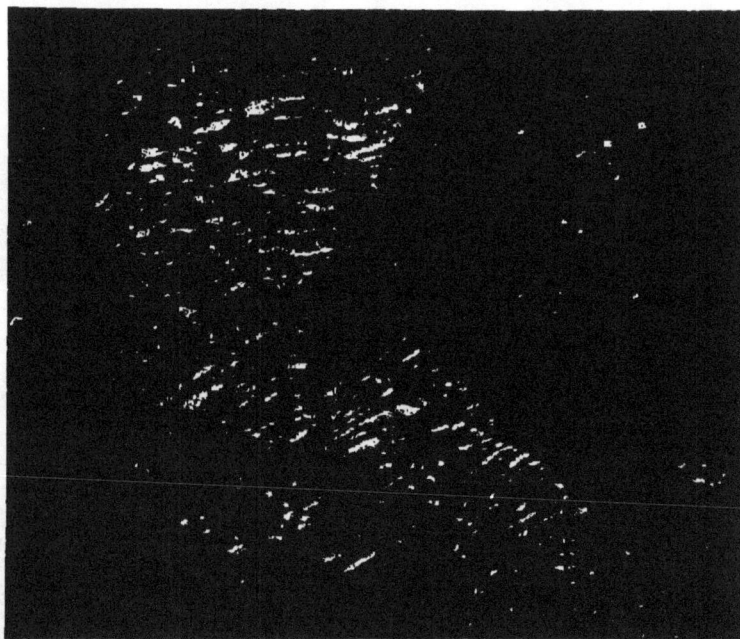

Fig. 27.—Steel containing 1.00 per cent. of carbon. Pearlite (laminated) passing into sorbite. Magnified 1500 diameters. (Osmond.)[1]

it is evident that a steel to be hardened, must be heated to above that temperature in order that the pearlite may be converted into austenite. If a steel containing 0.9 per cent. of carbon is heated to *just below* the point S in the diagram, no matter how rapidly it may be cooled it will be in the unhardened state. But if it is heated to above the point S, even though but slightly, austenite will be formed, and then if it is rapidly cooled past the point S, the complete change into pearlite will be prevented; that is, either martensite, troostite or sorbite will be preserved, and the steel will be more or less hard. (It is very difficult to

[1] Reproduced by permission from Sauveur's Metallography of Iron and Steel.

obtain the austenitic condition in ordinary carbon steel at atmospheric temperatures.) The degree of hardness for a steel containing 0.9 per cent. of carbon, is not dependent upon the distance above the point S to which it may be heated. *Maximum* hardness may be obtained in this case by cooling with the greatest rapidity from *just above* the point S. Beside a distinct ill effect arises from heating to a higher temperature, since the crystal size is thereby increased.

The proper temperature from which a steel should be quenched to secure best results in hardening depends upon the amount of carbon it contains. Thus, for steels containing less than 0.9 per cent. C, maximum hardness and best crystal size may be obtained by quenching from just above the line GOS. For steels containing over 0.9 per cent. of carbon, maximum hardness results from quenching from just above the line IS. But since heating to the line IS produces large crystal size, it has been found that for these steels best general results are secured by quenching from just above the line SK.

A very reliable check to prevent excessive overheating can be had by noting at what point in the heating the steel loses its magnetism. This occurs when alpha ferrite changes into the beta form, *i.e.*, at 750°C. This will not serve, however, for steels containing less than 0.4 per cent. of carbon since they must be heated above the line GO.

Steel in the excessively hardened state, such as is obtained by quenching in water, is also very brittle—too brittle for use—and, therefore, some of this brittleness must be removed. In lessening the brittleness some hardness is also unavoidably sacrificed. The process whereby the brittleness and hardness are lessened is known as tempering.

Tempering.—The austenite, martensite, troostite and sorbite of the hardened steels are not stable in the ordinary sense at atmospheric temperatures. However, these constituents are prevented from decomposing in the hardened steel by the rigidity or immobility of the particles of the mass. Therefore, to remove brittleness and soften the steel it is only necessary to impart a certain amount of mobility to the molecules. This can be accomplished by application of heat, but since the degree of softening must be regulated, the heat must be quite accurately controllable. In the treatment of edged tools by the water-quenching process, only the tip is quenched, and then the desired amount of heat is allowed to flow into the tip from the uncooled shaft. By noting the color of the oxide developed upon the cleaned tip the degree of heating may be judged, since the color of the oxide changes with the depth of the layer, and this in turn bears a definite relation to the amount of heat applied. When this secondary heating has proceeded to the desired point the influx of heat is stopped by quenching the whole of the hot portion.

The result of the tempering is, then, that the austenite or martensite of the steel is changed into some softer and less brittle form. The first heating causes the ferrite and pearlite of the steel to dissolve in each other, and the quenching keeps them dissolved, at least, in large measure, by not allowing time for the separation to occur. Then, as heat is again applied, this interrupted separation is resumed, and proceeds to an extent that depends upon the degree of heat and the time during which it is applied to the quenched steel.

It should be mentioned here, that because of the longer period during which the shaft of the tool is hot its crystal size may be much increased, and on this account, fractures not infrequently occur at this point.

Combined Quenching and Tempering.—Instead of preventing the decomposition of the solid solution as far as possible by the rapid cooling in water, and then allowing a secondary influx of heat so that the decomposition may be resumed, the same result may be obtained by so regulating the rate of cooling that the decomposition proceeds to just the desired point in the first cooling. By choosing the proper cooling medium, any degree of hardness may in this manner be secured. Cooling media used in this way are molten solder, fused salts, oils of varying viscosities, solutions of salts, etc. The results secured by the use of these materials depend only upon the rate at which they are able to remove the heat from the steel. The steels are cooled rather rapidly past their critical temperatures, but still not so rapidly but that the proper amount of austenite decomposition can take place.

These baths cool the steel less rapidly than water because

of various factors, such as: (1) the sensible heat of the bath, (2) the heat conductivity of the bath, (3) the viscosity of the bath (liquids carry heat largely by convection), (4) the specific heat of the bath, (5) the volatility of the liquid (a cushion of vapor, which is a poor conductor, may form about the steel).

Overheating (Commonly Called "Burning") of Steel.—Overheated steel may be really oxidized, *i.e.*, burned, or it may not. Overheating is recognized as existing in three stages. The first stage exists when drops of melted metal have begun to

FIG. 28.—Burnt steel. Carbon 1.24 per cent. Magnified 20 diameters. Quenched at a white heat. Unetched. (Osmond.)[1]

form in the interior of the mass. This melted metal segregates or collects between the crystal faces and causes weakness. The second stage is reached when the molten drops segregate as far as the exterior and leave behind cavities filled with gas. The third stage occurs when sufficient gas is formed to force small drops of liquid steel through the surface of the piece. This produces the well-known scintillating effect seen at this temperature. The gas which collects is probably CO, produced by the action of the occluded oxygen upon the carbon, or it may be the occluded N or H which steel almost always contains.

[1] Reproduced by permission from Sauveur's Metallography of Iron and Steel.

8

Into the cavities left by the forcing out of the molten portion, air goes, producing ferrous oxide on the surface of the cavity. The crystal faces are then in reality burned. Steel in this last condition is useless. Burned steel, because of the high heat to which it has been raised, is usually also coarsely granular in fracture.

Case Hardening.—When the manufactured steel article is required to have a tough core but a hard surface, it is case-hardened. The action in the case-hardening process is the same as in the production of cementation steel, except that the carbon enters to only a very slight depth, and the process requires but a few hours instead of several days.

The articles to be case-hardened are packed usually in nitrogen-bearing carbonaceous material, such as charred leather and other animal products, or in cyanides or ferrocyanides since best results are produced by such materials. Steels suitable for case hardening are those containing about 0.15 per cent. carbon and not above 0.25 per cent. manganese since this element has a tendency to produce brittleness in the case. In case hardening it is necessary to heat the steel to above its critical temperature, which means for 0.15 per cent. carbon, about 900°C. Also, the carbonizing seems to proceed best when the iron is in the gamma condition. The steel apparently acquires carbon chiefly from the gases, carbon monoxide, CO, and cyanogen, CN, probably according to these equations:

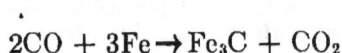

$$2CO + 3Fe \rightarrow Fe_3C + CO_2$$

and

$$2CN + 6Fe \rightarrow 2\ Fe_3C + N_2$$

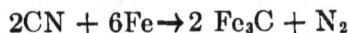

In this manner the carbon in the case will be raised to about 1 per cent.

ALLOY STEELS

The steels known as alloy steels are those that owe their properties, in a very large measure at least, to the presence of one or more elements other than carbon, although carbon is always present; whereas the properties of the older type of steels depend largely upon the percentages of carbon they contain.

For distinction then, in commercial terminology these latter steels are designated as *carbon steels*, and the former as *special or alloy* steels.

The alloy steels are divided into two classes. Those that are essentially made up of iron, carbon and one special element, as nickel, manganese, chromium, tungsten, and others, are called *ternary alloy steels*. Those that contain essentially iron, carbon and two special elements, as nickel-chromium, tungsten-manganese, tungsten-molybdenum or other combinations are known as *quaternary alloy steels*.

The results obtained by the addition of the special element are usually very marked and often very unexpected, so that the properties of the resultant steel cannot accurately be forecast, but must be determined by trial. This is especially true of the quaternary steels, for in some cases the effect of the two elements working in conjunction is directly opposed to the effect of each element alone in steel. The whole subject of special steels is still quite new and much remains to be done, both in preparing new combinations as well as in further investigating the properties of those already prepared.

The process of manufacture is practically the same as for the making of the more ordinary carbon steels. They are usually made in the openhearth, crucible or electric furnace. The special elements are added commonly at the time when the deoxidizer or recarburizer is added in the normal practice for each of the processes mentioned. An exception should probably be made in the case of nickel which is usually added in the early stages of the openhearth treatment since it has no great tendency to be lost through oxidation.

Structure and Properties.—Guillet has investigated the structure and properties of the alloy steels, and the following statements concerning ternary steels are based on his results, as quoted by Sauveur.[1]

If into a carbon steel, which when slowly cooled is pearlitic, a small amount of the special element is introduced, it will at first remain pearlitic, but, if the carbon remains constant, and the special element is increased in amount, the steel will become martensitic, then austenitic and finally cementitic, depending

[1] Metallography of Iron and Steel, Lesson xvii.

upon the amount introduced. These terms are used to signify that the steels naturally assume these structures as their stable condition when slowly cooled in air. *Cementitic steels* are those in which the austenitic condition does not appear, but in its stead there is a structure made up of particles of the double carbides of iron and the special element contained in a matrix of martensite, troostite, sorbite or pearlite. A similar result to that just described as being accomplished by increasing the amount of the special element may be obtained by keeping the special element at a constant percentage and increasing the quantity of carbon. Hence as either of these two constituents is increased in amount the less will be the quantity of the other needed to produce a different structure. In brief, the desired structure may be obtained by increasing either carbon or the special element, but it may be obtained more quickly by increasing both. As is to be expected, the alloy steels may be caused to assume, by suitable heat treatment, other structural forms than those which they acquire when slowly cooled in air. For example, a martensitic steel may be made austenitic by quenching and pearlitic by cooling slowly in the furnace.

The fact that the special elements are able to produce in steel the austenitic or martensitic condition on slow cooling depends upon an ability they have to lower the critical temperature to such an extent, that before this point is reached, the mass has become so rigid that the decomposition is either prevented altogether or else is inhibited to such a degree that only the martensitic stage is reached. The conclusion drawn from many observations seems to be, according to Sauveur, that when the critical temperature remains above 300°C. the steel when slowly cooled will be pearlitic, because at this temperature the particles are sufficiently mobile that the transformation can take place. When the critical temperature is below 300°C. the steel will be martensitic. When the critical point is lowered to atmospheric temperatures or lower, the structure will not be decomposed at all, and hence will be austenitic.

TYPES OF ALLOY STEELS

On the basis of the preceding explanation the special steels are divisible into at least four classes: (1) cementitic, (2) austenitic, (3) martensitic and (4) pearlitic.

Cementitic Steels.—Certain elements, for example, chromium, tungsten, molybdenum and vanadium when present in steel in sufficient quantity, cause crystals of the double carbide of iron and the special element to separate out when the steel is slowly cooled. These crystals may be set in a matrix of pearlite or some harder constituent and then exhibit a structure such as is shown in Fig. 29. The cementitic steels have less strength and ductil-

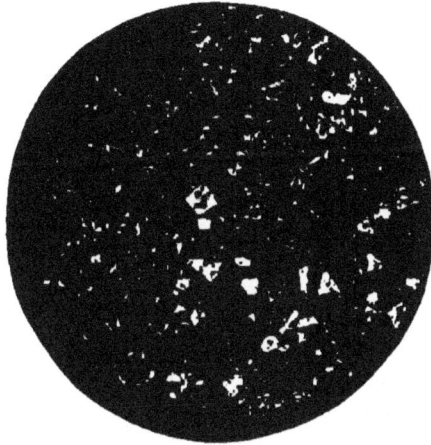

FIG. 29.—Showing the structure of cementitic tungsten steel. Tungsten 39.96 per cent. Carbon 0.867 per cent. The white crystals are the double carbide. Magnified 200 diameters. (Guillet.)[1]

ity and more brittleness than austenitic steels. However, when such steels are heated to about 1,000°C. so that these carbides go into solution, and are then cooled at a proper rate so that the carbides are held dissolved, the steel has the remarkable property of retaining its hardness at a red heat, and so may be used for "high-speed" tools.

Austenitic Steels.—The austenitic special steels have properties very similar to those of the austenitic carbon steels, *but unlike the quenched austenitic carbon steels, they are stable at all temperatures below their solidifying point,* and hence are only with difficulty affected by heat treatment. Austenitic steels are moderately tough, but are very ductile and hence have a low elastic limit. (By the term elastic limit is meant the number of pounds of stress per square inch needed to produce the first

[1] Reproduced by permission from Sauveur's Metallography of Iron and Steel.

permanent deformation.) They have a great resistance to wear by abrasion and to rupture by shock. They are very hard. Since austenite does not contain alpha iron these steels are *non-magnetic.* Their chief objectionable features are their low elastic limit and the extreme difficulty with which they may be machined. Some steels that are austenitic when cooled in air become martensitic or troostitic when slowly cooled in the furnace.

Martensitic Steels.—These steels generally contain less carbon and special elements than the austenitic steels. They are very hard—harder than the austenitic steels. Also they are brittle, which is one of their chief disadvantages. Beside, because of their hardness, it is almost impossible to machine them.

Some martensitic steels can be made pearlitic by slow cooling in the furnace and austenitic by quenching in water. This is a distinct advantage since they may be machined in the pearlitic form and hardened by cooling in air, which eliminates the risk of cracking during quenching and tempering. Because some alpha iron is present in the martensitic structure, these steels are at least mildly magnetic.

Pearlitic Steels.—The pearlitic special steels have a higher elastic limit and are frequently harder than the pearlitic carbon steels. Also, they are usually tougher and better able to resist rupture by shock. These superior qualities are probably due, partly at least, to the fact that the pearlitic structure is made up of finer layers in the special steels.

By increasing the amount of carbon in the exterior by case hardening, the pearlitic special steels may be given a hard martensitic or tough austenitic shell when cooled in air. In this way the risk of fracture which attends quenching from above the critical temperature is avoided.

Common Pearlitic Alloy Steels

Nickel Steel.—The nickel steels are the most used of all the alloy steels. They may contain as much as 42 per cent. of nickel for special uses, but the steels of the greatest commercial importance contain about 3.50 per cent. of nickel and about 0.5 per cent. of carbon. These are pearlitic. Within the range of 10 to about 28 per cent. the steels are martensitic and with more nickel than this they are austenitic.

The 3.50 per cent. nickel steels have the properties described under the general head of pearlitic steels.

According to Stoughton, it is probably due to the high elastic limit of nickel steel, that it so well resists that which is called "fatigue." This means that it resists rapidly applied, repeated and alternating stresses. All steels ultimately break down under such stresses even though the load is far less than the steel could bear indefinitely if constantly applied. The 3.50 per cent. of nickel increases the resistance of steel to such stresses about six times. Also it greatly increases the tensile strength.

The crystalline structure of nickel steel is much finer than that of carbon steel, and this is probably the reason for its increased toughness. Microscopic cracks, if started, develop much more slowly than when the crystals are larger, and the cleavage planes therefore greater.

Nickel steel is much harder than carbon steel, although it can be machined without difficulty. Because of this hardness, it is excellent for use on railways curves where abrasion is great. Nickel steel has also a low coefficient of friction, and on this account is well suited for making axles for automobiles, etc.

It has a very great shearing strength and so is well adapted for use in making rivets. Nickel segregates very little and seems also to lessen the segregation of other elements. Nickel steel has a lower melting point than the ordinary carbon steel and is easily cast, making exceptionally sound castings. It welds less readily than carbon steel.

When the amount of nickel present is about 36 per cent., the coefficient of expansion is practically zero. This metal is called "Invar" and is used for pendulums, surveyors tapes, etc. With about 42 per cent. of nickel, the steel has about the same coefficient of expansion as glass, so that it was at one time used in electric light bulbs, for fusing into the glass, where the joint must be kept air-tight. Such metal has been called "Platinite," because originally platinum only was used here. Platinite has not proven entirely suitable for lamps and now a compound wire, having a 38 per cent. nickel-steel core encased in copper and sometimes coated with platinum is employed. Nickel steel is used in the "armored glass," so much employed for fireproofing.

Nickel-steel corrodes very little. For this reason, with 30 per cent. of nickel, it has been found very serviceable in making boiler tubes.

In general, then, this steel is very suitable where exceptional strength, hardness, toughness, abrasion resistance, and corrosion resistance are required.

Chromium Steels.—Depending upon the amount of chromium present, the chrome steels may be pearlitic, martensitic, or cementitic; but those generally prepared contain under 3 per cent. of chromium, and from 0.8 to about 1.5 per cent. of carbon and are pearlitic upon slow cooling. Chromium seems to affect the position of the critical point but little. But a noteworthy effect that it does have, is that it causes the structural transformation to take place more slowly, so that the steel can be easily made martensitic by quenching.

Because of its hardness, chrome steel is used for files, safes, plows, etc. Also because chromium considerably increases the elastic limit and notably lessens the liability to rupture by shock, this steel is much used for armor plate and the wearing parts of crushing machinery. However, for armor plate it is used as a quaternary steel with nickel. The nickel-chrome alloy is also used for automobile parts, as crankshafts and gears.

Vanadium Steel.—Vanadium seems to have but little effect upon the critical temperature of steel. Any steel that contains less than 1 per cent. of vanadium will be pearlitic. In the commercial steels, the vanadium scarcely ever exceeds 0.2 per cent. and hence they are pearlitic when slowly cooled. With the single exception of carbon no element is so powerful in its effect on steel; even 0.1 per cent. will produce very noteworthy results. Ordinarily the amount ranges from 0.12 to 0.15 per cent., with carbon from 0.25 to 0.55 per cent. and manganese from 0.5 to 1 per cent. Usually, too, about 1 per cent. of chromium or nickel is used in the vanadium steels and so they may be considered quaternary.

A very noteworthy point is that the vanadium is an exceedingly active deoxidizer, so that it lessens blow-holes and apparently segregation as well. Because of the great readiness with which it combines with oxygen and perhaps with nitrogen also, it must be added to the steel only after the addition of manganese and carbon has been made. It would be lost otherwise.

The properties of the pearlitic vanadium steels are in many ways similar to the pearlitic nickel steels. They have a high combination of elastic limit and ductility, are very hard and tough and seem to be especially able to resist shock and vibratory stresses. Unlike nickel steel, vanadium steel welds readily.[1]

Silicon Steel.—Silicon has no very noticeable effect on critical temperature. It exists as a solid solution with the ferrite, probably first forming a silicide, $FeSi$, which in turn dissolves. If the silicon does not exceed 5 per cent., the steel is pearlitic, the carbon being combined, but with more than 5 per cent. silicon, some graphite is formed. Hadfield, who first produced this steel, recommends as the best alloy that which contains 2.75 per cent. silicon and the smallest possible amounts of carbon, manganese and other impurities.

The most important uses of this steel depend upon its magnetic properties. It has a higher degree of magnetic permeability than the purest iron. Its hysteresis is low; that is, it does not persist in a state of magnetism after the magnetizing force has been removed. For these two reasons it is very useful in constructing electromagnets and electric generating machinery. It also has a high electric resistance, and this lessens the effect of the "eddy currents" which detract from efficiency.

The chief *structural* use of silicon steel is in making the leaf type springs such as are used for automobiles.

SELF-HARDENING STEELS

Certain of the alloy steels remain austenitic upon being slowly cooled in air. They owe this property to the fact that the special element or elements they contain, lower the critical temperature to some point below the temperature of the atmosphere. On this account they cannot be very appreciably softened by heat treatment. *Being austenitic, these steels are all non-magnetic.*

Manganese Steel.—The usual manganese steel is that which contains about 12 to 14 per cent. of manganese and about 1.5

[1] For properties of vanadium steel, in detail, see pamphlets issued by the American Vanadium Co.

per cent. of carbon.   This steel when cast and slowly cooled is hard, brittle and non-ductile.   If it is heated to 1,000°C. or more and then is quenched in oil or in water, it will have pronounced ductility and toughness.   This toughening by quenching is interesting since it is the very opposite effect from that produced by similarly treating carbon steel.   Although tough, the steel

Fig. 30.—Manganese steel.   Austenitic.   Cast.   Magnified 50 diameters.[1]

when quenched is practically as hard as before, so that it cannot be machined or drilled and there seems to be no way of making it softer.

Manganese steel has a low heat conductivity and so must be heated slowly and uniformly lest it crack.   It is very difficult to forge and roll successfully.   This is particularly true of the high-carbon manganese steels.   When Hadfield first produced manganese steel it was impossible to produce a manganese steel low in carbon, since the only ferromanganese available was the product of the blast furnace and this was practically saturated with carbon.   But since manganese, practically carbon-free,

[1] Reproduced by permission from Sauveur's Metallography of Iron and Steel.

has been produced in quantity in the electric furnace and by the Thermit process, low-carbon manganese steels have been produced and these are relatively much more workable. Manganese steel has a rather low melting point, about 1,360°C.

It is used wherever extreme hardness and abrasion resistance is required, for example, the wearing parts of rock-crushing machinery, the teeth of steam shovels, railroad and trolley frogs, crossings and curves, mine-car wheels and burglar-proof safes. Its uses are somewhat limited by the fact that it must usually be cast into finished shape. Nevertheless, it can be rolled, and railroad rails are made of it.

Manganese-tungsten Steel.—This steel has been commonly known as "Mushet steel" since it was first produced by Robert Mushet. The proportions of the elements vary, but the following is a very good example: tungsten 9 per cent., manganese 2.50 per cent., and carbon 1.85 per cent.

This steel is very hard and durable. Tools made of it are used for cutting extra hard metals. It is considered economical, regardless of its higher cost, because it will take such deep cuts and last so long without regrinding. With these proportions it is not a high-speed steel.

Chrome-tungsten Steel.—This steel has a composition and properties very similar to the manganese-tungsten steel. In it the manganese has been eliminated and 1 to 2 per cent. of chromium is used in its stead.

Manganese-molybdenum Steel.—This is also similar to the manganese-tungsten combination. Here from 4 to 6 per cent. of molybdenum have been used in place of the tungsten.

HIGH-SPEED TOOL STEEL

The name "high-speed steel" is used to indicate those alloy steels that do not lose their hardness and toughness at a red heat. Hence in the form of cutting tools they can be used to do cutting work at a speed so great that the cutting edge is raised to and kept at a red heat by the frictional resistance. Tungsten, either alone or in combination with chromium, vanadium and cobalt, seems to be best able to confer this property.

The ability to maintain a cutting edge at a red heat seems to depend upon the manner in which tungsten, and molybdenum

also, influence the structural transformation. Carpenter[1] found
that when the steel was heated to and cooled from a temperature
of about 930°C., the structural transformation takes place between 700 and 750°C. But when the steel was cooled from about
1,250°C., a temperature that would ruin an ordinary carbon steel,
the critical changes were prevented altogether, the steel remaining in the austenitic condition. A peculiarity about this auste-

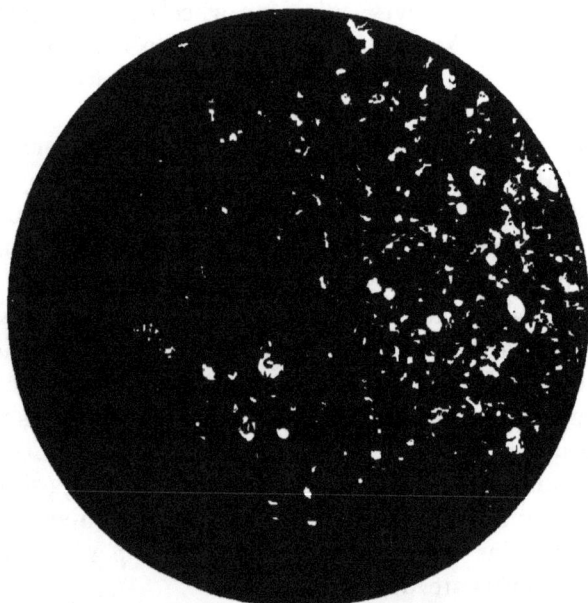

FIG. 31.—High-speed steel. Quenched at 1280°C. Magnified 1500
diameters. Tungsten 17.87 per cent.; chromium 3.46 per cent.; vanadium 0.81 per cent.; carbon 0.65 per cent. (Emmons.)[2]

nite is that it is stable and does not decompose when the steel is
heated again to temperatures that would transform austenitic
carbon steel into pearlite. It is stable at all temperatures up to
550°C., or somewhat higher. Metals have a visible glow at
temperatures a little below this, hence tools of high-speed steel
that have been heated to about 1,200° and quenched can be used
while red hot without losing their hardness and toughness, and,
therefore, their capacity for work.

[1] *Jour. Iron and Steel Inst.*, Part I, 1905, p. 433.
[2] Reproduced by permission from Sauveur's Metallography and Heat
Treatment of Iron and Steel, Ed. 2.

Generally the heat treatment given to tools designed for lathes and planers consists of heating to incipient fusion and quenching in oil. Cooling by air blast and double treatment, which were formerly recommended, are not now common, except that a second (drawing) heating is given to milling cutters and similar tools, the temperature employed depending upon the material to be cut.[1] High-speed steels may be forged at temperatures of 1,050 to 1,150°C. or higher, and may be annealed for machining by heating to 800°C., and cooling with extreme slowness in lime, ashes or in the furnace.

The composition of high-speed steels varies greatly. They usually contain 0.4 to 0.8 per cent. carbon, and seldom over 0.4 per cent. manganese. Sulfur and phosphorus, which are so deleterious in carbon steels, are considered less so in high-speed steels, their effect being either modified or masked by the other ingredients.

As has been said, tungsten is the most important alloying element. When used alone, the amount may be as high as 24 per cent., but it is usually between 10 and 20 per cent. Molybdenum is more active than tungsten in rendering steel capable of high-speed work, but it is much less employed now than formerly because of certain irregularities that attend its use. Chromium may in part replace tungsten, when used, ranging from 2 to 8 per cent. It seems to increase the hardness. Vanadium is used for a similar purpose, but in smaller amounts, generally not exceeding 1.5 per cent. Cobalt has become very important in the manufacture of high-speed steel. It increases the red hardness, thus allowing the tool to be used at a higher speed. Moreover, it has been observed[2] with cobalt steel that it seems to do its best work when run nearly if not quite red hot, rather than at lower temperatures.

References

A

Howe: Iron, Steel and Other Alloys, Boston, 1907.
Carr: Openhearth Steel Castings, Cleveland, 1907.
Campbell: The Manufacture and Properties of Iron and Steel, New York, 1907.

[1] Hibbard: Manufacture and Uses of Alloy Steels, U. S. Bureau of Mines, *Bull.* 100, p. 61 (1915).

[2] Hibbard: *Loc. cit.,* p. 60.

FORSYTHE: The Blast Furnace and Manufacture of Pig Iron, New York, 1908.
TIEMAN: Iron and Steel, A Pocket Encyclopedia, New York, 1910.
BECKER: High-speed Steel, New York, 1910.
STOUGHTON: The Metallurgy of Iron and Steel, New York, 1911.
LAKE: Composition and Heat Treatment of Steel, New York, 1911.
MOLDENKE: The Production of Malleable Castings, Cleveland, 1911.
HARBORD and HALL: The Metallurgy of Steel, London, 1911.
HATFIELD: Cast Iron in the Light of Recent Research, London, 1912.
SAUVEUR: Metallography and Heat Treatment of Iron and Steel, Cambridge, 1916.
HOWE: The Metallography of Steel and Cast Iron, New York, 1916.

B

GUILLETT: Special Steel Alloys, *Eng. Mag.*, **36**, 65 (1908).
BECKER: Manufacture and Use of High-speed Steel, *Cassier's Mag.*, **35**, 327 (1908) and **36**, 505 (1909).
GIBBS: Vanadium Steel, *Cassier's Mag.*, **38**, 174 (1910).
JOHNSON: Manganese Steel Castings, *Foundry*, **37**, 243 (1910).
SPRINGER: Manganese Steel, *Cassier's Mag.*, **39**, 99 (1910).
SMITH: High-speed Steel, *Eng. News*, **63**, 459 (1910).
EDWARDS: Theory of Hardening Steels, *Engineering*, **90**, 548 (1910).
SAVOIA: Production of Malleable Cast Iron, *Jour. Soc. Chem. Ind.*, **30**, 214 (1911).
NORRIS: Vanadium in Iron and Steel Castings, *Eng. News*, **67**, 1100 (1912).
WALKER: Metallurgy of Molybdenum, *Met. Chem. Eng.*, **10**, 110 (1912).
COE: The Influence of the Metalloids on the Properties of Cast Iron, *Engineering*, July 18, 1913.
CONE: The Annealing of Steel Castings, *Iron Age*, May 1, 1913.
HALL: Heat Treatment of Hypo-eutectoid Carbon Steels, Am. Soc. Test. Materials, June, 1913.
JOHNSON: Effect of Oxygen on Cast Iron, *Iron Age*, Feb., 1914.
STINNER: Manufacturing and Testing of Electric Steel, *Wis. Engr.*, December, 1914.
HALL: Manufacture and Utility of Manganese Steel, *Foundry*, April, 1915.
SEE: Manufacture and Use of Pure Iron, *Jour. Assoc. Eng. Socs.*, April, 1915.
HIBBARD: Manufacture and Uses of Alloy Steels, U. S. Bureau of Mines, *Bull.* 100 (1916).
JOHNSON: Blast Furnace Products, *Met. Chem. Eng.*, **15**, 325 (1916).
CARNELL: Acid-resisting Alloys, *Jour. Ind. Eng. Chem.*, **8**, 922 (1916).

CHAPTER V

THE CORROSION OF IRON AND STEEL

The five metals most commonly used in the industries, either alone or in alloys, are iron, copper, lead, tin and zinc. Of these, iron is for many reasons, by far the most important. Its various commercial forms together form the largest manufactured product of the world. But also, of all the metals, iron is the most perishable. When the other metals are exposed to the varying conditions of the weather, they become coated with a layer of insoluble compounds that in a large measure seems to protect the underlying metal, but this is not so with iron. When it is exposed it corrodes, and the compound that is formed serves to hasten the corrosion, rather than to retard it, so that as a result, the metal is quite rapidly converted into a loosely coherent compound called *rust*. The term *corrosion* is usually applied to the process by which any of the metallic elements are, by natural agencies, converted into compound forms, but the term *rusting* is applied more specifically to the corrosion of iron. The compounds formed are as a rule insoluble, and are produced by the addition to the metal of some negative ion derived usually from the surrounding atmosphere or water. In the case of iron, the compound that is the most commonly formed is the red oxide, which, disregarding the water it contains, very closely resembles one of the most important of the ores of iron. In fact, the corrosion of iron may be considered to be nothing more or less than a reversion of the metal to the state in which it existed before it was smelted or reduced in the blast furnace. It is not at all unreasonable to suppose that the deposits of ore now found were at one time deposits of iron in the metallic state.

Composition, Weight and Bulk of Rust.—Rust is the hydrated red oxide of iron, of the composition $Fe_2O_3 \cdot xH_2O$. It is impossible to ascribe a definite numerical value to x, indicating the number of molecules of combined water that rust contains, because the amount varies. However, in the following discussion

127

it will be considered as ferric hydroxide, $Fe(OH)_3$, the equivalent of which is, when written so that the anhydride is expressed, $Fe_2O_3 \cdot 3H_2O$. In addition to having an indefinite amount of water, the degree of oxidation varies slightly, that nearest to the iron being somewhat more ferrous, or in a lower state of oxidation.

If we consider rust as the ferric hydroxide, the weight of the rust is nearly twice (1.91 times) the weight of the iron from which it was formed. This, of course, must not be considered to apply with mathematical accuracy to the commercial forms of iron, since they are not pure.

The bulk of the rust is even more increased. Bauermann[1] says that one volume of wrought iron when completely rusted, produces ten volumes of rust. By thus increasing in bulk during rusting, iron has been known to part heavy stones and walls.

The Theories of Corrosion.—There have been various theories advanced to explain the exact method by which iron rust is formed. Of these, but two will be mentioned here. They are: (1) the carbon dioxide theory, and (2) the electrolytic theory. These two agree in respect to the statement that both water and oxygen must necessarily be present in order that rusting may go on, but the former differs essentially from the latter in that it postulates the necessity for the presence of carbon dioxide gas as well. It should be stated also that the water must be present in the liquid form. Iron will not rust even though it is in contact with moisture-laden air, if the moisture does not condense upon it in the liquid form or, of course, if it does not become wetted in any other way.

The Carbonic Acid Theory.—This theory denies the solubility of iron in pure water. It assumes that the following reactions are necessary in order that rust may be formed:

$$4H_2O + 4CO_2 \rightarrow 4H_2CO_3$$

Then the carbonic acid thus formed, produces ferrous carbonate, as:

$$4Fe + 4H_2CO_3 \rightarrow 4FeCO_3 + 4H_2$$

which, by means of water and oxygen is converted into the hydrated ferric oxide, as:

$$4FeCO_3 + O_2 + 6H_2O \rightarrow 2(Fe_2O_3 \cdot 3H_2O) + 4CO_2$$

[1] *Jour. Iron and Steel Inst.*, 2, 135.

It will be seen that all of the carbon dioxide used in the beginning is set free at the end of these reactions, ready to dissolve in water and attack fresh iron. Consequently, a very little carbon dioxide could, in time, bring about a great deal of rust formation. There is no doubt that rust is formed according to the reactions expressed in these equations,[1] but judging from the evidence that has been presented in favor of the electrolytic theory, it seems that the intermediate acid formation is not essential.

The Electrolytic Theory.—This theory explains the solubility of iron in pure water.[2] The explanation of the manner in which the solution is effected is based on electrochemical action such as occurs in a simple form of primary cell. In order that this theory may be more easily comprehended, a brief discussion of the electric cell will be inserted here. The cell action depends upon the solution pressure of the metals used, hence this will be mentioned first.

The Solution Pressure of Metals.[3]—Whenever a metal is immersed in water, a dilute acid, or in a solution of one of its salts, it exerts a pressure that tends to drive ions[4] from the surface of the metal into solution. This is known as the *solution tension* or *pressure* of the metal. The solution of a metal in this manner is illustrated in the familiar experiment in which a piece of metallic sodium is dropped upon the surface of water. The sodium ions leave the metal and enter the water with the

---

[1] For views in support of this theory see J. NEWTON FRIEND: The Corrosion of Iron and Steel, p. 56.

[2] It has been rather difficult to either prove or disprove that iron is soluble in pure water, since it is not definitely known that water has ever been prepared in the pure state. That which is considered pure may contain residual traces of dissolved gases or it may dissolve traces of material from the sides of its container. Even when treated most carefully by repeated distillation and boiling, it will still contain, according to LEDUC (quoted by FRIEND, *loc. cit.*, p. 16) 1 c.c. of gases per liter, and since carbon dioxide is more soluble than the other gases of the atmosphere, it is assumed that this 1 c.c. contains not a small proportion of carbon dioxide. And this residual CO_2 is claimed by some to initiate the dissolving action in the experiments designed to show its solubility in pure water. (See CUSHMAN and GARDNER: Corrosion and Preservation of Iron and Steel, p. 43.)

[3] For a more complete explanation of this subject, see p. 393.

[4] An ion is an atom, or a group of atoms, charged with electricity. The number of charges carried is the same as the valence of the element or group.

9

resultant formation of sodium hydroxide and the liberation of hydrogen. Practically all of the metals possess this solution pressure, but some to a much greater degree than others.

The Electrochemical Action in a Simple Form of Cell.—In building up a primary cell it is necessary to choose two metals, one of which has a high and the other a low solution pressure. Here we shall consider zinc and copper, zinc having a high, and copper a low pressure. If we insert a plate of copper in dilute sulfuric acid for a few minutes no action will be noticed. Also, if we repeat the operation using a plate of *chemically pure* zinc no action will be visible. If we now insert both metals in the acid at the same time and connect their outer ends with a piece of wire, as shown in Fig. 32, we find numerous small bubbles arising from the copper but still none from the zinc. But if we allow the metals to remain in the acid with their outer ends connected for an hour or more, it will be noticed that the zinc is considerably wasted away. Also, if we hold a magnetic needle beneath the connecting wire it will be found to be deflected, showing that a current is passing.

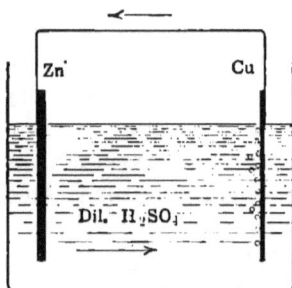

FIG. 32.—A simple form of electric cell.

The explanation of the action that occurs in the preceding experiment is given more fully in Chapter XVIII, which see, but the essential facts are these. Zinc ions enter the dilute acid and the following reaction occurs:

$$Zn + H_2SO_4 \rightarrow ZnSO_4 + 2H$$

Since the zinc ions carry positive charges with them into solution, the zinc plate is left negatively charged. On this account, when it was inserted alone, the action could not proceed because the positive charges could not be returned to the metal. But when the zinc plate was connected with the copper, hydrogen ions precipitated upon the copper, gave up their charges and became atoms of hydrogen, which gas was seen collecting in the form of bubbles. The charges given up by the hydrogen ions flowed through the connecting wire back to the zinc, the deflection of the needle indicating their passage. The zinc being replenished with positive charges was then able to send off more ions; in other words, it continued to dissolve. This action of the hydrogen ion should be carefully noted, since it serves in the same way in the corrosion of iron where its action is very important.

The following terms are used in describing the primary cell. The electrochemical action occurring in the cell is frequently called *galvanic or voltaic action*, so named in honor of Galvani and Volta, who were pioneer investigators in this field. The liquid in the cell is called the *electrolyte*. The surfaces of metal plates are called *electrodes*, or in distinguishing them from each other, the active, dissolving electrode is called the *anode* and the other the *kathode*, as in the preceding experiment the zinc was the anode, and the copper the kathode. Also in a cell system such as this, the metal that sends its metallic ions into solution is said to have a high potential and the other a low potential, or it is said that a *difference of potential* (difference in pressure) exists between them.

"**Local Action.**"—Electrochemical action, of the sort just described, is not limited to such cases where two metals are joined by means of a wire. It may readily occur within a single piece of metal. It will be recalled that in the experiments in elementary chemistry, zinc was considered a metal that was very soluble in the dilute mineral acids, so much so that it was most commonly chosen as the metal that was most conveniently able to liberate hydrogen from the acids. This ready solubility of zinc depends upon the fact that it is impure, for as was seen in the preceding experiment, when pure zinc was inserted alone in the acid it did not dissolve. The ordinary or commercial zinc used in the generation of hydrogen contained many impurities, such as lead, iron and cadmium, which are segregated so that the surface is by no means uniform. The presence of these impurities is quite readily shown, for when a sheet of commercial zinc is dipped into acid its surface soon becomes blackened. The impurities are not soluble while in contact with the zinc, but separate in a fine sponge form, in which condition they appear like a layer of soot on the zinc. The various parts of the piece of zinc act like dissimilar metals in contact, little currents being produced that run from part to part of the piece itself. The spots where the zinc is more nearly pure, have the higher solution pressure and act as anodes, while the relatively impure spots become kathodes.

Iron Corrosion a Case of "Local Action."—Now according to the electrolytic theory, iron corrosion is a case of "local action." From the standpoint of this theory, all iron and steel must be thought of as a composite structure, as though it were made up of strands and patches of more or less dissimilar material. It is between these somewhat dissimilar parts that the electro-chemical action occurs. It was shown in the discussion of iron and steel, that their important properties either depended upon, or were controlled by the silicon, sulfur, phosphorus, manganese (and sometimes other metals), they contained. Although these elements are quite essential in order to produce these desirable properties, they are a decided detriment in their influence on corrosion, because they cannot, by the present methods of manufacture, be distributed uniformly throughout. The unavoidable segregation of these impurities on the surface will cause the solution pressure to be greater at one point than at another. The points where the solution pressure is greater become anodic to those points where it is less. ' Hence, when the surface of a piece of iron or steel becomes coated with a conducting film of water, a current flows through the film from the anode to the kathode point. Iron ions enter the solution at the anode point and hydrogen ions are deposited at the kathode point. Thus, it is apparent that if the iron were pure, or even if its impurities were uniformly distributed, it would not corrode. Moreover, the results of observations indicate the same thing, for although absolutely pure iron and perhaps perfectly uniform iron has never been obtained, the more nearly these conditions are approached, the less the iron corrodes. And another observed fact is, that when it does corrode, the corrosion does not begin or take place evenly; some spots are more liable to attack than others, although as the corrosion proceeds, layers having a different composition are exposed, so that the position of the anode and kathode spots may change, and eventually the whole of the surface may become corroded. However, in any case, the iron dissolves only at those spots that are for the time being at least anode points, and this leads to the formation of hollows which is described as *pitting.* It is very unfortunate that this pitting occurs, for the iron or steel may be entirely rusted through at one point and the remainder of the metal be but little affected. In some cases, as in a boiler tube, the value of the article is thus entirely destroyed.

Essential Reactions.—The chemical reactions that occur during the corrosion of iron may be summed up in the following manner. Water ionizes or dissociates to a certain slight degree, producing hydrogen (H) and hydroxyl (OH) ions. The iron ions that enter the water at the anode points on the metal, react with the ionized water, producing ferrous hydroxide, thus:

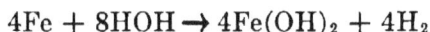

$$4Fe + 8HOH \rightarrow 4Fe(OH)_2 + 4H_2$$

Then by the action of oxygen and water, the ferrous hydroxide is converted into the ferric state and precipitates out as rust, as:

$$4Fe(OH)_2 + O_2 + 2H_2O \rightarrow 4Fe(OH)_3 \text{ or } 2 (Fe_2O_3 \cdot 3H_2O)$$

Evidence tending to show that the ferrous hydroxide is formed in this way is furnished by the fact that when polished pieces of iron or steel are sealed up with the purest water obtainable and allowed to stand for several weeks, the water develops a basic reaction, turning phenolphthalein pink, when tested with this indicator. Ferrous hydroxide is a weak base. Also when oxygen is admitted to the flask in which the iron and water was kept, there is immediately a rather copious precipitation of rust from all parts of the water in the flask, showing that the iron must have previously been disseminated in the dissolved state.[1]

Factors that Affect the Corrosion Rate.—It has been observed that there is a variety of factors that may either hasten or retard the rate at which corrosion may proceed. These factors have to do both with the condition of the iron and the medium that surrounds it.

Corrosion Stimulated by Contact with Other Materials.— When some material capable of assuming the kathode relationship to iron touches it at some point, and a film of moisture connects the two at another point, corrosion proceeds much more rapidly. A small current is established, flowing from the iron

[1] For experiments showing the solubility of iron in water see CUSHMAN and GARDNER: The Corrosion and Preservation of Iron and Steel, p. 42.

through the moisture toward the kathode substance, which aids in dissolving the iron, in that it serves as a point where the hydrogen ions may be precipitated and their positive charges returned to the corroding spot. The action of "mill scale" is very pronounced in this respect, since it has a low potential as compared to iron. If it were continuous over the whole surface of the iron, it would be a good protecting coat, because it has a very low solution tendency, but because it is very brittle it readily becomes cracked and flaked off in spots, and wherever iron is exposed, it is rapidly corroded by the electrochemical action set up.

Rust itself assumes the kathode relationship to iron. It has been noticed that the rate of rusting during the second year is about twice as fast as during the first, this being due to the action of the accumulated rust. Railroad rails in use, where the vibration constantly detaches the rust, do not rust nearly so rapidly as those on unused switches. Of course, beside its negative voltaic action, the rust is porous and retains moisture, which helps the corrosion. Similar galvanic action occurs when metals of different kinds, or even the different forms of iron are brought into contact. Iron railings fixed in stone copings by means of lead are most rapidly corroded where the iron joins the lead, because lead is kathodic to iron. For this reason it would be better to use spelter for this purpose, since zinc is anodic to iron and protects it, as will be explained later in the discussion of the protective metal coatings. Other examples of corrosion brought about by galvanic action occur where the malleable cast-iron connecting rings and T's join wrought-iron or steel pipes, or about the soft rivets in steel structures.

There is also a difference of potential between strained and unstrained portions of iron and steel. Corrosion in the neighborhood of punched holes is greater than in the neighborhood of drilled holes. Scratches and indentations made by tools are almost always anodic to the surrounding areas.

The Effect of Acids and Alkalis.—The solvent action of water on iron is very slow, because the ionization of water is so slight that the number of hydrogen ions is very limited. Consequently, the introduction of any substance into the water that will increase the number of hydrogen ions will allow the corrosion to proceed more rapidly. As is well known, acids are substances that ionize in water with the production of hydrogen ions; therefore, when the water in contact with the iron is acidified, the return of the positive charges to the metal can go on more rapidly because of the greater number of carriers. Carbon dioxide and sulfur dioxide gases in the presence of water produce acids, and since these gases are quite common in the atmosphere, much corrosion is caused by them.

On the other hand, the presence of alkaline solutions of suitable strength retards corrosion. But this is not true of the exceedingly dilute solutions. Experimental results[1] indicate that protection is afforded by alkaline solutions having the following concentrations:

Sodium hydroxide..... about   1 to   10 grams NaOH per liter.
Potassium hydroxide.. about   1 to   10 grams KOH per liter.
Calcium hydroxide.... about 0.5 to 1.35 (saturated) grams CaO per liter.

Alkalis, when dissolved in water, ionize with the production of hydroxyl ions. Hydroxyl ions and hydrogen ions have a strong tendency to combine and form molecules of unionized water; therefore, when the hydroxyl ions are in excess in the proper degree, they prevent the existence of free hydrogen ions.

The Quantity of Dissolved Oxygen.—In the air, the proportion of oxygen to nitrogen is approximately one volume of oxygen to four of nitrogen; but when dissolved air is boiled out from water, it is found that the proportion is one of oxygen to two of nitrogen. This is accounted for by the fact that oxygen is twice as soluble in water as nitrogen. It has been seen that dissolved oxygen is a very important factor in the formation of iron rust. It is active in two ways. First, it hastens the corrosion by its ability to remove the hydrogen that precipitates out at the negative poles and adheres there in the form of a film that would produce a so-called polarized condition of the iron and tends to stop further action. The dissolved oxygen removes this hydrogen by uniting with it to form water. This depolarizing action of oxygen is of great importance, so much so that it is frequently considered to be the cause of corrosion.[2] Secondly, the dissolved oxygen converts the ferrous hydroxide into insoluble

[1] FRIEND: *Loc. cit.*, pp. 120–122.
[2] *Met. Chem. Eng.*, **14**, 389, April, 1916.

rust, so that it is removed from solution. This removal of the dissolved iron compound allows more iron ions to enter the solution, it being a well-known fact that any metal dissolves less readily in a solution that already contains ions of that metal, the solubility proceeding less rapidly as the concentration increases.

Because of this action of oxygen, when iron is immersed in water, the more deeply it is immersed, the less rapidly it will corrode, other things being equal. The lower strata of the water are not so well supplied with dissolved oxygen, because when the supply is used up, a further supply is obtained only as fresh oxygen slowly diffuses in from the surface that is exposed to the air. Consequently, iron that is deeply immersed in water as in deep wells or deeply buried in the ground corrodes less than that which is near the surface. Also tanks, pipes and other containers, in which the water is frequently changed, corrode more rapidly than those in which the water is allowed to stand. Of course, it should be remembered that those conditions that allow fresh supplies of oxygen to be readily obtained, also allow carbon dioxide to enter more readily, which, as has been shown, hastens corrosion by forming an acid and so increasing the number of hydrogen ions. Rain water is very highly corrosive to iron. This is dependent upon the fact that during its passage through the air it has become saturated with these gases.

Because fresh charcoal has a high power for absorbing gases, it has been found that when blocks of it are put into water or floated in powdered form on the surface of the water, in which iron is immersed, the corrosion is very materially lessened.

Partial Immersion.—It is quite well known that iron that is partly immersed corrodes most rapidly at the surface of the water. The reason is obvious, for at this point there is a plentiful supply of both water and gases. Also, the action is hastened at this point by the slight catalytic effect of light, and the slightly higher temperature here than at lower levels. For similar reasons, iron that is alternately wet and dry corrodes more rapidly than that which is permanently wet.

The Action of Cinders.—Cinders are found to be extremely corrosive in their action on iron as well as on other metals buried in them. This is, of course, because they are porous and filled with gases. The two gases that are contained in greatest abundance are carbon dioxide and sulfur dioxide, the latter being formed by the burning of the iron pyrite, FeS_2, that exists as an impurity in the coal, and the former by the burning of the coal itself. Both of these gases are acid anhydrides, and as the water from the rains or other source percolates through the cinders, it unites with them to produce acids. Although not strong acids, they nevertheless ionize to a certain extent, and bring about accelerated corrosion.

Contact with cinders will bring about the corrosion of metals that are far more resistant than iron. For example, even lead

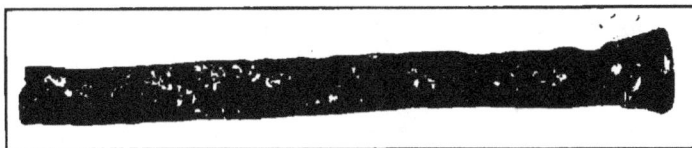

FIG. 33.—A section of lead pipe showing the corroding action of cinders.

is quite rapidly acted upon. In Fig. 33 is shown a section taken from a line of lead pipe that passed through a cinder fill.

This corroding influence must be remembered in laying pipe lines for any purpose through cinders, as for example, railroad embankments, etc. The metal must be encased in some way that it will be kept perfectly free from contact with the water that percolates through the cinders. If it is possible to use such, it would be better to substitute entirely some fiber or wooden pipe or conduit.

Corrosion in Pipes and Boilers.—Here, as in corrosion elsewhere, the effect of the dissolved oxygen is most objectionable. Since gases are less soluble in hot water than in cold, open feed-water heaters are often used to remove the oxygen, as well as other gases. Often, also, quantities of scrap iron are placed in the tank, which by their rusting aid in depriving the water of its oxygen. The addition of tannic acid, tannin and tannin extracts, as well as pyrogallic acid and the alkaline pyrogallates is very good practice because these substances are all excellent oxygen absorbents. In addition, if all of the "mill scale" is removed from the interior of boiler tubes and boilers, their life is immeasurably prolonged. ǁ

Passivity of Iron.—When iron is for a certain length of time immersed in solutions of suitable strength of nitric acid, chromic acid, potassium bichromate, or certain other strong oxidizing agents, its solubility in acids is almost destroyed and it will remain free from a tendency to corrode for a long time. It is then, said to be in the passive state. When in this state, no change can be seen on the surface of the metal even with a high-power microscope, and it is not known to just what cause this passivity is due. Various theories have been advanced to explain it. One is, that a very thin, invisible layer of iron oxide, Fe_3O_4, is produced on the surface. Another is, that the oxygen is not chemically combined, but merely occluded, that is, in a sense, condensed as a film on the surface. Other theories have been advanced but consideration cannot be given them here.[1]

If we assume that the surface of the iron is oxidized in one way or another, a possible explanation of the action of the passive state in preventing corrosion is that the oxidized state of the surface prevents the hydrogen ions from giving up their free positive charges to the metal, which in turn would prevent it from sending positively charged metallic ions into solution.

Electrolysis an Aid to Corrosion.—When a current of electricity is passed through water that contains a little acid so that the resistance of the water to the flow of the current is diminished, the water is broken up into its two elements, hydrogen and oxygen. The process of thus bringing about the decomposition of any substance by means of the electric current is known as *electrolysis.*[2]

Now, when iron is buried in the ground, or is placed in some other situation where it is in contact with moisture, if a current passes through the iron, then the moisture, and then to some neighboring substance, as for example the ground, electrolysis occurs. Corrosion of the iron at the point where the current leaves it is very much accelerated. The moisture being decomposed, oxygen is liberated on the surface of the iron, and this, in the ways that have been shown, is an exceedingly active corroding agent, especially so in this case because it is nascent. Corro-

[1] For a discussion of the various theories see *Met. Chem. Eng.*, December, 1913, p. 679, and FRIEND: Corrosion of Iron and Steel, p. 181.
[2] See Chapter XIX.

sion by electrolysis is very rapid, and it occurs wherever a current, for example, a stray current that leaks from some electric system, passes from one metal through moisture in contact with it, to some other metal or substance. It may occur in the seams of boilers, about rivets and bolts in metal, in the joints in pipe lines, between the pipe and the ground, between metal and moist wood and masonry, between structural metal and the earth, or even between the reinforcing rods and the concrete encasing them. An example of electrolysis is shown in Fig. 34. This is a section

FIG. 34.—Showing a section of brass pipe corroded by electrolysis.

cut from a brass pipe at a point where it passed through a board partition in a damp basement. An electric system having been grounded on the pipe, the electrolysis was produced where a portion of the current passed to the earth by way of the damp boards.

A very troublesome point about this sort of corrosion is, that it usually goes on, in some unexposed situation where an inspector cannot see it. Therefore, great care must be exercised to prevent currents from passing through such metallic systems. Of course, in some cases, electrolysis may be prevented by making the metallic connection continuous, as in the case of a soldered joint.

The Relative Corrodibility of Iron and Steel.—There is a wide difference of opinion relative to the respective abilities of wrought iron and steel in resisting corrosion. However, it seems that the best conclusion that can be drawn from all the various observations to date, is that given by Stoughton.[1]

"It is now generally understood that the difference between these two materials is not great, if both are well-made. There is, however, a great superiority in well-made over badly made material of either class. Steel which contains entrained slag, manganese oxide, iron oxide, much occluded gases or blow-holes, or in which the impurities

[1] ROGERS and AUBERT: Industrial Chemistry, p. 362.

are segregated will yield more rapidly to rusting; and wrought iron, which contains an excessive amount of slag, not thoroughly worked, or which is irregular in composition, due to incomplete puddling or to the mixture with it of steel scrap during the welding process, or which has been made from low-grade scrap without remelting, will rust faster than well-made material. In short, it is now recognized that the quality of manufacture has a much greater effect than the process of manufacture. Any irregularity in the composition of the metal which produces a difference in potential will hasten the rate of corrosion."

It is often the case that the difference of opinion as to the relative corrodibility of these metals is traceable to inherent differences in the methods of investigation employed.

As regards the resistance of cast iron to corrosion, when compared with wrought iron and steel, results seem to show that if the skin on the casting produced by the sand of the mould is allowed to remain, it will resist corrosion better than the other forms, but that if this skin is removed it will corrode more rapidly. The more rapid corrosion in this case may be due to the fact that gray foundry iron is quite porous, which allows an entrance of air and water to points beneath the surface. Also the electronegative graphite imbedded in it forms a couple with the iron and hastens corrosion by galvanic action.

PROTECTION OF IRON AND STEEL FROM CORROSION

The methods employed for the protection of iron and steel from corrosion nearly always consist of the application of some coating. The coatings applied may be divided into three classes: (1) other metals or alloys, (2) coatings of magnetic oxide, (3) paints and similar materials.

1. Metal Coatings.—The metals that are usually applied as protective coatings to iron are zinc, nickel and tin, although occasionally copper and lead are used. The manner in which these metals protect is not the same in all cases. The difference in their action depends upon the difference in their solution pressure, and before proceeding further it will be necessary to explain the manner in which this acts.

The Solution Pressure Series.[1]—It is quite well known that if a piece of clean iron is put into a solution of copper sulfate,

[1] For a more complete discussion see p. 395.

the iron receives a coating of metallic copper. If the copper sulfate solution is now examined, it will be found to contain also some iron sulfate. In fact, as many ions of iron entered the solution as there were atoms of copper that came out of it into the metallic state. The reaction is merely a case of substitution as is shown in the following equation:

$$Fe + CuSO_4 \rightarrow FeSO_4 + Cu$$

The solution pressure of iron is quite high, while that of copper is relatively low. Hence iron enters the solution and copper assumes the metallic state. It has been found that all the metals bear a similar relation to each other. In the following list some of the common metals are arranged in the order of their diminishing solution pressures:

1.	Magnesium	5.	Lead
2.	Zinc	6.	Silver
3.	Iron	7.	Tin
4.	Nickel	8.	Copper

Any metal in this series is able to throw out any metal lower in the series from a solution of a salt of the latter. Thus, for example, zinc will precipitate iron, and iron will precipitate lead, copper, etc.

Zinc-coated ("Galvanized") Iron.—Suppose now that a piece of zinc-coated iron, exposed to the atmosphere, becomes scratched or cracked, laying bare the iron. As the corroding agents of the atmosphere find their way to this point, because of its higher solution pressure, the zinc reacts with them producing zinc compounds, which the iron is unable to decompose. Thus zinc protects iron by itself entering into solution. However, if we assume that some iron compounds are formed at first, the zinc decomposes them, causing the iron to again assume the metallic condition. Thus, zinc can protect iron even though the zinc coating is broken. Of course, as this protective action continues, the zinc is gradually used up and so retreats from the point where the original break occurred. But it still continues to exert a protective action over a zone or space of iron that it does not actually cover, although it does so only within limits.

Zinc, Nickel and Tin Coatings Compared.—If the iron is coated with nickel or tin, when the coating is broken, the action is just the reverse of the action with zinc. Nickel and tin have low solution pressures as compared to iron, and they do not enter solution in preference to iron, nor do they deposit iron from iron

Fig. 35.—Steel plates containing buttons of copper, zinc, tin and lead.

salts. As long as the nickel and tin coatings are unbroken they have excellent protective values and last almost indefinitely. Because of their low solution pressures, they do not dissolve in

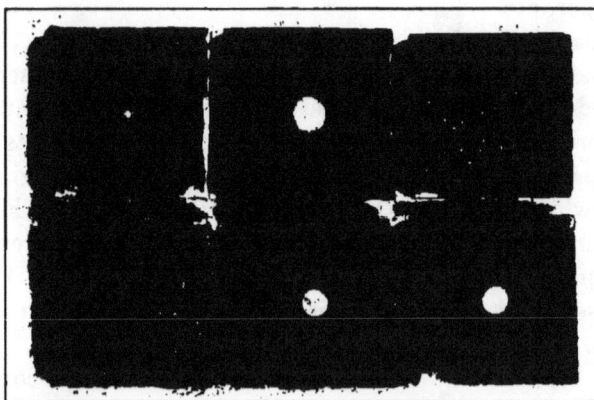

Fig. 36.—Same plates as shown in Fig. 35, after corrosion. Showing the inhibiting action of zinc and the accelerating action of other metals.

the presence of the natural corroding agents. In this respect they are superior to zinc, for zinc protects iron at the expense of its own destruction. Most of the early failure of zinc coatings is due to the fact that they are made so very thin. This is done partly, of course, to prevent the cracking of the coat during the bending of the coated article, but as can be seen, the cracking is not nearly so serious as the thinness of the coating.

All coated articles that are subjected to rough usage are almost certain to be scratched, and besides, small "pin holes" exist even in the original coating. And when openings are present in the nickel or tin coatings, it may be said that these metals become stimulators of corrosion, because they act as kathodes and enable iron to go into solution more rapidly.

The action of different metals in contact with iron is shown to some extent in Figs. 35 and 36. Fig. 35 shows 3-in. plates cut from low-carbon steel, $\frac{1}{4}$ in. thick. In the center of these plates are buttons of copper, zinc, tin, and lead, as indicated. Two plates were left blank for comparison. Fig. 36 shows these same plates after they had been wetted with faucet water and allowed to corrode. Copper, tin, and lead showed corrosion accelerated at the point of contact; particularly was this true in the case of copper, about which the rust was built up in the form of a ridge. That the acceleration is greater in the case of copper is readily explained, since copper is farthest removed from iron in the solution pressure series. Zinc shows a well-defined protective action. There is a zone about $\frac{1}{8}$ in. wide about the zinc button, that has been entirely protected from corrosion, and beyond this a much wider zone, in which the action seems to be largely inhibited although not entirely prevented. Further, the button of zinc itself is free from rust spots, while the others show them. These spots resulted from the corrosion of the iron particles drawn into the button when the plate was ground to secure a clean surface.

Methods of Applying Zinc Coatings.—There are at least three methods by which zinc is applied as a coating to iron. These are known as the hot dip, in which the cleaned article is dipped into molten zinc; the "cold" or electrolytic, in which zinc is plated on the article from a solution of zinc salts; and the "dry" or Sherardizing process, in which the article is coated by heating it in contact with zinc dust. In all cases the value of the coating depends upon the purity of the metal, the thickness of the coating and its continuity, and these factors may vary as much between coatings applied by the same process as between coatings applied by different processes. In exposure tests carried out by the author on samples of galvanized electric conduits, those coated by the hot-dip process were found to have a longer life, but in all these cases the coatings applied by this process were much thicker than those applied by the other processes.

Iron Oxide Coatings.—There are various processes for giving the iron article a coating of the black magnetic oxide, Fe_3O_4. A familiar example is the Bower-Barf process, in which the article is heated in a closed retort to a high temperature, after which superheated steam is injected. The so-called Russia iron, or black sheet iron, is produced by a similar process.

Like the coatings of nickel and tin, these black oxide coatings are efficient as long as entirely continuous and not in any way broken or scratched; but when the coating is broken and moisture is present, a galvanic action is set up that brings about rapid corrosion of the iron.

Paint Coatings.—It has been found that certain substances used as paint pigments have an influence on the rate of corrosion when used in paints applied to iron. Some have been found to have a retarding or inhibiting effect, while others have a stimulating or accelerating action. The inhibitive value of the pigment seems to be due to two causes. Some pigments, as basic white lead, have an alkaline reaction and so retard corrosion for this cause, while others like the chromates, may have an effect in causing iron to assume the passive state. On the other hand, a pigment like gypsum, or any other form of calcium sulfate, which dissolves slightly and ionizes in water producing an acid reaction, hastens the formation of rust. The properties of the pigments in this respect are given in a later chapter under the subject of "Paint Pigments," which see.

Protection of Iron by Concrete. Reinforced Concrete.—It is a subject of much discussion whether or not the iron or steel that is imbedded in concrete will corrode. There are several factors that have a bearing on this question. In the first place, Portland cement when set, contains considerable free calcium hydroxide which is produced during setting when the tribasic calcium silicate, Ca_3SiO_5, and the dibasic, Ca_2SiO_4, revert to the mono- form, $CaSiO_3$. It has been found by experiment that a

saturated solution of calcium hydroxide has a very high protective value. In a concrete mixture such as should be used for
reinforced work there is much more than enough free calcium
hydroxide to produce this saturated solution. Although the
calcium hydroxide in the exterior layers of the concrete structure
soon becomes carbonated by the carbon dioxide in the air, the
formation of the more dense calcium carbonate makes the concrete less porous and the further penetration of the carbon
dioxide more difficult. Still, it is very obvious that the metal
should not be allowed to be too near the surface of the concrete.

Much can be done to insure the protection of the encased
iron by properly proportioning the concrete so that it is as free
from voids as possible, since these allow an easier penetration of
the corroding agents. Also, round rods are better in this respect
than square and twisted rods, since the chance for the formation
of voids is less, and the concrete should be of a sufficiently wet mix
for the same reason. It is said[1] that the rods, before being imbedded in the concrete, need not be cleaned of rust, since the
calcium hydroxide set free in the cement appears to dissolve the
iron oxide with the formation of calcium ferrites or ferrates,
leaving the metal clean and bright. It should be noted that
cinder concrete should never be used for reinforced work, since
it without fail causes corrosion, as was shown in the preceding
discussion of the action of cinders.

It seems, then, reasonable to believe that if the reinforced
concrete is properly made, the reinforcing rods will not corrode.
And, in general, this belief has been thus far confirmed by the
facts. In concrete structures that have been torn down after
many years, the metal has been found clean and free from rust.

However, there are two factors that may accomplish the
destruction of reinforced-concrete structures. These are the
presence of cracks and voids, and the action of electrolysis. If
cracks and voids of sufficient magnitude exist, corrosion of the
encased metal without doubt will occur. And since the bulk of
the rust produced is much greater than that of the iron, the
increase will widen the opening so that the final result will be
disastrous. Also, if conditions should happen to be such that
electrolysis occurs about the encased rods, corrosion will certainly

[1] FRIEND: *Loc. cit.*, p. 159.

10

take place, for under these conditions iron corrodes even in the presence of the alkali. This is especially true of the rods or portions of rods that are in the anode position, that is, so placed that the current passes from them into the surrounding concrete[1]

References

A

STERN: Rust Prevention, 1907.

SANG: Corrosion of Iron and Steel, New York, 1910.

CUSHMAN and GARDNER: Corrosion and Preservation of Iron and Steel, New York, 1910.

FRIEND: Corrosion of Iron and Steel, New York, 1911.

FLANDERS: Galvanizing and Tinning, New York, 1916.

B

CUSHMAN: Corrosion of Fence Wire, U. S. Department of Agriculture, *F. Bull.* 239 (1905).

CUSHMAN: Corrosion of Iron, U. S. Office Public Roads, *Bull.* 30.

CUSHMAN: Preservation of Iron and Steel, *Engineering*, **87**, 710. Report of Committee Upon the Corrosion of Iron and Steel, *Proc. Am. Soc. Test. Matls.*, **7**, 209 and **8**, 231.

WALKER: Function of Oxygen in the Corrosion of Metals, *Trans. Am. Elec. Chem. Soc.*, **14**, 175 (1908).

HOWE and STOUGHTON: Relative Corrosion of Steel and Wrought Iron Tubing, *Proc. Am. Soc. Test. Matls.*, **8**, 244 (1908).

WALKER: Protection of Iron and Steel from Corrosion, *Eng. Mag.*, **37**, 198 (1909).

WALKER: Detection of Pin-holes in Tin Plate, *Jour. Ind. Eng. Chem.*, **1**, 295 (1909).

CLEMENT and WALKER: An Electrolytic Method for the Prevention of Corrosion, U. S. Bureau of Mines, *Tech. Paper* 15 (1913).

GARDNER: Paints to Prevent Electrolysis in Concrete, *Jour. Ind. Eng. Chem.*, **7**, 504 (1915).

ROSA and McCULLOM: Electrolysis and Its Mitigation, U. S. Bureau of Standards, *Technologic Paper* 52 (1915).

BUCK and HANDY: Research on the Corrosion Resistance of Copper Steel, *Jour. Ind. Eng. Chem.*, **8**, 209 (1916).

WALKER: Corrosion and the Engineer, *Met. Chem. Eng.*, **14**, 388 (1916).

RICHARDSON and RICHARDSON: Atmospheric Corrosion of Commercial Sheet Iron, *Met. Chem. Eng.* 15, 450 (1916).

[1] For a more complete discussion of electrolysis in concrete, see ROSA and McCULLOM: Electrolysis and Its Mitigation, U. S. Bureau of Standards, *Technologic Paper* 52 (1915).

CHAPTER VI

THE NON-FERROUS METALS

There are about 15 metals that are generally well known. Of these, iron, copper, lead, tin, zinc, and aluminum are quite common, and of great industrial importance. Of course, iron, whose properties were discussed in a preceding chapter, is by far the most important and useful, and on this account is classed by itself. In contradistinction to it, the other metals are commonly grouped under the name of the non-ferrous metals. In the present chapter, the properties of some of the non-ferrous metals will be discussed.

ALUMINUM

Occurrence.—Of all the metals, aluminum exists in the earth in the greatest quantity. It makes up nearly 8 per cent. of the earth's crust, while the total quantity of iron amounts to less than 4.5 per cent. The other common metals, as copper, lead, zinc, and tin, taken together, make up only a small fractional part of 1 per cent. of the solid portion of the earth. Aluminum oxide, called *alumina*, which when pure is white in color, is an essential ingredient of such minerals as clays, slates, mica, feldspar and granite. Alumina forms several precious stones.

Corundum consists of anhydrous alumina. The ruby is made up of crystals of corundum colored red by a small amount of chromium. The sapphire is another variety colored blue by traces of cobalt. **Emery** is an impure variety of corundum, containing ferric oxide. It has a granular structure and a degree of hardness only a little less than that of the diamond. Because of its hardness it is used as an abrasive material.

Reduction of Aluminum Ores.—Although aluminum is very abundant in the earth it is never found free, nor can it by present methods be obtained commercially from clay in which it exists in such great quantities. It is obtained from a hydrated oxide, called *bauxite*, $Al_2O_3 \cdot 3H_2O$. The bauxite is dissolved in molten

147

cryolite, $AlF_3·3NaF$, and the metal is then separated from the bauxite by electrolysis.

Commercial aluminum contains commonly, silicon and iron, the former both in combination and in the free state, like carbon in cast iron. The amounts of impurities range from 0.3 to 2 per cent. of silicon and 0.15 to 2 per cent. of iron.

Physical Properties.—Aluminum melts at 659°C. (1,218°F.). Its specific heat is 0.2185. Because of its high specific heat, aluminum melts, cools and thus solidifies very slowly.

Pure aluminum shrinks on solidifying from the molten state 0.2031 in. per foot. Its casting alloys, which contain about 10 per cent. of copper, shrink rather less than this, about 0.156 in. per foot. In heat conductivity it rates at 31 with silver at 100. With pure copper taken as a standard at 100, pure aluminum has an electrical conductivity of 62.

The specific gravity of aluminum is very low, being about 2.56 cast; and about 2.68 wrought, about one-third that of iron. It is most malleable at about 150°C.; above 200°C. it becomes brittle; with frequent annealing it may be rolled cold. It is about as hard as silver, but for commercial uses its hardness is increased by alloying with it a small amount of copper. Its tensile strength when cast is about 6 or 7 tons per square inch.

Pure aluminum does not cast very well. When molten it absorbs nitrogen, carbon dioxide, etc., which are expelled again upon cooling. Small additions of copper, zinc, nickel, chromium, etc., increase its efficiency for casting. In melting aluminum, sodium chloride is used as a flux.

At proper temperatures aluminum can be welded, and if suitably cleaned it may be soldered without a flux.

Chemical Properties.—When strongly heated in the air, aluminum oxidizes quite considerably. Thin pieces will burn in the air with a brilliant light, resembling magnesium, aluminum oxide and some nitride being formed.

On exposure to the air at ordinary temperatures it oxidizes very slowly, for a thin film of the oxide is formed which is very closely adherent and therefore protective. Hydrogen sulfide has no action on aluminum.

In dilute sulfuric acid, aluminum is practically unaffected. But in the hot, concentrated acid it dissolves in a manner similar to copper, sulfur dioxide being evolved. Nitric acid also affects it very little. Hydrochloric acid dissolves it forming the soluble aluminum chloride. In solutions of the caustic alkalies it dissolves very readily, developing great heat. The products formed are hydrogen and aluminates—as for example, sodium aluminate, Na_3O_3Al. Hence solutions of the alkalies, such as caustic soda, should not be used in cleaning articles made of aluminum.

Uses.—Aluminum is used in making a great variety of alloys. Aside from the aluminum bronze, mentioned on page 184, which see, it is used in making magnalium, an alloy consisting of 10 to 25 per cent. of magnesium and 90 to 75 per cent. of aluminum. It is lighter and much harder than aluminum and is susceptible to a high polish. With cadmium, aluminum forms a very tough alloy. With lead and antimony it alloys only with difficulty.

It is also used in making cooking utensils, electric wires and cables. In very finely divided form it is used for paint and in the form of leaf for stamping titles on the covers of books.

It is used in deoxidizing molten iron and steel, especially in the top of the ingot when steel is poured in the ingot mould. In a similar manner it is used to prepare the metals from their oxides, by heating a mixture of powdered aluminum and the oxide of the metal to be reduced. The mixture is known as thermit.

Thermit.—Finely divided aluminum is mixed with the metallic oxide in about equal proportions by volume. When the reaction between the two is started, it proceeds of itself and develops great heat, producing molten aluminum oxide as slag, and the molten metal whose oxide was reduced. The reaction is started in the thermit by means of a small quantity of ignition powder placed on top of it. The ignition powder is made up of a mixture of barium peroxide and finely divided magnesium, and reacts according to the following equation:

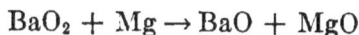

$$BaO_2 + Mg \rightarrow BaO + MgO$$

In the thermit itself, if the sesqui-oxide of iron is used, the reaction is as follows:

$$Fe_2O_3 + 2Al \rightarrow Al_2O_3 + 2Fe$$

The mixture of iron sesqui-oxide with aluminum is known as "red thermit" and the magnetic oxide mixture as "black thermit," the latter being, probably, the most used. The reaction with the magnetic oxide occurs thus:

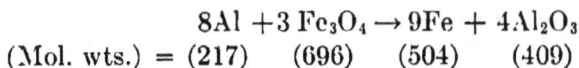

$$8Al + 3 Fe_3O_4 \rightarrow 9Fe + 4Al_2O_3$$

(Mol. wts.) = (217)   (696)   (504)   (409)

In this reaction, the weights used are approximately 3 of aluminum and 10 of magnetic oxide, which produce about 7 of metallic iron and 6 of slag. But in actual practice the proportions are usually 1 of aluminum and 3 of magnetic oxide. It is generally estimated that the molten metal produced is about one-half the original mixture by weight and one-quarter by volume.

The temperature produced by the heat of the thermit reaction is very great. It has been calculated by Richards[1] as 2,694°C. Féry using the radiation pyrometer found the temperature of the stream of metal flowing from the crucible to be 2,300°C.

The exceeding hot metal has proven to be very useful in a certain form of *welding*. For the purpose of welding, the parts to be joined are cleaned, butted together, and the joint surrounded by a suitable mould, and the ends are then heated by a gas flame playing through an opening provided in the mould for this purpose. The hot metal is allowed to flow from the reaction crucible into the mould, where the ends being joined are fused by the great heat, and upon cooling, the mass solidifies as a whole.[2] The process is applicable to wrought iron, steel, cast iron, and other metals whose oxides are capable of reduction by aluminum, as for example, nickel, manganese, and chromium.

Many metallic sulfides also react with aluminum powder, producing aluminum sulfide and free metal. Hence the thermit process is much used in obtaining metals from sulfide, as well as oxide ores.

COPPER

Ores.—Copper has been found in the metallic state in large quantities in the vicinity of Lake Superior. It is also found in the Ural Mountains, in Sweden, in China and Japan. In Montana are found great ore deposits, such as copper glance,

[1] Elec. Chem. and Met. Ind., January, 1905.

[2] For methods of making welds by thermit see Welding by R. H. HART; also literature issued by the Goldschmit company.

Cu_2S; the chalcopyrite, $Cu_2S \cdot Fe_2S_3$; and bornite, $Fe_2S_3 \cdot 3Cu_2S$. Other ores are ruby copper, Cu_2O; azurite, $Cu(OH)_2 \cdot 2CuCO_3$; and malchite, $CuCO_3 \cdot Cu(OH)_2$, a great deal of which is found in Arizona and Siberia.

Reduction of Ores.—If the copper ore contains no sulfide the process of extracting the metal is comparatively simple, for then it is necessary only to heat the ore with coal or coke in a blast furnace.

If sulfides are present as they very commonly are, the process is a great deal more difficult, for such ores contain iron, lead, arsenic, antimony, together with the sulfur which must be eliminated.

The methods employed are rather varied and complex, the particular method used depending upon the character of the ore being treated. Only one of the more common methods can be outlined here.[1] This will be considered under three heads: roasting, smelting, and converting.

The object of the *roasting* is to convert a part of the sulfide into the oxide form, sulfur dioxide gas being evolved. This partially oxidized ore is then *smelted* to separate the earthy material from the metallic oxides and sulfides. The smelting may be carried out in a blast furnace or in a special form of reverberatory furnace, but in either case, the product is known as *matte*. The matte is not metallic as is the product of the iron blast furnace, but consists chiefly of the sulfides of copper and iron, together with other metals that may have been present in the ore. This matte is then "*blown*" in a Bessemer converter, which resembles the steel Bessemer, except that the blast is introduced at the sides rather than at the bottom. The result of this "Bessemerizing" is the production of a certain amount of copper oxide, which in turn reacts with the copper sulfide present. In this way the sulfur and oxygen are eliminated as sulfur dioxide gas, and metallic copper, known as "blister copper," is produced.

Refining.—The impure, or blister copper, must be refined, and this may be accomplished in one of two ways, fire refining and electrolytic refining.

In fire refining, the blister copper is melted in a reverberatory

---

[1] For a full treatment of the metallurgy of copper, see PETERS: Principles of Copper Smelting.

furnace and air is blown over the charge. Cuprous oxide is formed which acts as an oxidizing agent for the impurities, these being then removed as a slag. An excess of cuprous oxide is left, and must be reduced, since if allowed to remain, it causes the metal to be in a condition known as "dry," meaning that it is brittle, breaking with a granular fracture, instead of being tough with a fibrous fracture. Reduction is accomplished by covering the molten charge with charcoal or coke and stirring with a green pole. Over-reduction must be avoided, since the copper almost always contains traces of the oxides of arsenic, antimony, and bismuth, which in their oxide state are not so objectionable; but if the reducing action of the carbon and the gases from the pole is too great, these oxides are reduced, and the resultant metals are more active in producing brittleness than the cuprous oxide itself.

Fire-refined copper is not sufficiently pure for electrical work, and since over half that produced is used for this purpose, the electrolytic method is usually employed.

For electrolytic refining the impure copper is cast in the form of thick plates, known as anode plates. These are suspended in an electrolyte consisting of an acid solution of copper sulfate. The kathodes are thin sheets of pure copper. As metal is deposited upon the kathode from the solution an equal quantity is dissolved from the anode, the process ultimately amounting to the transference of copper from the anode to the kathode. The impurities are not soluble in the electrolyte and fall to the bottom as a "slime" which is later treated for the recovery of gold, silver, etc. Copper refined in this way is known as "electrolytic copper," and possesses a very high degree of purity. It is about 99.95 per cent. copper, the remainder consisting chiefly of hydrogen with small amounts of metallic impurities. Because the kathode plates do not stand shipment well, small pieces being easily broken off, the copper is usually not marketed in this form, but is first melted and cast in bars. The melting lessens the purity slightly.

Physical Properties.—Copper melts at 1,083°C. (1,981.5°F.). At a white heat it is sufficiently volatile to color a flame green, but loss on melting is not noticeable. Just below its melting point it becomes so brittle it may be pulverized. At a red heat it may be welded. Its specific gravity is 8.94.

Like aluminum, pure copper does not cast well. When molten it absorbs gases, such as carbon monoxide, hydrogen and sulfur dioxide, which separate out on cooling and cause "blow-holes."

To produce sound castings having a high electrical conductivity, Weintraub[1] adds about 1 per cent. by weight of boron suboxide to the molten copper, which has been heated to about 1,350°C. To secure high conductivity, pure copper must be used, since the suboxide acts only as a deoxidizer. Sound castings can be made by this treatment regardless of the purity, but the conductivity is reduced in proportion to the amount and kind of impurity present. Electrical conductivity as high as 97 per cent. has been obtained in castings made in this way, but usually it is about 88 to 90 per cent.

The following percentages show the extent to which small amounts of impurities affect the conductivity of copper. Traces of arsenic will lower the conductivity about half; 0.13 per cent. of phosphorus will lower it 30 per cent.; and 0.4 per cent. of iron, 64 per cent.

Copper is malleable both at ordinary temperatures and at a red heat. Lead is added to copper in quantities less than 0.5 per cent. that it may be able to be worked more easily, but if more than 0.5 per cent. is added it causes the copper to be brittle. In copper alloys for casting, as much as 10 to 20 per cent. may be added to cheapen the product, but this lead largely separates out in globular masses on cooling. Lead will not alloy with copper in amounts greater than 3 per cent.

Chemical Properties.—Copper is not affected by dilute sulfuric acid, but in the presence of hot concentrated acid it dissolves with the formation of copper sulfate, water and sulfur dioxide. It is much more readily attacked by sulfuric acid in the presence of air. Cold hydrochloric acid has but little action, but the hot concentrated acid attacks it with the formation of hydrogen and cuprous chloride. It is very soluble in nitric acid both dilute · and concentrated. It is important to remember that many organic acids, even when dilute, act upon copper. Such acids are often found in food products, and copper is used in making utensils for cooking. All copper compounds are poisonous.

Ammonia water slowly dissolves copper in the presence of air,

[1] *Gen. Elec. Rev.*, **15**, 459 (1912).

with the formation of a double compound, $Cu(NH_3)_4(OH)_2$, a blue-colored solution being produced.

ZINC

Ores.—The chief ores of zinc are the carbonate, $ZnCO_3$, called smithsonite or zinc spar, and the sulfide, ZnS, called zinc blende, or black jack. Other ores of less importance are also found, but in whatever state the zinc may occur it is necessary to convert it into the oxide before metallic zinc is obtained.

Smelting.—The oxide is produced by roasting the other forms and then it is reduced by carbon, usually finely divided coal, at high temperatures. The temperature necessary is about 1,200 to 1,300°C., and since this is above the temperature at which zinc volatilizes, the reduced metal is always obtained as a vapor. This vapor must be condensed only at some temperature above the melting point of zinc, so that the metal may be obtained in the form of a liquid. If the condensation should occur below the melting point, so that the zinc condenses to the solid state, a very finely divided product known as zinc dust is obtained. This dust is useful in certain ways, but it evaporates at about 200°C., and oxidizes very readily so that it cannot be melted to form a solid mass of metal. The vaporizing point of the dust is about 700° lower than that of the ordinary metal, the difference being due to the peculiar physical state of the particle. There seem to be present internal strains such as exist in the laboratory curiosity known as "Prince Rupert's drops." In ordinary smelting practice, some dust is generally formed, and this will be mixed with about 5 to 10 per cent. of zinc oxide. However, the greater part of the metal collects in the liquid form and is drawn off and cast in moulds, thus producing the form of zinc known as *spelter*. In the United States the spelter constitutes the commercial zinc, although in European countries, especially in those localities where the ore contains considerable lead ore, it is necessary to further refine the spelter.

In American spelter the chief impurities are lead, iron and cadmium, but small amounts of arsenic, antimony and sulfur are usually present also.

According to the amount of impurities present, the American

Society for Testing Materials[1] divides the spelters into four grades, as shown by the following specifications:

STANDARD SPECIFICATION FOR SPELTER

	Lead shall not exceed, per cent.	Iron shall not exceed, per cent.	Cadmium shall not exceed, per cent.	The sum of the Pb, Fe, and Cd shall not exceed, per cent.
A High-grade......	0.07	0.03	0.05	0.10
B Intermediate....	0.20	0.03	0.50	0.50
C Brass special....	0.75	0.04	0.75	1.20
D Prime western...	1.50	0.08		

Grades A, B, and C shall be free from aluminum.

Chemically pure zinc is obtained by distilling pure precipitated zinc carbonate with pure carbon. Also, it is prepared by the electrolysis of pure zinc salts.

Physical Properties.—Zinc melts at 420°C. (787°F.), and boils at about 918°C. under atmospheric pressure. Consequently it distils readily and considerable loss must be allowed for in making alloys. Upon solidifying from the molten state it becomes very brittle. Jones says[2] that it manifests the following remarkable behavior on warming. When heated above 100°C. it becomes malleable and can be rolled into sheets, drawn into wire, etc. Having acquired this property, it retains it even at ordinary temperatures. When heated still higher—to somewhat above 200°C.—it becomes brittle again, and can be readily pulverized. Having become brittle, it retains this property even when cooled again to the atmospheric temperature.

Cast zinc has a specific gravity of 7.08, but the wire and hammered zinc have a specific gravity of 7.19.

Because of its great fluidity when molten, zinc casts well. It shrinks but little on solidifying. It is much used for casting small statues and other ornamental work, which are then electroplated with bronze or brass. So-called French bronzes are thus made.

Chemical Properties.—Zinc is extremely soluble in practically all of the acids. The rate of solution becomes less as the degree

[1] Year Book, 1915, p. 344.
[2] JONES: Principles of Inorganic Chemistry, p. 387.

of purity increases. Absolutely *pure zinc* will not dissolve in any of the acids except nitric. In contact with hot concentrated sulfuric acids, zinc sulfate, water and sulfur dioxide are formed. In hot caustic alkalies zinc is soluble, as:

$Zn + 2NaOH \rightarrow Na_2O_2Zn + H_2$ (compare aluminum).

Zinc and water at ordinary temperatures do not noticeably react, but when steam is passed over heated zinc, zinc oxide and hydrogen are formed. In the air, it burns readily with a bluish-white flame forming dense clouds of zinc oxide.

CADMIUM

This less common metal is obtained usually in small quantities from zinc-bearing ores. It is silver-white, malleable and ductile even at ordinary temperatures, resembling tin in appearance. In the air it is quite stable, although it does not long retain a bright surface due to the formation of a thin film of closely adherent oxide. Unless considerable care is exercised in melting it in making alloys, much will be lost, since it burns readily with the formation of a brown oxide, CdO.

Cadmium has a specific gravity of 8.6. It melts at 321°C. and is much used in making alloys requiring a low melting point. Cadmium is quite readily soluble in nitric acid, but less so in hydrochloric and sulfuric.

Cadmium sulfide, CdS, is a bright yellow compound used as a pigment in paints.

TIN

The chief tin ore is cassiterite, SnO_2, called also tin stone. It was formerly found in great quantities in Cornwall and Devon in the south of England, but now the world's supply is obtained chiefly from the Malay Peninsula and the island of Banca in the Indian Ocean. Tin is the only metal of commercial importance that has not been found in considerable quantities in the United States.

In the metallurgy of tin, the ores are first roasted to expel the sulfur and arsenic present. The iron, copper and other metals are then removed by treatment with crude hydrochloric acid. After this, the mass is agitated with water to remove the soluble constituents, and the finely divided ore is then mixed with carbon and heated in a furnace. The molten tin resulting from the reduction is drawn off and cast into bars.

Commercial tin is never absolutely pure but nearly so, being rarely below 99.9 per cent. Traces of lead, iron, copper and antimony may be present. The terms *block tin* and *sheet tin* indicate pure tin, while *tin plate* has reference to sheet iron or steel coated with tin.

Physical Properties.—Tin melts at 232°C. (449°F.) and boils at about 2275°C[1].

The specific gravity of pure tin is 7.3; of commercial tin it is somewhat higher. Tin is of a decidedly crystalline character, and when bars of it are bent a peculiar crackling noise known as the "tin cry" is noticeable. This is due to friction of the crystals as they move over one another.

Tin is rather soft and so malleable that it can be rolled into foil $\frac{1}{1,000}$ in. thick. However, it is not ductile, since its tensile strength is only about 2 tons per square inch. At about 200°C. tin is brittle and may be powdered. When pure tin is kept at low temperatures for some time it changes to a gray, extremely brittle mass—an amorphous form of tin. At −48°C. the change takes place most rapidly, but it occurs quite noticeably at even −15°. This is a cause of great annoyance in Russia, where the winters are extremely cold. The disintegration is there known as the "tin pest." This conversion of tin into the amorphous variety has also been known to occur in cold-storage warehouses. At above 20°C. the ordinary modification of tin is stable. Tin is a poor conductor of both heat and electricity.

Chemical Properties.—Tin is only very slightly affected by cold sulphuric acid. In hot concentrated acid it acts in the same way as the other metals, stannous sulfate, water and sulfur dioxide being produced. Nitric acid when diluted attacks tin slowly with the formation of stannous nitrate, $Sn(NO_3)_2$. Concentrated nitric converts it into insoluble, hydrated stannic oxide, known as metastannic acid, the composition of which varies according to the temperature and other conditions of preparation, but probably is most commonly $H_{10}Sn_5O_{15}$. This mass adheres to the metal, and the action of the acid is soon checked.

[1] *Proc. Roy. Soc.*, 82, 396.

Hydrochloric acid dissolves tin quite readily with the formation of stannous chloride, $SnCl_2$.

Above its melting point it oxidizes readily to stannic oxide, SnO_2, commonly known as putty powder. At very high temperatures it burns in a manner similiar to magnesium.

Tin does not ordinarily corrode or even tarnish very much when exposed to the atmosphere.

LEAD

Lead is sometimes, but not often, found in the free state in nature. Galena is the chief ore, and it occurs in considerable quantities in the United States, Spain, Great Britain, Australia and Germany. The United States produces more lead than any other country. Spain produces about one-fourth of the world's supply.[1] Sometimes lead is found in small quantities in the form of the carbonate, chromate and sulfate.

The galena ore is roasted so that it is converted partly into the oxide and partly into the sulfate, and then these compounds are heated with more sulfide, in the absence of air, sulfur dioxide and lead being produced. The lead is then further treated to recover from it the silver which it frequently contains to some extent.

Commercial lead is quite pure, but small amounts of copper, arsenic, antimony, tin, zinc, iron, sulfur, etc., may occur. With the exception of copper, these elements are more readily oxidized than the lead and can be removed from it by some oxidizing treatment. High temperatures, with air and stirring will remove the most of them. Antimony, arsenic, zinc, and copper make the lead hard, and zinc lowers its resistance to acid attack.

Physical Properties.—Lead is the softest of the metals used in the trades. The tensile strength is low, about 1 ton per square inch, and it is very malleable. Just before it melts it becomes brittle, but at a slightly lower temperature it is so malleable that it can be squeezed or squirted into tubes or rods.

Lead melts at 327°C. (621°F.), and boils at about 1,525°C. Its specific gravity is 11.34.

Chemical Properties.—Lead is not much affected by dilute acids, being especially indifferent to sulfuric. The sulfuric, and

[1] PRESCOTT and JOHNSON: Qualitative Chemical Analysis, p. 29.

hydrochloric as well, act but slightly because their compounds with lead are so sparingly soluble. Nitric, when moderately concentrated, acts the most readily. In acetic acid, lead is also quite soluble. Many other weak organic acids act upon lead quite noticeably, hence it cannot be safely allowed to remain in contact with any food products. All compounds of lead are poisonous, so care must be exercised to prevent it from finding its way into the system, since small amounts taken daily seem to accumulate and in time may cause serious trouble. Stains and splashes of paints containing lead compounds or pigments should not be allowed to remain upon the skin, since lead compounds from this source are rather readily absorbed through the skin.

Uses.—Lead is used for the making of lead pipes for plumbing; in fact, from the Latin word for lead—*plumbum*—is derived the name "plumber." In the form of sheet lead it is used for lining acid chambers, and for chemical-proof plumbing in general. Sheet lead is also much used for roofs and gutters. Until recently lead has been used almost solely in the making of the plates for storage batteries, and the lead plate-sulfuric acid type of battery is still most common. It is used in making shot, which contain also from 0.2 to 0.4 per cent. arsenic, and some antimony.

Lead enters into a great variety of alloys, as solder, Babbitt's metal, type metal, pewter, etc. It is used also in making the low melting-point alloys, such as Rose's and Wood's metal.

Molten lead will dissolve gold, silver and platinum, hence it must not be allowed to come into contact with apparatus or ornaments made of these metals.

BISMUTH

Bismuth occurs in nature usually in the free state, and is commonly quite pure. It is sometimes found also as the sulfide and the oxide. It is refined by fusing it with a flux of sodium nitrate or a mixture of sodium carbonate and potassium chlorate. Thus arsenic, antimony, iron, lead, sulfur, etc., are converted into their oxides and removed.

Bismuth remains practically unaffected in the air, and this no doubt accounts for the fact that it is found in the metallic state.

Physical Properties.—Bismuth melts at 271°C. (520°F.). At about 1,700°C. it volatilizes. The specific gravity of bismuth is 9.82. It has the greatest atomic weight of all the common metals.

Bismuth is brittle and very crystalline and, therefore, not malleable or ductile. The crystal faces are gray with a reddish sheen. On account of the latter characteristic bismuth may be readily distinguished from antimony, which it resembles in crystalline structure.

Unlike most metals, bismuth expands upon cooling, increasing over 2 per cent. upon solidifying from the molten state. It is a poor conductor of electricity.

Alloys of bismuth are very readily fusible and are used chiefly where this property is desired. Also it is used sometimes for type metal where the casting must have a sharp outline, which is produced by the expanding power of cooling bismuth.

Chemical Properties.—Upon being heated in the air above its melting point, it burns with the formation of the trioxide, a yellow powder, Bi_2O_3. Considerable loss is encountered in this manner in making alloys.

Bismuth is only very slightly attacked by hydrochloric acid, but nitric and sulfuric acids dissolve it.

ANTIMONY

Antimony is found chiefly as the mineral stibnite, Sb_2S_3, which occurs mainly in Hungary and Japan. However, a considerable amount of the metal is obtained as a by-product in the metallurgy of lead and silver. It is obtained from the sulfide by heating it with iron. Also the sulfide is converted into the oxide by roasting and then reduced with carbon. The impure metal thus obtained is known as *"regulus."*

Antimony is silvery-white, hard and so brittle that it can be powdered under the hammer.

It melts at 630°C. (1,166°F.), and volatilizes at about 1,600°C. It has a specific gravity of 6.62. Like bismuth it expands considerably upon cooling and hence is used in alloys to be cast with sharp detail, especially in type metal.

Antimony does not tarnish upon exposure. When heated it burns readily forming the white trioxide, Sb_2O_3.

MERCURY

Mercury occurs chiefly in the form of the sulfide, HgS, known as cinnabar. The metal can be obtained by simply roasting this sulfide in the presence of air. The sulfur oxidizes, and vapors of mercury distill off and are condensed. Sometimes also the sulfide is heated with lime. Calcium sulfide is formed and the mercury distills off.

Physical Properties.—Mercury is the only metal that exists in the liquid state at ordinary temperatures. It melts at $-38.9°$C. ($-38.0°$F.) and boils at $357°$C. The vapor conducts electricity. The specific gravity of mercury is 13.59.

The alloys of mercury are known as *amalgams*. The amalgam assumes the liquid form when the mercury is in excess, and the solid form if the other metal is present in the greater proportion. Practically all of the metals excepting iron and platinum form amalgams, and even platinum will amalgamate under certain conditions. Tin amalgam is used in making mirrors. The zinc electrodes of primary batteries are amalgamated on the surface to prevent undue action of the electrolyte in which they are placed.

Chemical Properties.—Mercury is practically unaffected in the air. At ordinary temperatures hydrochloric and sulfuric acid have scarcely any action upon it. However, nitric and hot concentrated sulfuric acid dissolve it quite readily.

NICKEL

Nickel is obtained chiefly from New Caledonia and the Province of Ontario. The most important ores are nicollite, NiAs, and nickel glance, NiAsS.

Nickel is a silver-white, lustrous metal, is very malleable and ductile, and has a high tensile strength. At temperatures below $350°$C. it is magnetic. It is very permanent upon exposure to the air and is, therefore, much used to plate other metals. It is not much affected by any aid excepct nitric in which it dissolves readily.

It has a specific gravity of 8.9 and melts at $1,452°$C. ($2,646°$F.).

It is used in German silver, in nickel steel, and in an alloy called manganine consisting of nickel, copper and manganese, which is used for resistance wires. It is used in plastic bronze for bearings and in many other alloys.

11

COBALT

Cobalt is obtained chiefly from Canada, and from various localities on the European continent. Important ores are smaltite, $CoAs_2$, and cobaltite, $CoAsS$.

Cobalt resembles nickel in many of its properties, but it is considerably harder than nickel. It is silver-white, malleable and very tenacious. Cobalt can be made magnetic, but unlike iron and nickel it retains its magnetism even at a white heat. It is growing to be of considerable importance in the manufacture of alloy steels.

Cobalt is permanent in the air, and also resists quite well the action of all acids except nitric. It has a specific gravity of 8.95, and melts at 1,480°C. (2,696°F.).

MANGANESE

There is a variety of ores of manganese. It occurs in the form of three of its oxides, as well as the sulfide and carbonate. Chief of these is the dioxide, MnO_2, known as pyrolusite, the major amount of which comes from the Caucasus Mountains.

Two iron alloys of manganese are much used in the steel-making industry. They are speigel-eisen, containing not over 21 per cent. of manganese, and ferromanganese, containing in the neighborhood of 81 per cent. manganese. These alloys are produced direct by reducing mixtures of iron and manganese ores in the ordinary type of blast furnace.

By the thermit process, metallic manganese can be obtained almost entirely pure. The absence of carbon in manganese so obtained is a point worthy of comment. By using this metal manganese steel can be made which is much softer and far more workable than that produced by the use of manganese reduced by carbon.

Manganese is hard, steel-gray and brittle, resembling cast iron a great deal in appearance. Upon exposure to the air it oxidizes in a manner similar to iron but more readily. It dissolves even in dilute acids. It reacts with boiling water forming the hydrated oxide and free hydrogen.

Its specific gravity is 7.39 and it melts at 1,260°C. (2,300°F.).

CHROMIUM

Chromium is obtained chiefly from the ore chromite, which is otherwise called chrome iron ore, $FeO \cdot Cr_2O_3$. Chromium may be obtained by the Goldschmidt process or by reducing the oxide with carbon in the electric furnace. It is steel gray in color, hard and brittle. It is unaffected upon exposure to the atmosphere. Chromium resists the action of nitric acid, but dissolves in warm dilute hydrochloric or sulfuric acids.

It has a specific gravity of 6.9, and melts at 1,520°C. (2,768°F.).

An alloy of chromium with iron—ferrochrome—is much used in making a special steel, known as chrome steel. The ferrochrome is obtained directly from the chromite ore by heating it with carbon in an electric furnace.

MOLYBDENUM

Molybdenum is obtained from the ore molybdenite, MoS_2, which very closely resembles graphite in appearance, and from wulfenite, $PbMoO_4$. Molybdenum is a hard, lustrous metal resembling iron in its physical properties. It may be welded and tempered.

It has a specific gravity of 9.01, and melts at 2,500°?C. (4,500°F.).

It is used in making molybdenum steel.

TUNGSTEN

Tungsten is obtained from the ores scheelite, $CaWO_4$, wolframite, $(FeMn)WO_4$, and stolzite, $PbWO_4$. The metal is hard, brittle and steel gray in color. It has a specific gravity of 18.73, and melts at 3,000°C. (5,430°F.).

It is used in making filaments for incandescent lamps and in an alloy steel known as tungsten steel.

TITANIUM

Titanium is never found native. It occurs chiefly as the dioxide, TiO_2, which is found in two mineral forms, brookite and anatase. It is also found in the polymeric form, Ti_2O_4, known as rutile. The oxide may be reduced with carbon by means of the electric arc, or by aluminum in the thermit reaction, although the element obtained by these methods contains carbon and aluminum in the combined and alloyed form.   It is these impure forms that are used in the steel industry as deoxidizers.

The metal often exists as a dark gray, lustrous, amorphous powder.   But these particles are shown to be distinctly metallic when examined under the microscope.   Also it may exist as steel-gray metallic mass, which is fairly hard but may be worked at a red heat.

It melts at 1,800°C. (3,272°F.).

When titanium is heated in the air it burns with a very brilliant incandescence, forming the dioxide.   It is one of the very few metals that combine directly with nitrogen when heated in it. It decomposes boiling water with the evolution of hydrogen. Also, it dissolves in hydrochloric and sulfuric acids in the usual manner.

Because of the extreme readiness with which titanium unites with oxygen and nitrogen it is used in molten steel, subsequent to the addition of manganese, to remove the last traces of these gases.

VANADIUM

Vanadium is found in nature chiefly as the lead, zinc or bismuth salt of vanadic acid.   For example, the ore vanadinite is $3Pb_3(VO_4)_2 + PbCl_2$.   Also, it is found in some iron and copper ores.

The vanadium metal is a light gray powder with a silver-white luster.   The microscope shows the particles to be crystal-line.   They do not become coherent when compressed.

Vanadium has a specific gravity of 5.8; melts at 1,720°C. (3,128°F.), and is non-magnetic.

The metal is only slowly oxidized in the air at ordinary temperatures, but burns when ignited.   Vanadium also combines with nitrogen when burning in the air, forming a mononitride.

It is insoluble in hydrochloric acid or in solutions of the alkalies. It is somewhat attacked by concentrated sulfuric acid, but is readily soluble in nitric acid both dilute and concentrated.

Vanadium is used in the steel bath in the same manner as titanium, since it also is an excellent deoxidizing agent.

CHAPTER VII

THE NON-FERROUS ALLOYS •

The *common alloys* are solidified solutions of metals in each other; but it must be understood that when the molten solution solidifies, it does not necessarily form that which is called a solid solution. The most of the common alloys undergo a selective crystallization as they turn from the molten into the solid state and consequently may be described as mixtures of crystals of the constituent metals, or of definite substances formed from them, the structure being less homogeneous than in the case of solid solutions. In order that the physical structure of alloys may be comprehended, it is necessary that the formation of solid solutions and eutectics be discussed.

Solid Solution.—When a solution, upon solidifying, retains

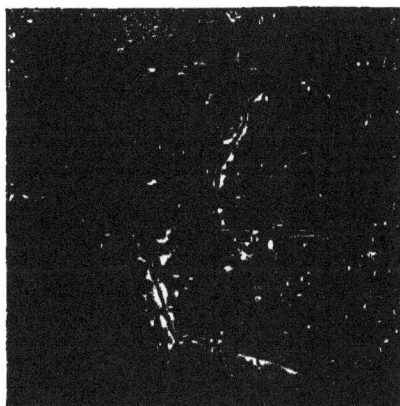

Fig. 37.—Copper-nickel alloy, containing 30 per cent. of nickel, annealed, etched with acid ferric chloride. Uniform solid solution. Magnified 60 diameters. (Kurnakow and Zemczuzny.)[1]

in the solid state its essential characteristics, it is described as a solid solution. It will be remembered that an essential feature of the ordinary liquid solution is that the constituents are completely merged; that is, the union is much more intimate than

[1] Reproduced by permission from Gulliver's Metallic Alloys.

165

mere mixture, so intimate, in fact, that the separate existence of the constituents cannot be determined by any physical means, such as, for example, by the examination with a microscope even with the greatest magnification. It appears like a simple uniform body. In this respect it differs from a mixture, for the constituents of a mixture can always be detected by microscopic examination. A solution, in brief, resembles a chemical compound in respect to the intimate association of its constituents; but differs from it in that the proportions are not fixed and definite, but may vary through a rather wide range. Glass is a familiar example of a solid solution.

A solid solution is not necessarily homogeneous throughout. The portion remaining fluid longest, for example in the center of the mass, becomes enriched in the dissolved substances, so that the solid solution may be more concentrated in one part than in another.

Eutectics.—The term eutectic is taken from the Greek and signifies "well melting." It is used most commonly to designate the lowest-melting-point alloy that can be formed of the constituents being used. The composition of the eutectic alloy is determined automatically, as shown in the following discussion.

Very often it happens that two metals are soluble in each other through a wide range of proportions in the liquid state, but are entirely insoluble in the solid state. Consequently, as the molten solution, or alloy in this case, cools, a separation of the constituents occurs. The temperature at which the separation begins, depends upon the percentages of the various constituents in the alloy, but the temperature at which it ends is fixed and definite. It has been found by careful observation, that when the two metals are present in the molten solution, in proportions other than that required to make the eutectic alloy, the excess constituent precipitates out and solidifies or crystallizes as the alloy cools, so that the eutectic composition is automatically formed in the molten remainder. As has been said, the temperature at which the excess constituent begins to separate depends upon the amount present, or conversely, the degree of solubility of the one metal in the other depends upon the temperature, just as is the case in salt solutions where a high temperature causes more of the salt to be dissolved.

This formation of eutectics can be illustrated by the use of tin and lead. We will consider first an alloy containing 80 per cent. of lead, at a temperature of 300°C. This will be at the point a in Fig. 38, and represents a liquid solution of lead in tin. If cooled, the solution will remain liquid until a temperature of about 265° is reached, which is a point on the line AB. This is the lowest temperature to which the solution can cool and retain 80 per cent. of lead in solution. If cooled further, lead will precipitate out and solidify, because the temperature is quite well below its melting point. Thus it becomes no longer a part

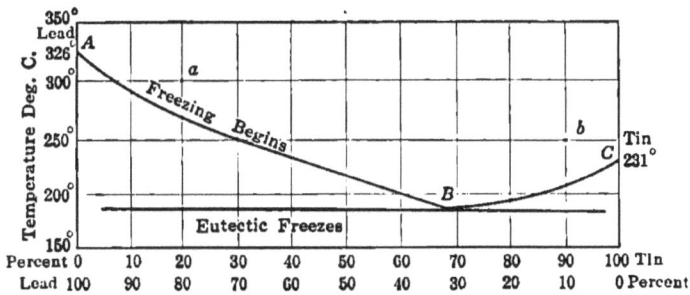

FIG. 38.—Freezing-point curve of the lead-tin alloys.

of the solution, although the crystals will be mechanically mixed with the remainder that is still in the molten state. In other words, the portion that is still molten serves as a medium in which these crystals are carried. The arrangement of crystals that have been separated out in this way is shown in Fig. 46, which see. These crystals of lead that separate out do not consist of absolutely pure lead, but contain some tin. Still at the maximum they will carry no more than 4 per cent. of tin held in solid solution, and usually it is considerably less than this.

Now as the still molten portion is further cooled, the solubility of lead becomes still less. Consequently, more lead separates out, and it will continue to do so as the temperature falls until the point B is reached, when the relatively small amount that still remains molten will contain 31 per cent. of lead and 69 per cent. of tin. So every point on the line AB represents the maximum amount of lead that can remain in solution at that temperature. The point B represents a temperature of 180°, and at this temperature only 31 per cent. of lead can remain in solution. Then, if the original alloy contained any percentage of lead more than 31 per cent., the excess will be intermingled with the solution in the form of lead crystals when the alloy has cooled to 180°.

If, on the other hand, we should start with an alloy containing 90 per cent. of tin, at a temperature of 250°, which would be represented in the figure by the point *b*, this would cool without any change in composition until a temperature of about 210° is reached as is shown by the line *CB*. If the alloy cools further,

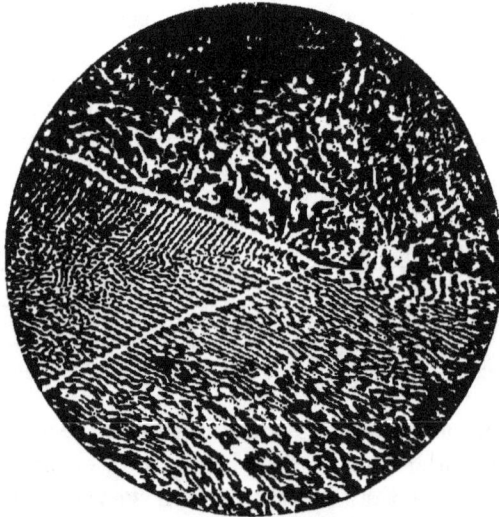

Fig. 39.—Eutectic alloy of bismuth and tin. Magnified 200 diameters. (Desch.)[1]

tin will precipitate out and solidify in crystal form. This separated tin will not be entirely pure but practically so. It will contain at a maximum no more than 2 per cent. of lead. As the cooling continues, the separation of tin will continue, the composition at the various temperatures being shown by the line *CB*. At *B* the liquid portion will contain in solution 69 per cent. of tin and 31 per cent. of lead.

Thus it is seen that regardless of the composition of the alloy with which we start, if it contains other than 69 per cent. of tin and 31 per cent. of lead, a sufficient quantity of the excess constituent will be thrown out of solution so that finally a certain

[1] Reproduced by permission from Sauveur's Metallography of Iron and Steel.

amount of solution containing the metals in the 69 to 31 ratio will be formed, mixed, of course, with crystals of either tin or lead, depending upon which metal was in excess.[1] The solution, the composition of which is determined in this way, is known as the *eutectic* solution.

Now as this eutectic solution cools below 180° it decomposes or separates entirely into its constituents, tin and lead. It is obvious that it cannot cool below 180° without crossing the point *B*, and to do this it must cross both the line *AB* and *BC* at once. And as has been seen, it cannot cross these lines without precipitating either lead or tin. As a matter of fact, this is exactly what happens. The lead and tin are both precipitated. They separate completely and crystallize in the form of fine crystals, which arrange themselves in the form of a parallel, banded structure, which is now no longer a solution, but partakes more of the nature of a very intimate mixture. The structure of a eutectic of this sort is shown in Fig. 39.

Liquation.—Many alloys, either when solidifying from the molten state, or when being heated while in the solid state, allow certain of their constituents that have a lower melting point than the remainder, to noticeably separate out. This separation is termed liquation, or sometimes if not pronounced, it may be termed segregation. The material that separates out in this manner, may be either the eutectic or at times almost a pure metal, a good example of the latter being the almost complete separation of lead from zinc or copper.

The desirable qualities of an alloy increase as the liquation becomes less. Slow cooling tends to increase, and rapid cooling tends to lessen it, hence the separation is more pronounced the thicker the casting. In some cases after the exterior has solidified, its contraction brings such a pressure to bear on the interior of the casting, that the more fusible constituents are forced into the pores of the solidified exterior, and in some cases they may pass entirely through and form globular masses on the surface of the casting.

---

[1] Exception to this statement must be made in the case of the alloys containing less than 4 per cent. of tin and less than 2 per cent. of lead. All alloys that contain no more than 4 per cent. of tin crystallize as solid solutions as do also those that contain no more than 2 per cent. of lead.

Reasons for Alloying Metals.—By forming alloys of metals, undesirable properties may be lessened and desirable properties increased. For example, the alloy may be harder, tougher, more tenacious, more ductile, or more fusible than the constituent metals. Also, the cost of production may be lessened. Aside from the introduction of cheaper metals into the alloy, it may be rendered less expensive to work. For example, some metals as copper and aluminum can only with difficulty be made to produce sound castings, and they can be machined and filed only with difficulty as well, while their alloys are much more conveniently manipulated. The effect of the formation of alloys on the physical properties of the metals is stated more in detail in the following paragraphs.

Specific Gravity.—In only a few cases does the specific gravity correspond to that which would result from the calculation of the mean of the constituents. In most cases it is greater than the calculated mean, contraction having taken place upon alloying. Specific gravity is in all cases dependent upon the density, or upon the closeness of the contact between the molecules, and this seems to be increased in most cases by the solution of one metal in another. The specific gravity of metals is increased by wire-drawing, hammering and pressure which closes the pores, but this is not permanent. Annealing removes it.

Hardness is the resistance a substance offers to the separation of its molecules by the penetrating action of another substance. By alloying, hardness is scarcely ever reduced, while in many cases it is greater than that of the separate constituents. Frequently, the increased hardness may be the only reason for forming the alloy. Pure copper is harder than pure tin, but an alloy of copper and tin containing only 5 per cent. of tin is almost twice as hard as pure copper.[1] The hardness of copper and copper-tin alloys is also much increased by small amounts of iron and manganese. Zinc hardens copper, but its action is much less pronounced. The greatest hardness is produced by about equal parts of copper and zinc, when it is about twice that of pure copper.[1] Lead-tin alloys are harder than pure lead.

Toughness is the property which enables a metal to resist breaking after sufficient stress has been applied to produce a

[1] Brannt: Metallic Alloys, p. 102.

permanent deformation. Tough metals are very ductile. And those metals lacking in ductility are very brittle. For the effect on ductility produced by alloying copper and zinc, see **Fig. 45, page 177.**

Tensile Strength. Tenacity.—This is the resistance a substance is able to oppose to a force tending to pull it apart. The tensile strength of copper is considerably increased by the addition to it of a small quantity of tin, although tin by itself has a comparatively low tensile strength. For the effect on tenacity produced by alloying copper and zinc, see Fig. 44, page 176. The tenacity of metals depends much upon the mechanical treatment to which they have been subjected.

Conductivity for Heat and Electricity.—Good conductors of heat are also as a rule good conductors of electricity, and *vice versa.* Pure metals are the best conductors and alloying them almost always lessens this property. Fractions of 1 per cent. of such metals as arsenic, bismuth, antimony and lead lower the conductivity of copper very greatly. Oxides contained in the metals also lower their conductivity.

Casting Qualities.—Metals that soften only gradually when heated and assume an intermediate pasty state, as copper and wrought iron, are more thickly fluid when molten than those that melt suddenly, as bronze and cast iron. Alloys pass more sharply into the fluid state than pure metals, and are, as a rule, more thinly fluid when molten. On this account, they are more easily cast, both because they fill the mould more readily and because they allow the gases to escape from them more quickly. Copper and bronze that contain cuprous oxide are rendered more thickly fluid by it, just as is iron by ferrous sulfide.

Gases in Metals.—Gases that separate out from molten metals are very objectionable because they produce flaws in the casting. The gas may separate because of a lessened solubility as the metal solidifies, or because they are being generated by a reaction that proceeds in the molten metal, as is the case with copper. Commercial copper contains small quantities of cuprous oxide, Cu_2O, and cuprous sulfide, Cu_2S. Since they are present only in small quantities they may be considered as being in very dilute solution in copper. These compounds react with each other, as:

$$2Cu_2O + Cu_2S \rightarrow 6Cu + SO_2$$

but being in dilute solution, the reaction is very slow and quite continuous. Hence as the molten copper remains in the mould, gases are formed, making it very difficult to produce sound copper castings. A similar action may occur between carbon and metallic oxides, producing carbon monoxide gas. This difficulty is eliminated by introducing some element, as phosphorus, silicon, manganese, aluminum, or other metals that use up the oxygen with the formation of difficultly reducible oxides.

The Preparation of Alloys.—If the metals being alloyed have much tendency to be destroyed by oxidation, they should be protected by a covering of small lumps of charcoal (no dust), borax or ground glass as soon as they are put into the crucible. When melted the alloy should be thoroughly stirred, preferably with a green stick, since the gases and vapors liberated from this, aid in producing a more thorough mixing. Also since the gases are mainly of a hydrocarbon nature they have a certain reducing effect.

If a large amount of one metal is being alloyed with a small amount of another, greater uniformity may be obtained by first preparing an alloy containing approximately equal parts of the two metals and then adding the remainder of the excess metal to this alloy in a second melting.

If the constituent metals have widely separated melting points, the one with the higher melting point is melted first, being covered well to prevent oxidation, and the lower-melting-point metal is then added. If the lower-melting-point metal were melted first, a great deal of it would be lost through oxidation while the temperature was being raised sufficiently to melt the other metal. However, if the lower-melting-point metal is much in excess, and is not very volatile, time and heat energy may be saved by melting it first, and then adding the higher-melting-point metal in small pieces. The molten metal usually dissolves the added metal at a temperature much below the melting point of the latter.

When three or more metals with widely differing melting points are being alloyed, less loss will occur if the constituents are first formed into two independent alloys, having melting point lying not far apart. The final alloy may then be made by combining the two intermediate alloys.

FIG. 40.—Alloy containing 79.7 per cent. of lead, 14.5 per cent. of antimony and 5.8 per cent. of tin, slowly cooled, etched with nitric acid. The white crystals of antimony-tin solution have floated to the top of the ingot. Magnified 100 diameters. (Gulliver.)

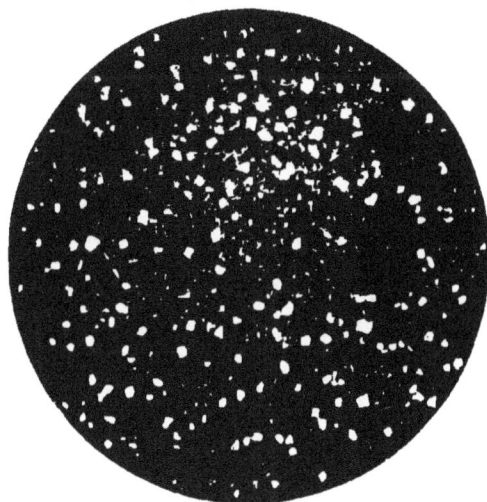

FIG. 41.—Same alloy as that of Fig. 40. Quickly cooled, etched with nitric acid. The crystals are much smaller and are distributed with fair uniformity. Magnified 100 diameters. (Gulliver.)

The Crystalline Structure of Alloys.—The natural state of all metals is crystalline. In alloys, when eutectics form, the physical structure consists of well-defined crystals imbedded in a matrix of smaller ones. And the more slowly the crystals are formed the larger they become, as shown in Figs. 40 and 41. Slow cooling allows a larger crystal growth, as is shown in Fig. 40, where the cooling has been sufficiently prolonged to allow the crystals that were first formed to float in that which was still melted. But large crystals are not usually desirable because they produce a weaker alloy. Rapid cooling produces fine crystals, and makes a harder, stronger and more uniform alloy, although one that is much more brittle and less ductile. The center of a mass of alloy, cooling more slowly, will consist of coarser crystals than the outside. Also, it should be noted that crystal growth does not cease with the solidification of the alloy. While it is still hot the molecules still have considerable freedom and can redistribute themselves. Then the larger crystals grow, absorbing the smaller ones. It is said that even when at atmospheric temperatures, vibration will also cause this crystal growth.

The Effect of Hot and Cold Working.—When an alloy is subjected to hot working, the crystal structure is broken down, but being hot there is molecular vibration and the crystals form again, although they are small. The effect of this is to harden and to some degree strengthen the alloy. But cold working produces a different effect. When cold, molecular activity has largely ceased, and the broken crystals remain broken, so that the metal is left in a permanent condition of strain, causing decided brittleness and weakness.

Annealing of Alloys.—The object of annealing an alloy is to relieve the strains that may exist in it, for example, those produced by working it. This result is accomplished almost entirely by re-formation or growth of the crystals, which is illustrated in Figs. 42 and 43. Annealing lessens the strength of the alloy because it increases the crystal size and for the same reason softens it. But if the annealing is carried too far, the crystals may become too much enlarged and brittleness also may result.

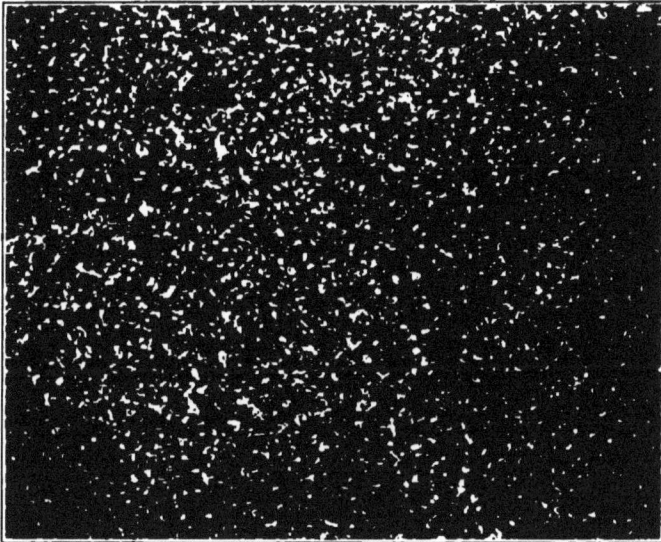

Fig. 42.—Rolled aluminum bronze containing 7.5 per cent. of aluminum. Magnified 10 diameters. (Law.)

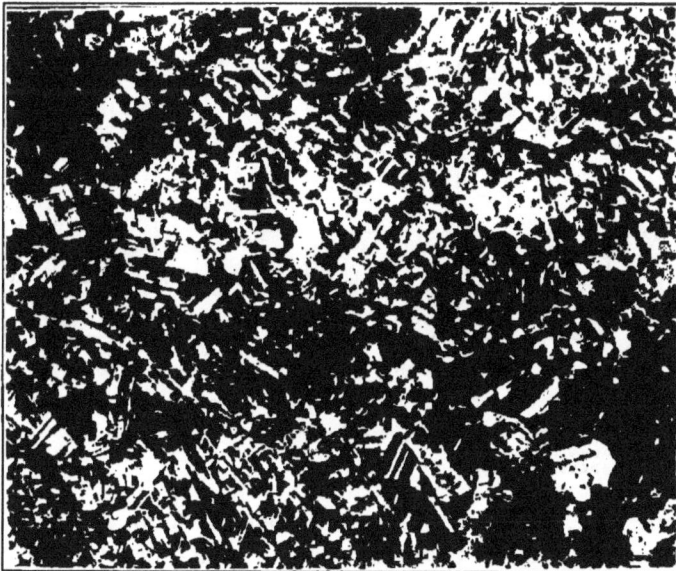

Fig. 43.—Same alloy as shown in Fig. 42, after prolonged annealing. Magnified 10 diameters. (Law.)

Brass

Structure.—Brass is an alloy of copper and zinc. The commercial brasses may contain from 5 to 60 per cent. zinc, but the most important are those that do not contain above about 35

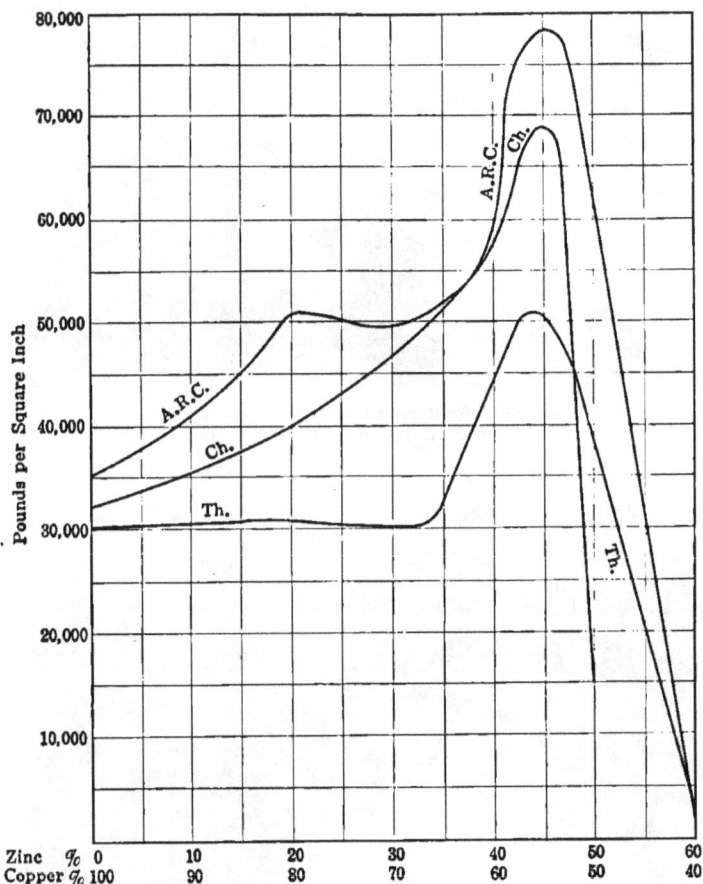

Fig. 44.—Tensile strength of copper-zinc alloys.   (After Sexton.)

A.R.C.—Alloys Research Committee (worked rods).
Th.—Thurston (castings).
Ch.—Charpy (annealed brass).

per cent. zinc.   The alloy exists as a simple solid solution if the zinc does not exceed this amount and consequently any brass containing less than 35 per cent. of zinc is somewhat more homogeneous than those that contain more, because in these latter alloys eutectics are formed. Still even in these high zinc alloys, the uniformity is very fair because the brasses show practically no tendency toward liquation.

Effect of Composition on Physical Properties.—If a brass were wanted for strength only and if hardness and brittleness were not objectionable, it should contain about 40 to 45 per cent. of zinc, but if ductility and toughness are required, as usually is the case, the zinc should not exceed 35 per cent. The curves

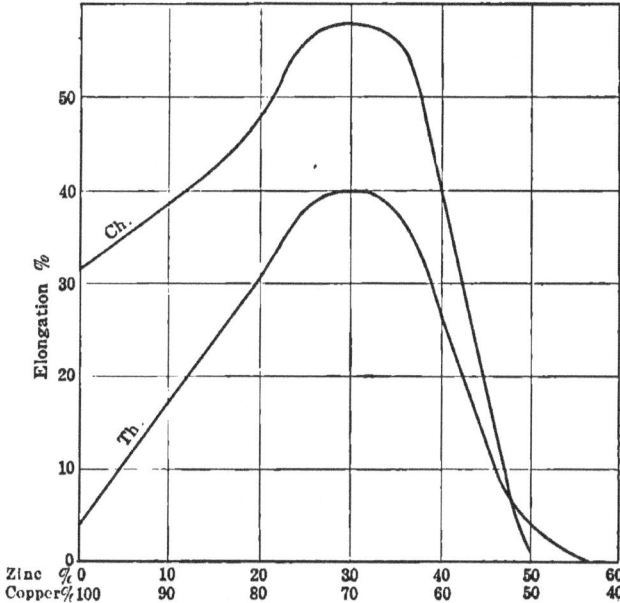

FIG. 45.—Ductility of copper-zinc alloys. (After Sexton.)
Th.—Thurston. Ch.—Charpy.

shown in Fig. 44 serve to roughly indicate the relation between the composition and tensile strength of brass. It should be noted that the strength varies greatly with the mechanical and heat treatment that the alloy receives.

However, brass so proportioned that its tensile strength is at the maximum is very brittle as is shown by the low ductility with this composition in Fig. 45. Ductility reaches a maximum with about 30 per cent. zinc, and since those alloys that are ductile are also tough, the most serviceable brass for general use is that which contains zinc in the neighborhood of this amount. However,

12

many of the brasses contain more zinc than this. Those that are often included under the name *Muntz's metal* may contain from 35 to 50 per cent. zinc. These alloys are harder and stronger, but also more brittle, than those that contain less than 35 per cent. of zinc. Brazing solder also contains a high amount of zinc, from 40 to 50 per cent., because the high zinc alloys have lower melting points.

Any brass that contains under 35 per cent. of zinc may be converted into wire and sheets, but usually these products contain between 20 and 30 per cent. of zinc. As a rule, cast brass contains more zinc than that which is worked into sheet and wire.

In general, brass has less strength than bronze, but the introduction of a small quantity of a third metal often increases the strength of brass, and modifies other of its properties as is illustrated by the following examples.

Effect of Other Metals.—Sometimes *aluminum* in quantities up to 3 per cent. is added to brass. The product has a deep golden color and is called aluminum brass. It casts well. Also, a small quantity of *tin* is often introduced into brass. As much as 2 per cent. noticeably hardens the alloy and increases the tensile strength. *Manganese* in brass acts as a deoxidizer. It also hardens and strengthens the alloy. For brass intended for filing and turning, 1 to 2 per cent. of *lead* is added to prevent fouling of the tools. Lead increases the softness and is sometimes added to brass that is to be worked, but care must be taken to add but a small amount. Only about 3 per cent. will alloy with the brass and if more than this is added, it has a strong tendency to liquate. Sometimes lead, in small amount, is introduced accidentally from the spelter. *Antimony* in brass is very objectionable. When present it is accidental. It makes the brass exceedingly brittle. The action of *bismuth* is very similar. The effect of *arsenic* is also similar to that of antimony but its action is not so decided.

Preparation of Brass.[1]—According to the general principles already given for making alloys, the copper, having the higher melting point and being in excess, is melted first, preferably under granular charcoal. Then the zinc, previously warmed, is added to the melt in pieces small enough that they do not chill

[1] For further information see SEXTON: Alloys, p. 248.

the copper and cause it to solidify. If this should happen, zinc would be lost while awaiting the remelting of the copper. When all the zinc has been added, the melt should be well stirred.

Since the melting point of copper is above the boiling point of zinc, considerable of the latter will be lost through vaporization and oxidation in any case, the amount varying from about 5 to 10 per cent. of that added, depending upon the care exercised. This loss must be taken into account in making brass to conform to given specifications.

Also, in remelting brass, zinc is lost, so that remelted brass is always richer in copper than before remelting.

Delta metal is essentially a brass containing a small quantity of iron. The alloy was first produced by A. H. Dick, who assigned to it the name "delta," this being the Greek equivalent of the initial of his own name. The alloy contains usually 55 to 60 per cent. of copper, 40 to 43 per cent. of zinc and 1 to 2 per cent. of iron, with a fraction of a per cent. of manganese or aluminum. This alloy is stronger, harder, and tougher than brass and is easily cast. The tensile strength is about two-fifths greater than a brass of similar composition with the iron omitted. In making delta metal the iron and zinc are first alloyed, producing an alloy containing about 8 to 10 per cent. of iron. This is then combined in the proper ratio with the copper.

German silver may be considered a brass that by the addition of nickel has acquired a white color and much increased hardness. It resists chemical action far better than ordinary brass and many of its uses depend upon this property. It has been found that iron further whitens and hardens the alloy, and practically all of the commercial varieties contain some iron, in amounts varying from 1 to 3 per cent. The composition of German silver is not definite, but it varies usually within the following limits: copper, 50 to 65 per cent.; zinc, 19 to 30 per cent.; nickel, 13 to 20 per cent. Disregarding the small amount of iron, and sometimes manganese also which it may contain, the following may be regarded as a typical alloy: copper, 55 per cent.; zinc, 25 per cent.; nickel, 20 per cent. In making this alloy, common practice is to melt together the nickel and part of the copper and then add the zinc and the remainder of the copper in the form of brass.

Monel Metal

Monel metal is an alloy containing approximately 68 to 70 per cent. nickel, 26 to 28 per cent. copper, and 2 to 3 per cent. iron.[1] The alloy is not made by melting together the pure metals but is produced directly from a copper-nickel matte. Copper-nickel alloys exist as solid solutions in all proportions.

Monel metal melts at about 1,360°C., and has a specific gravity (cast) of 8.87. Its tensile strength when cast is about 85,000 lbs. per square inch. The color of the metal very closely resembles that of nickel but it has a slightly darker tinge. It is susceptible to a high polish and does not tarnish readily. It is very resistant to corrosion. It is difficult to produce sound castings of monel metal because of the pronounced tendency to form blow-holes. To correct this, ferromanganese and magnesium are sometimes added in small amounts. It may be rolled into sheets and drawn into wire. Because of its high resistance to corrosion, it has been found very useful in making piston rods, roofing sheets, in marine work, or in other situations where resistance to corrosion is essential.

Stellite

Stellite is an alloy consisting essentially of cobalt and chromium,[2] but may also contain considerable molybdenum as shown by the following analysis:[3]

	Per cent.
Co	59.500
Cr	10.770
Mo	22.500
C	0.870
Si	0.770
Mn	2.040
S	0.084
P	0.040
Fe	3.110
W	0.000
Ni	0.000

[1] *Trans.* Canad. Min. Inst., **16**, 241 (1913).
[2] *Chem. Eng.*, **22**, 170 (1915).
[3] HIBBARD: High-speed Tool Steels, U. S. Bureau of Mines, *Bull.* 100, p. 61.

Stellite is grayish in color, is very strong, tough, hard, and resistant to abrasion and corrosion. It retains its hardness at a full red heat and is used for high-speed tools. It cannot be forged but must be cast and ground.

BRONZE

Structure.—Bronze is an alloy of copper and tin. When the tin does not exceed 5 per cent. the alloy solidifies as a solid solution, but with more tin than this, eutectics are formed. Bronzes that form eutectics have quite a tendency to liquate, and this tendency increases with an increase of tin. Liquation is said to be overcome by the addition of a small quantity of zinc.

The Effect of Composition on Physical Properties.—The tensile strength of bronze increases gradually with the amount of tin, reaching a maximum with about 20 per cent. of tin, but as the tin increases beyond this amount the tensile strength very rapidly diminishes. Bronze is most ductile when it contains about 5 per cent. of tin. With this amount it may be rolled satisfactorily at a red heat. However, bronze is used chiefly for casting. As the amount of tin is increased above 5 per cent., the ductility gradually lessens and practically disappears with about 20 per cent. of tin; and since ductility is coördinate with toughness, these alloys are very brittle. They are also very hard. The most useful of the bronzes are those that contain from 8 to 10 per cent. of tin, since the maximum *combined* strength and toughness is secured with about these amounts. The bronze containing about this amount of tin was formerly known as *gun metal*, since because of its strength it was used for making guns, but now steel has entirely replaced it for this purpose. At the present time, the term gun metal is very loosely used, so that it cannot be said to have any definite significance.

Bell metal is a bronze containing from 15 to 25 per cent. tin, and the remainder copper. It approaches the lower limit of the useful bronzes, being very hard and brittle, but is also resonant, hence its use for bells. The amount of tin increases within the range shown, according to the size of the bell, the larger bells containing the more tin. It must not be understood that all bells are made of this copper-tin alloy only; silver and other elements are sometimes added, or also bells may be cast of brass.

The Effect of Other Metals on Bronze.—*Zinc* is added to increase the fluidity of the molten bronze and so make the alloy cast better, as well as to increase the ease of its machining. But if more than 2 per cent. is added, the hardness and tenacity are decreased. Color is said to be improved by zinc. The introduction of as much as 2 per cent. of *lead* into bronze causes the strength and ductility to be very noticeably lessened. If cooled quickly, the lead is distributed in fine grains, but if allowed to cool slowly it liquates. Leaded bronzes containing considerable lead are used for bearing metals and will be discussed later under that head. *Iron* in bronze confers great hardness, and for this reason it is sometimes added to bronze bearings. It whitens the bronze and lessens the fusibility. *Ancient bronze* was hardened largely by hammering, but also it contained elements beside the copper and tin that hardened it, as for example about 1 per cent. of iron.

Preparation of Bronze.—In making bronze the copper is first melted and then the tin is introduced and the mass is stirred at once in order that the tin may be quickly mixed and the oxidation lessened. The loss of tin by oxidation is liable to be very great, being as much as 10 per cent. sometimes. If the loss is no greater than 2 per cent., it is considered excellent.

Bronze, like brass, has a very much lowered tenacity while at a temperature of 200°C. or above. It has been found that the tensile strength of any bronze is greatly increased by quenching from about 600°C.

SPECIAL BRONZES

With perhaps the single exception of aluminum bronze, the special bronzes are common bronzes to which one other special element has been added. The special elements chosen are those that possess more or less pronounced deoxidizing powers, and the superiority of the special bronzes seems to depend upon the fact that their entrained oxides have been so thoroughly reduced. Greater additions are made of those elements having the lesser deoxidizing power, but in some cases the whole of the amount added may be used up in removing oxygen from the alloy.

Karr and Rawdon[1] have pointed out in their report of their

[1] U. S. Bureau of Standards, *Technologic Paper* 59, p. 63 (1916).

investigations on the common bronze known as Government bronze, which consists of copper 88 per cent., tin 10 per cent., and zinc 2 per cent., that a microscopic examination of the fractured test specimens showed that the most common source of weakness was entrained or occluded oxides in the metal. Such oxides frequently occur as thin films in otherwise sound castings, producing brittleness and low ductility. By the reduction of these oxides with a suitable agent, the desirable properties of the alloy are much increased.

Phosphor Bronze.—Since phosphorus may be added to a bronze of any composition, the significance of this term is quite indefinite. It is certain, however, that whatever good qualities the bronze might have, they are greatly increased by the addition of phosphorus. The tensile strength, the elasticity, and the power to resist repeated stresses—pulls, twists, and bendings—is enormously increased, so much so that it may almost be considered as an entirely new metal.

Phosphor bronze of proper composition can be forged, drawn cold, rolled and cast. It seems to resist corrosion better than other bronze, especially of sea water, so that it is much used for propeller blades. Phosphorus is best added to the bronze in the form of phosphor copper, a hard brittle substance of white fracture, containing about 16 per cent. of phosphorus; or as phosphor tin, a white brittle crystalline material containing about 6 per cent. of phosphorus. Upon being added to molten bronze nearly all of the phosphorus oxidizes and escapes, so that most phosphor bronzes contain no more than a tenth of 1 per cent. The crystalline structure of two bronzes, one with, and the other without phosphorus, as shown by the microscope, seems to be the same in both cases.

Although phosphorus produces very beneficial results through its activity as a deoxidizing agent, if there is much of it left in the alloy, it may be a decided detriment. This residual phosphorus increases the hardness and brittleness of the bronze, so the amount left must be kept within low limits. When special hardness is desired there may be 1 per cent., but as much as 4 per cent. makes the alloy useless. Phosphorus greatly decreases the electrical conductivity of the bronze.

Silicon bronze is a bronze containing only a fraction of a per cent. of silicon. The silicon is added to the bronze in the same manner as phosphorus, but nearly all that is introduced unites with oxygen and is removed as a slag. Like phosphorus its use is as a deoxidizer. This bronze has the excellent properties of phosphor bronze but to a somewhat less degree. It has a much greater electrical conductivity than phosphor bronze, so that it is much used for telephone and other wires, particularly in those situations where high tensile strength is required. Castings for electrical purposes may be made of bronze to which has been added about 0.75 per cent. of silicon.

Manganese Bronze.—Manganese also serves as a deoxidizing agent and is frequently all oxidized out. Hence some manganese bronzes may, when finished, contain no manganese at all, but also some may contain as much as 4 per cent. Zinc is usually added to manganese bronzes to increase the fluidity of the molten alloy and thus produce sounder castings. Manganese bronze seems to contract considerably upon solidification and this must be considered in casting. It has a very high tenacity, and is much used for gear wheels and other parts of machinery. It has a high corrosion resistance and because of this is much used for screw propellers, etc.

Aluminum Bronze.—Although this alloy is really not a bronze, having no tin in it, this name will be retained because it is customarily so called. Copper and aluminum alloy in all proportions, but it is only those alloys containing 5 to 11 per cent. of aluminum that are of commercial importance. With more aluminum than this, the alloy is highly crystalline and brittle. Exception should be made, however, to the aluminum casting alloys, which are about 90 to 93 per cent. aluminum and the remainder copper. This combination produces also a very tough and useful alloy. All of the alloys containing less than 11 per cent. of aluminum solidify as solid solutions, and the structure is therefore quite uniform.

The aluminum bronzes are of a fine yellow color resembling gold, and are much used for imitation jewelry and other ornamental work. With about 5 per cent. aluminum, the color is much paler than that with 10 per cent., but beyond 10 per cent. the color pales with increase of aluminum.

The specific gravity of these alloys is greater than the calculated mean of the specific gravities of copper and aluminum, hence there is a contraction upon alloying. With the 7.5 per cent. alloy, the contraction is about 5 per cent.

The tensile strength is high, and increases with the amount of aluminum up to 11 per cent. Also its strength is much modified by heat treatment. The Alloys Research Committee[1] found that the tensile strength of a 10.78 per cent. alloy cast in sand and slowly cooled was 29.5 tons per square inch. When this same casting was quenched from 800°C., the strength was 50 tons. For a chilled casting containing 10 per cent. aluminum, the strength was found to be 36.93 tons, and when this same casting was annealed the strength was 27.7 tons per square inch.

Aluminum bronze shrinks much upon cooling from the molten state. The 10 per cent. alloy shrinks about twice as much as brass. It resists corrosion very well and when exposed at a red heat for some time it oxidizes but little. Because of its fine color and permanent brightness it is useful for ornamental parts of machinery. Because of its malleability and toughness it has been found very useful as an antifriction metal.

BEARING ALLOYS

The friction between two metals is directly proportional to the pressure applied. Of course, this is true only when both surfaces are sufficiently hard that one does not gouge or rub into the other. The amount of the load that can be borne before gouging begins is greater the harder the metals. Arguing from this point, to reduce friction and avoid cutting, hard surfaces should be used for bearings. And this is true, but only under perfect adjustment. In an ordinary bearing there are points of greater and lesser pressure, and therefore of friction, hence the bearing must be sufficiently soft to mould itself about the shaft. But also it must be able to resist wear as much as possible. The best fulfillment of these requirements has been obtained by the use of an alloy containing hard grains in a soft matrix. The soft matrix allows an adjustment, so that local pressures, which are generally responsible for trouble, are avoided, and the hard grains on the surface resist the wear. This structure is found in

[1] SEXTON: Alloys, p. 157.

the so-called white alloys, consisting of lead, tin, antimony, bismuth, copper, etc.

The White Alloys.—1. *The Lead-antimony Alloys.*—These two metals alloy readily in all proportions, the hardness and brittleness of the alloy increasing with the amount of antimony. The eutectic contains 13 per cent. of antimony. With less antimony than this the structure of the alloy is made up of crystals of soft lead in a harder eutectic. But with more than this amount, the alloy consists of hard crystals of antimony in a softer eutectic, as shown in Fig. 46. This fulfills the requirements

Fig. 46.—Lead-antimony alloy, containing 40 per cent. of lead, etched with picric acid. Bright crystals of antimony surrounded by eutectic. Magnified 100 diameters. (Gulliver.)

of a good antifriction metal. As much as 15 to 20 per cent. of antimony is advantageous in some cases, as for very heavy pressures.

2. *The tin-antimony-copper alloys* are called Babbitt metal. They are more expensive than the preceding, because they contain a large amount of tin, but for the same reason they are considerably better. The tin becomes a constituent of the hard grains, diminishing their hardness it is true, but also their brittleness. The tin enters the eutectic also and increases its compression strength.

The structure of the Babbitt metal is made up of a mass of tin, containing cubical crystals of tin-antimony alloys, and needle-like crystals of an antimony-copper alloy. The use of the three metals is advantageous, because to secure enough hard crystals by the use of one metal alone, would require so much of it that brittleness would result.

An alloy having the same composition as Babbitt's original alloy is made by first melting together the following: copper,

FIG. 47.—Tin-copper-antimony bearing metal. (Law.)

4 parts; tin, 12 parts; antimony, 8 parts, and then adding 12 parts of tin after fusion. The antimony and the first portion of the tin are melted together, then the copper is added, and finally the remainder of the tin. This is called the hardening alloy. The bearing metal proper is made by melting 1 part of this alloy with twice its weight of tin, making a final alloy having the following composition: copper, 3.7 per cent.; antimony, 7.4 per cent.; and tin, 88.9 per cent.[1]

The Leaded Bronzes.—Bronzes to which a considerable amount of lead has been added are also used as bearing metals, but their structure is the reverse of the white alloys, since they consist of soft grains in a hard matrix. And they are inferior

[1] For detailed directions for making Babbitt alloy, see *Foundry Mag.*, January, 1911, data sheet 81.

to the white alloys as bearing metals, because they are less plastic and do not mould about the shaft so well. And although they are stronger than the white alloys this does not allow them to bear heavier loads, because they have a greater tendency to "cut."

A leaded bronze that is used to a considerable extent for bearings has the following composition: copper, 64 per cent.; tin, 5 per cent.; lead, 30 per cent.; and nickel, 1 per cent. This is known as plastic bronze, for although it is not so plastic as the white alloys, it is more plastic than an ordinary bronze.

The structure of the plastic bronze is as follows. Lead does not alloy much with copper nor with copper-tin alloys, and hence to a large extent remains as an independent constituent in the bronze. The ratio between the copper and tin is such that a eutectic is formed. This eutectic becomes the matrix and in it are held the crystals of the excess copper, together with the grains of lead or of a soft lead-tin alloy. The small amount of nickel probably lessens the segregation or liquation of the lead by causing the alloy to solidify more quickly; that is, at a higher temperature.

TYPE METAL

This alloy must be capable of being readily cast and must expand somewhat upon cooling, so that it will take a sharp impression of the mould. Also it must be sufficiently hard that it can withstand the pressure of the press without crushing. These properties are possessed by alloys of lead and antimony, and type metal either consists of these two metals alone or else contains them as a base. Tin is frequently added to increase the strength of the alloy.

Other metals, such as bismuth and copper, are also sometimes used in type metal. The following percentages[1] will serve to show the composition of alloys that may be employed: lead 75 per cent. and antimony 25 per cent.; or lead 60 per cent., antimony 20 per cent., and tin 20 per cent.; or lead 55 per cent., antimony 30 per cent., and tin 15 per cent.; or lead 77 per cent., antimony 15.4 per cent., and bismuth 7.6 per cent.

[1] SEXTON: Alloys, p. 183.

SOLDER

Solder is a readily fusible alloy used to join the surfaces of metals. Its joining ability depends upon the fact that a surface alloy is formed between the solder and the parts soldered. Soldering may be accomplished by the means of various alloys, the composition of which depends upon the melting point desired and the metals to be joined. However, the most common solder is an alloy of tin and lead, known as *soft or tinner's solder*. This is used for a great variety of purposes.

Physical Structure of Soft Solder.—The freezing-point curve for the lead-tin alloys is shown on page 167. As is indicated by the limited extent of the line that shows the freezing of the eutectic, those alloys containing no more than 4 per cent. of tin or no more than 2 per cent. of lead solidify as solid solutions. All of the other alloys of these two metals solidify with the formation of eutectics as is shown in this figure, the eutectic containing approximately 31 per cent. of tin.[1] The tin content of the commercial lead-tin alloys ranges from about 25 to 75 per cent., hence the structure of these alloys consists of more or less pure crystals of either tin or lead, depending upon the proportions used, held in a matrix of smaller crystals of the eutectic.

Composition and Properties.—The tensile strength of solder is greatest with 72.5 per cent. of lead, but this alloy is not sufficiently fusible to be used for general soldering. Any of the alloys that contain above 70 per cent. of lead have a melting point too high to be used in ordinary soldering with a copper tool, although they may be used with a steel one. The chief use to which such alloys are put is for coating iron or steel sheets for roofing, for filling hollow castings, etc.

The alloy that contains about 67 per cent. of lead is used for plumber's solder and will be discussed later under that head. The alloys containing from 55 to 60 per cent. of lead melt from about 215° to 230°C., and are sufficiently fusible and freely flowing for ordinary soldering. That which contains 58 per cent. of lead is used to a considerable extent for soldering joints in electric wiring since it is considered to flow quite well.

But by far the favorite alloy with the consumer is that which

[1] For a discussion of this figure see p. 167.

contains 50 per cent. lead and 50 per cent. tin, known commonly as "half and half." This melts readily, flows freely, and presents a bright surface when the joint is finished. It can be used for every purpose to which soft solder is applicable, except wiping joints, as it is technically called, in plumbing. But all kinds of work do not require a solder as high in tin as this, even though, because of its many good properties, it is much desired by the user. Its high price is a very great objection. And since the consumer desires to buy solder cheaply, it happens that many grades of solder appear on the market designated as "half and half" that contain much less than 50 per cent. of tin. Very often the upper limit of the amount of tin contained in the "half and half" solder is determined by the price the customer is willing to pay.

For some purposes, as for manufacturing gas meters, electrical instruments, etc., where a bright, freely flowing solder is desired in order that all joints may be completely filled with the solder, greater amounts than 50 per cent. of tin are used. As is shown by the freezing-point curve in Fig. 38, the melting point of the alloy falls gradually with increase of tin up to the amount of 69 per cent. This is the eutectic alloy and melts at 180°C., the lowest temperature at which any alloy of lead and tin will melt. A disadvantage in using a solder containing so much tin is that it forms an alloy with the copper of the soldering tool and etches it away, leaving the surface pitted. And this is somewhat difficult to file smooth, since the surface alloy formed is rather hard. The difficulty may be avoided by using a steel tool.

In soldering work that has been galvanized, it is necessary to use solder containing an excess of tin. Galvanizing consists of a surface coating of zinc, and since lead does not alloy well with zinc, a good joint cannot be made if an excess of lead is present.

Schultz says[1] that solder that contains at least 50 per cent. of tin may be known by the character of its surface when it is allowed to cool in an open mould. It will be uniformly bright. If any frosted lines appear it may be known that it contains less than 50 per cent. tin. Solder that contains more than 50 per cent. tin will have the same bright surface, but when lead is in excess, the frosted surface appears and increases with the increase

[1] F. W. SCHULTZ: Tin, Lead and Solder, Chapter XIII (1908).

of lead until when the alloy contains 67 per cent. of lead the surface will be frosted all over. Workmen make use of this fact in preparing solder containing a preponderance of lead, as for wiping joints in plumbing. Lead is added to the molten solder until the desired proportions are obtained as is shown by the surface when it cools.

Plumber's solder contains usually 67 per cent. of lead and 33 per cent. of tin. As this alloy cools through its solidification range, as shown in Fig. 38, it acquires a plastic condition, in which it has about the consistency of baker's dough, before it cools enough to grow hard. While in this plastic state it may be moulded into shape in the so-called wiping of joints. Solders may be used in this way that contain as low as 60 per cent. of lead, but the more the composition varies from 67 per cent. lead and 33 per cent. tin the less easily are the alloys manipulated, since they either do not have sufficient plasticity or else do not retain it sufficiently long. That which contains 67 per cent. lead begins to assume the plastic state at about 245°C. and finally solidifies at 180°C.; that is, at the eutectic point. This makes a range of about 65°C. in which it is plastic. The plasticity is due to the presence of the lead crystals that separate out as the temperature falls. The portion that still remains fluid serves as a vehicle in which the crystals are carried. The crystals increase the consistency of the fluid portion in much the same manner as additions of sand increase the consistency of a thin mortar. It is these lead crystals also that give to the wiped joint its frosted appearance, and they are responsible for the frosted appearance of the solder when poured in the open mould as was mentioned in a preceding paragraph.

Occasionally it is necessary to use block tin[1] for certain purposes, as for making receptacles and conductors for distilled water, and for syrups and liquids charged with carbon dioxide. Now since plumber's solder begins to assume the pasty state at about 245°C. and tin melts at 231°C., the usual solder cannot be used for tin. The tin would melt before the joint could be made. Therefore, an alloy consisting of 2 parts tin, 2 parts lead, and 1 part of bismuth is employed. This melts at 145°C.

[1] The terms *block tin* and *sheet tin* are used to indicate pure tin, while *tin plate* means only sheet iron or steel coated with tin.

Care of Solder.—When using solder it is highly essential that it be kept absolutely free from oxides and foreign material of every kind. It is bad practice to gather scraps and droppings from the floor and put them back into the melting pot, since foreign metals are then likely to be introduced. More than ordinary care must be exercised when soldering upon galvanized articles or brass, since the solder that has been brought into contact with these zinc-bearing alloys is almost certain to acquire a small quantity of zinc. And when lead is added to the solder by the consumer there is also a likelihood that foreign metals will be introduced. Lead is very likely to contain at least traces of such metals as arsenic, antimony, zinc, etc. In the manufacture of solder these impurities are removed, usually by a process of oxidation.

The Effect of Foreign Elements in Solder.—Zinc, sulfur, arsenic, antimony, iron, and copper all have detrimental effects on solder. Users of solder object especially to the presence of *zinc*, saying that if even a trace of it is present, it can be detected in the working qualities of the solder. In the manufacture of solder, zinc is removed by oxidation and a similar process may be employed by the user.

Removal of Zinc.—Zinc boils at 918°C., and if the solder is heated to this temperature and then is stirred and ladled so that the air may gain access to it, the zinc will distill out and oxidize. After the solder has cooled somewhat, if a small amount of ammonium chloride is added and the mass is again ladled so that the salt is mixed with the solder as much as possible, the entangled products of oxidation will be enabled to rise to the surface, whence they may be skimmed.

Users of solder have been taught to employ sulfur as a medium for the removal of the real or fancied presence of zinc. But it should be remembered that when sulfur is brought into contact with any of the common metals in the molten state, it immediately combines with them with the formation of sulfides. By the introduction of sulfur into molten solder, a great amount of lead and tin are wasted, because their sulfides which are formed are skimmed off and thrown away. Still, small amounts of these sulfides remain occluded in the solder and are in themselves objectionable.

Arsenic causes the solder to be hard and brittle. Fortunately it oxidizes very readily and but little of it remains for very long. The presence of arsenic may be detected by its garlicy odor as it oxidizes.

Antimony has an effect similar to that of arsenic, but it is not so readily removable. When as much as 2 per cent. is present, small, bright crystal faces appear on the surface of the solder when the alloy is poured in an open mould.[1]

Iron, even in amounts much less than 1 per cent., causes the solder to have a noticeably higher melting point and to be sluggish when molten. When solder containing iron is poured in an open mould the surface will have readily perceptible dark streaks upon it.[1] Iron in solder is difficult to remove.

Copper finds its way into solder perhaps more frequently than any other foreign metal. This results from the practice of filing the copper tool to brighten it before tinning and then allowing the filings to find their way into the molten solder with solder scrap. When poured into an open mould the presence of copper in solder is indicated by the presence of a slight iridescence or pale blue color where the solder comes into contact with the iron. It is difficult to remove copper from solder because of the readiness with which it alloys with tin.

Aluminum Solder.—Richards[2] describes as a satisfactory solder for aluminum, an alloy containing the following: tin, 29 parts; zinc, 11 parts; aluminum, 1 part; and phosphor tin, 1 part. Brannt states that this alloy has a fusing point sufficiently low that it can be melted with a copper soldering tool. It is tough and has nearly the same color as aluminum. It darkens somewhat upon exposure, but is readily brightened by polishing.

In using this solder, the edges to be joined are filed or scraped clean, and if the article will allow it, it is heated with a torch or similar device to a temperature at which the solder will melt and then the edges are rubbed briskly with a stick of solder. If the object cannot be heated in the manner stated, then the edges are heated with a tinned copper tool and the solder is rubbed on as before described. No flux of any kind is used.

Another solder that has been found to produce very satisfac-

---

[1] Schultz: Tin, Lead and Solder, Chapter XX.
[2] Aluminum; Its Properties, Metallurgy and Alloys, p. 467.

13

tory results is made up as follows: tin, 32 parts; lead, 12 parts; zinc, 12 parts; and aluminum, 1 part. This is applied without flux in the manner described.

The Use of Fluxes in Soldering.—The value of a flux depends upon its ability to gather up, by dissolving or otherwise, the oxide or other adherent matter on the surface of the metal. Also, it must stay in place and keep away the oxygen of the air so that further oxidation cannot take place before the solder can be caused to flow over the cleaned surface and alloy with it. If the metal were cleaned mechanically, as with a file, a film of oxide, imperceptible to the eye, would spread over the surface of the metal before the solder could be applied. Solder cannot alloy or combine with metallic oxides. In the method described for soldering aluminum, although the metal becomes oxidized after it is cleaned by filing or scraping, this relatively thin oxide layer is removed at the moment the melted solder is applied, by rubbing with the stick of solder itself.

Kinds of Fluxes.—Acids alone are not universally good as fluxes, for although they readily dissolve the oxide and keep away the air, they possess an objectionable feature in that they attack the metal as well. A solution of zinc chloride is suitable for many kinds of soldering. This is most cheaply prepared by diluting commercial hydrochloric acid with water and then neutralizing it with zinc in excess. If the acid is not sufficiently diluted it will not be all neutralized, since after the zinc ions in solution reach a certain concentration more are prevented from entering it. .After action has ceased the solution may be tested for complete neutralization of acid by taking a sample and diluting it with about half its volume of water and then inserting a piece of zinc. If effervescence occurs, the whole of the stock solution must be further diluted and allowed to stand longer with the zinc.

Ammonium chloride is also a good flux, being the best obtainable for copper and brass. And, as is well known, rosin and rosin-bearing preparations are much used. Paste fluxes commonly contain fats, as olive oil or tallow, and rosin, together with zinc and ammonium chlorides or similar salts.[1] The fats and rosin

---

[1] For a comparison of the values of hydrochloric acid and zinc and ammonium chlorides, see LIPPMANN: The Chemistry of Soldering Agents, *Sci. Am.*, **101**, 389 (1909).

are melted together, and then a concentrated solution of the salt is introduced and the whole is stirred until cold so that the salt solution is retained in the form of an emulsion.

Borax is used as a flux in brazing.

Cleaning Metal Surfaces[1]

Grease and old lacquer may be removed by boiling in about a 15 per cent. solution of caustic soda. As water evaporates, more must be added. Aluminum articles must never be treated in this way, and zinc and tin should be allowed to remain in this bath for only a short time, because strong alkaline solutions attack these metals, aluminum very actively. After removal from the bath, the article should be washed and dried as in the same manner as described later for articles removed from acid pickling baths.

Or grease may be removed also by dipping in benzol or petroleum spirit. A succession of at least three baths should be used, for if only one is employed, it becomes contaminated with the grease and a film of this is left on the article when it is removed from the bath and allowed to dry.

Oxide layers and salts produced by corrosion may be removed by immersion in suitable acid baths. This process is known as "pickling." *Rust and scale* may be removed from iron articles by immersion in a solution made up of 25 parts of concentrated, commercial hydrochloric acid and 75 parts of water. The black scale is not soluble in acid but is removed by acid treatment because the underlying layer of iron is dissolved. .

To prepare a pickle that will leave the iron article very bright, pour 800 c.c. of concentrated sulfuric acid into 10 liters of water (not the water into the acid) with stirring, and dissolve 60 grams of zinc in the mixture. Then add 350 c.c. of concentrated nitric acid. After removal from these baths the article must be washed and dried as described later.

Copper, brass, bronze and German silver may be cleaned by dipping in the following: water 40 parts, commercial concentrated sulfuric acid 40 parts, concentrated nitric acid 20 parts, and concentrated hydrochloric acid 1 part. Articles of the

[1] For a more complete treatment of this subject, see HIORNS: Mixed Metals, pp. 201 to 212.

same sort may be cleaned and brightened by dipping in a bath composed as follows: water 25 c.c., concentrated sulfuric acid 30 c.c., concentrated nitric acid 50 c.c., sodium chloride 1 gram, lamp-black 1 gram. The articles are dipped in this and washed alternately until the desired surface is obtained.

Washing and Drying.—After the removal of an article from any of the preceding acid or alkaline solutions, it should be washed thoroughly, preferably in running water, and then immersed in boiling water and allowed to remain until the metal has been fully heated to the temperature of the water. Then it should be removed from the hot water, dried as much as possible with a clean cloth, or most of the adherent water may be removed by shaking it, after which it should be suspended in a warm place until drying is complete. Since the metal is hot, the water will evaporate quickly. · Rapid drying is necessary in the case of iron articles to prevent corrosion.

References

A

HIORNS: Metal Coloring and Bronzing, London, 1892.
RICHARDS: Aluminum; Its Properties, Metallurgy and Alloys, Philadelphia, 1895.
BRANNT: The Metallic Alloys, Philadelphia, 1908.
SEXTON: Alloys (Non-ferrous), Manchester, 1908.
LAW: Alloys, and Their Industrial Applications, London, 1909.
HIORNS: Mixed Metals, London, 1912.
GULLIVER: Metallic Alloys, London, 1913.

B

THURSTON: Report of Committee on Metallic Alloys, of U. S. Board Appointed to Test Iron, Steel, etc., 1879 to 1881.
CARPENTER and EDWARDS: Alloys of Aluminum and Copper, *Engineering*, Jan. 25, 1907, p. 127 and February, 1907, p. 158.
HOWE: An Introduction to the Study of Alloys, *Engineering*, **75**, 647 (1903). Theory of the Constitution of Alloys, *Castings*, December, 1908, p. 82.
EDWARDS and ANDREW: Aluminum, Copper, Tin Alloys, *Engineering*, **88**, 664 (1909).
LIPPMANN: The Chemistry of Soldering Agents, *Sci. Am.*, **101**, 389 (1909).
CAMPBELL: Monel Metal. *Trans. Canad. Min. Inst.*, **16**, 241 (1913).
HAYNES: Stellite, *Chem. Eng.*, **22**, 170 (1915).
KARR and RAWDON: Government Bronze, U. S. Bureau of Standards, *Technologic Paper* 59 (1916).
MERCIA: The Effect of Corrosion on the Ductility and Strength of Brass, *Met. Chem. Eng.*, **15**, 321 (1916).

CHAPTER VIII

FOUNDRY SANDS[1]

Foundry sands, as a rule, consist of a mixture of sharp sand and clay, the latter being necessary to make the parts adhere. Sand is called sharp when its grains are angular; dull, when round; strong, when it retains a shape given it; weak, when it does not. If the amount of clay and all other factors are equal, the sharper the sand the stronger it will be, but the sharpest sand is weak without some cementing material.

In general, foundry sands may be divided into two classes, moulding sands and core sands, the former being used to make the mould proper, and the latter to make the core with which the hollow parts of the casting are formed.

Moulding Sands.—Usually moulding sands are of a finer grain and contain more clayey matter than the core sands, but this distinction is not sharp, since both classes vary greatly in this respect, and often overlap.

Grades of Moulding Sands.—Moulding sands cannot be of fixed composition and grain size, since they must be selected with respect to the kind of casting to be made. For light castings with smooth surfaces, the sand should be fine; for heavy ones, an open, coarse-grained sand is best. If sand suited for heavy castings is used for light work, the casting will have a rough skin; the indentations caused by the larger grains are more noticeable on the smaller casting. If the fine sands suitable for light castings are used in making heavy ones, there is danger that the sand will be fused and form a coating of scale on the casting. In making the small casting the heat is quickly disseminated, but with large masses of melted metal, the temperature may be kept elevated sufficiently long to cause a vitreous coating to form. This coating is removable from the casting only with difficulty, as

[1] This discussion of foundry sands is based on the report of RIES and ROSEN on Foundry Sands, published by the Board of Geological Survey of the State of Michigan, 1907.

197

with hammer and chisel, sand blast or "pickling," which adds considerably to the cost. Further, in using fine sand for large castings, there is danger of producing blow-holes because of the sand being insufficiently permeable to allow the larger quantities of gases to escape.

In choosing a moulding sand, it is necessary to consider not only the size of the casting but also the kind of metal to be poured. Thus, for a steel casting, a highly siliceous sand containing 97 per cent. of silica, or over, is used. For iron castings, a clayey sand is employed, its fineness depending upon the size of the casting as has been explained. Fine-grained, clayey sands are used for brass and aluminum, the grades used being very similar to those employed for stove plate. Bronze is cast in an imported sand of very fine grain and good bonding power.

Cohesiveness or Bonding Power.—Sands used for moulds must possess sufficient cohesiveness, when slightly moistened, to retain a shape and to resist the action of the flowing metal when the casting is poured. Ries says that the cohesiveness probably depends in part, not only on the clay, but on the *character* of the clay and its texture. Fineness tends to increase cohesiveness, while coarse-grained sands with little or no clay are almost lacking in this property.

Refractoriness.—The refractoriness of foundry sands should be sufficiently high to prevent the pores from being closed by fusion during casting. Several factors, such as fineness of grain, chemical composition, and uniformity of distribution of the components, influence refractoriness, but none of these alone is determinative. Highly siliceous sands are more refractory than those containing much clay. The former are used for making steel castings, while the latter are used for making castings of metals having lower melting temperatures. Iron oxide, lime, magnesia, and the alkalies are known as fluxes, and sands containing considerable amounts of these are less refractory than those in which such materials are relatively absent. The fluxes react with the silica of the sand and form silicates having rather low melting points.

Texture has reference to the degree of fineness and is a very important factor in the choice of sands. Cohesiveness and permeability are in a large measure dependent upon it. Sands are commonly rated in respect to texture by determining the percentages retained on sieves of different mesh, but other methods, such as the decantation and aspirator methods, which are considered by some to be more satisfactory, have been devised.

Permeability and Porosity.—In discussing these properties, Ries says:

"The permeability of a sand may be defined as the property which it possesses of allowing liquids and gases to filter through it. The porosity can be defined as the volume of pore space between the grains. These two properties are therefore different and should not be confused.

"Two sands might have exactly the same percentage of pore space, but vary in their permeability, and other things being equal, the one with the larger pores should be more permeable.

"Again, a sand might have a large total pore space, but owing to the smallness of the pores its permeability might be low.

"The permeability of the sand will be influenced by several things, such as the tightness of packing or tamping, size of grains, water content and fluxing impurities in the sand. The first two factors operate throughout the period of use in the mould, the third only while the sand is moist, and the fourth only after it has been exposed to the heat of the molten metal."

Concerning tamping, he says:

"It would seem that the degree of tamping is perhaps too little considered in the use of sands. For example, one sand may yield excellent results with a given amount of tamping, but another one to yield good results may require less tamping. In trying the new sand this point may be overlooked, and the material turned down as worthless because it was not properly handled.

"Porosity alone cannot be used as a gage of permeability, although it is probable that coarse texture combined with high porosity indicates good permeability."

The permeability of a sand is also affected by the amount of water used to moisten it. Although it would seem on first thought that the addition of water to sand would lessen its permeability to gases, just the opposite has been found to be true up to a certain point. In the following tables[1] the rates of air flow through two samples of sand containing known percentages of moisture are shown:

[1] RIES and ROSEN: *Loc. cit.*, p. 59.

Sample A		Sample B	
Per cent. of moisture	Time required to draw 1 liter of air, seconds	Per cent. of moisture	Time required to draw 1 liter of air, seconds
Air dry.........	3,300	Air dry.........	840
5...............	2,880	5...............	560
7...............	1,080	8...............	240
9...............	640	10...............	280
11...............	240	12...............	320
13...............	305	15...............	375
15...............	560	20...............	3,000
25...............	4,200		

The moist sand was rammed as nearly alike as possible in each case.

"The data obtained would seem to indicate that there is for each sand an *optimum water content* when the permeability is highest. After this optimum is passed the permeability again decreases. The explanation of the fact is that when the sand is dry the particles pack close together, and smaller ones fitting between the larger ones, the porosity is reduced to a minimum. When water is added the particles readjust themselves, swell up and adhere to each other, forming a *spongy* mass with numerous passages, and the sand cannot be packed so close as before. When an excess of water is added, the sponges of grains fall apart, pack close, and the pore spaces fill with water. The behavior of the sands can be easily observed under a microscope with a stereoscopic arrangement."

The effect of fluxes in diminishing permeability by closing the pores with fused material can be readily understood.

Durability or Life of Moulding Sands.—The following discussion on this subject is quoted from the same source as the preceding:

"Practically all moulding sands after being exposed to the full heat of the molten metal lose some of their desirable qualities, and become 'dead' as the foundryman calls it. A dead sand has had its cohesion and texture destroyed, but may have changed but little in its ultimate composition. In casting, it is the layer of sand next to the metal which is most affected, and the thickness of this will depend upon the size of the casting and the temperature of the metal. On removing the flask from the casting it is impracticable to separate all this burned sand from the unburned, and moreover there is no sharp line of division between the two, so that much of it gets mixed up with the unaltered material. Since it is deficient in bonding power a small quantity of fresh sand has to be added to counteract it.

"The deadness of the sand is no doubt due to several causes. In the first place, the heat brings about a dehydration of the clayey particles of the sand, and thus destroys its plastic and bonding qualities. Secondly, the heat may be sufficient to cause some or many of the grains to agglomerate by fusion, thus altering the mixture. Thirdly, the iron may be reduced largely or in part to the ferrous condition, but this change need not necessarily affect the physical properties of the material. From the first of these causes it would appear that a clayey sand (loam) would become dead sooner than a more siliceous one.

"The length of a sand's life is a matter of some importance. Some sands can be used over several times without the admixture of fresh sand, while others are easily 'burned.'

"In this connection the following analyses are of interest: No. I is an unused moulding sand from Richmond, Va. It is much used for general work in the foundries of that locality and is known as Redford sand. No. II is some of the 'dead' sand from the layer next the metal.

MECHANICAL ANALYSIS

Number of mesh	Per cent. retained on mesh	
	I	II
20	1.51	5.34
40	1.26	14.73
60	1.27	10.41
80	0.56	1.28
100	6.27	14.61
250	71.69	59.37
Clay	16.52	3.52

CHEMICAL ANALYSIS

	Per cent.	
Silica (SiO_2)	83.49	82.32
Alumina (Al_2O_3)	7.25	7.80
Ferric oxide (Fe_2O_3)	4.74	3.98
Ferrous oxide (FeO)	....	2.38
Lime (CaO)	0.36	0.54
Magnesia (MgO)	0.35	0.41
Potash (K_2O)	1.30	1.64
Soda (Na_2O)	0.41	0.80
Titanium oxide (TiO_2)	0.30	0.22
Water (H_2O)	1.66	0.19
Total	99.85	100.28

"Comparison of these two sets of analyses indicates that there is a decided increase in coarseness of texture, due to the fusing together of the particles. There was, in fact, a greater agglomeration even than is represented by the mechanical analysis, because some of the coarse grains have been screened out.

"The chemical analysis shows little difference between the fresh and the used sand, except in its water percentage and in the iron contents, which have been in part reduced to ferrous oxide, and moreover the sand has apparently absorbed some iron from the metal during casting."

Chemical Composition.—The ultimate chemical analysis of a moulding sand seems to have very little value, since sands used for the same purpose may bear no chemical relation to each other. It is generally very difficult, if not impossible, to predict even the refractoriness from a chemical analysis, since texture plays such an important part, a coarse-grained sand being much more refractory than a fine-grained one of the same composition. And there is no relation between the amount of alumina and the degree of plasticity. In summing up, Ries says that it appears that a chemical analysis furnishes no information regarding the cohesiveness, degree of plasticity, texture or use, and that in his judgment it is of little direct aid except possibly in selecting sand for steel casting.

Mineral Composition.—This is determined by the so-called rational analysis, the usual object of which is to ascertain the amount of quartz, clay and feldspar present.

Quartz is always present in moulding sands in large amounts and may be regarded as the shrinkage-resisting and refractory element. The grains may be white but are generally colored by iron oxide, which frequently forms a film on their surfaces. The size of the grain may vary from fine silt to particles as large as a pin head or larger, the size determining in a large degree the use for which the sand is suitable. Since quartz confers refractoriness, the amount of it present should be as high as possible without displacing necessary bonding material. Quartz itself is practically lacking in cohesiveness.

Clay.—In moulding sand, clay is next in abundance to quartz. Since it furnishes the bonding power a certain amount is necessary, but an excess is undesirable because it lessens porosity, and shrinks and cracks when heated. It should be understood that the amount of bonding power is not dependent alone on the amount of clay, since different clays possess this property in different degrees. It is also the clay that carries most of the fluxing constituents of the sand. Especially the alkalies will be present in it if it contains much unweathered feldspar. Iron oxide has a noticeable fluxing action especially if present in the ferrous condition, or if present in the ferric state in conjunction with reducing material, as carbonaceous matter.

Core Sands.—In general, core sands resemble medium- or coarse-grained moulding sands, although as a rule they contain much less clayey matter. Clay is so little necessary that even crushed quartz rock may be used very satisfactorily. The necessary bond is obtained by the use of oils, resins, and other organic substances called binders or core compounds. The presence of much clay would cause the core to bake too hard and would prevent it from disintegrating in the desired manner when the casting is cleaned.

Parting Sand.—Parting sand is usually either sharp or burned sand used to prevent two surfaces from adhering. For brass founding the ordinary parting sand is considered too coarse and rough to allow it to be introduced into the moulding sand, so pulverized rosin, charcoal, or lycopodium powder are used. Lycopodium powder is a very fine dust consisting of the spores of a moss-like plant of the genus Lycopodia. The powder is very inflammable. Sometimes also, in place of a parting sand, the surface of the mould is covered with soot from a lamp or burning piece of rosin, which serves the purpose also of drying it somewhat.

FOUNDRY FACINGS

With the exception of sea coal the substances included under the head of foundry facings are those that are used to line the mould to prevent the formation of a scale by a reaction between the silica of the sand and the ferrous oxide developed at the surface of the molten metal. The materials used as surface coatings must spread easily, have great heat-resisting power, and give the casting a fine blue surface. Sea coal is not a facing in the true sense but is used to mix with the sand to make the mould more porous or open. It serves as a vent and so prevents

the formation of blow-holes, and indirectly prevents the forma- ·
tion of the ferrous silicate mentioned previously. Because of
this latter effect, which will be explained in the following paragraphs, it is usually classed among the facings.

Sea Coal or Bituminous Facing.—Sea coal derives its name
from the fact that in England it was shipped from New Castle-
on-Tyne by the sea, to London, and was so named to distinguish
it from charcoal. In the eastern United States the coal used in
this manner as a so-called facing, comes from the Westmoreland
region in Pennsylvania. It is crushed, screened to remove slate
and is then ground. It is a high-grade gas coal. It should never
contain less than 35 per cent. of volatile matter and never more
than 8 per cent. of ash and 1.5 per cent. of sulfur. The reason
for the need of the high volatile matter is that when the molten
metal is introduced into the mould, the heat drives off that portion of the coal that is volatile, thus leaving the mould sufficiently porous to allow the escape of the heated gases. If the
gases escape freely, blow-holes are prevented. Also, these
gases in escaping carry away a considerable amount of heat from
the surface where the metal is in contact with the mould. Thus
the heat intensity at this point is sufficiently lowered that the
ferrous oxide formed at the surface of the molten metal is prevented from fusing with the sand. In other words, it prevents
the formation of "scabs," which consist of ferrous silicate and
entangled sand. Sand itself is not a good conductor of heat.

For heavy castings, the coal dust is mixed in about the proportion of 1 part of dust to 8 or 10 parts of sand; for light castings, about 1 of dust to 12 or 14 of sand. The fineness of the
dust for heavy work is about 10 mesh and for light work about
20 to 40 mesh. For very light work and where extreme smoothness of surface is desired, the dust is bolted. If the coal dust
were not of proper fineness, upon burning away it would leave a
pit-marked casting.

Graphite.—Apparently the best of the foundry facings is
graphite, the finest grade of which is obtained from the island of
Ceylon. The composition and general properties of graphite
have been given on page 58, which see. As a facing it spreads
easily and has a high heat-resisting power. To the foundry trade,
graphite is known by several names which are really synonymous.

However, custom has designated the highest quality as "graphite," the second grade or blends as "plumbago," and the very cheapest grade as "black lead" or sometimes also as plumbago. Plumbago is often mixed with talc and can be sold at any price.

Coke Blacking.—This is made from the best grades of 72-hr. coke, finely ground. It can be used only with dry sand work, or as a wet blacking since it will not slick with a tool. It is used as a roll or ingot mould blacking.

Soapstone or talc makes a very satisfactory facing for gray iron castings. It can be used on either dry or green sand moulds since it "slicks" easily. It is said that there is objection to it in that it causes the casting to have a reddish color, this often being mistaken for burnt iron. Finely ground mica can also be used as a facing.

CORE BINDERS

The ideal core binder is one that will bind the sand when green, become a hard, solid mass when dried, and after casting will break up and allow the sand of the core to readily run from the interior. Core binders may be divided into two major classes, liquid and dry.

Dry Binders.—*Rosin* is the best of the dry core binders. It makes a solid core and readily burns out during casting. Rosin-bound cores can be allowed to remain in a green sand mould without absorbing moisture. Also they are said to preserve sharp corners well and to allow free escape of gases in cores difficult to vent. However, rosin melts at a low temperature and thus remains soft so that the core cannot be handled until comparatively cold. Also, the core loses shape if heated too much in the core oven. For use the rosin must be very finely ground.

Flour.—Commonly flour of a poorer grade, really a fine grade of bran is used as a core binder, but since it is only the starch and gluten of the flour that possess any binding power, it is in reality more economical to use the better grades, which consist almost wholly of these substances. The bran itself has no value as a binder. There is an objection to the use of flour, since cores made of it cannot be left in a green sand mould for very long because they absorb moisture and become soft. Also, during pouring, the burning flour produces considerable quantities of stifling gases.

Dextrine is a colorless, uncrystallizable substance having strong adhesive properties. It is in some respects similar to starch, having the same ultimate composition, and can be made from starch without much difficulty. It is quite readily soluble in water. It is the substance used on postage stamps and the flaps of envelopes.

As a core binder it is very apt to absorb moisture and like flour cannot be left long in a green sand mould. Also, like flour, it produces stifling gases during casting.

Liquid Core Binders.—*Linseed oil* is the best liquid core binder. It can be used in proportions of 1 part of oil to 75 or 80 parts of sand and can be dried at extreme heat. Linseed oil is used for making thin intricate cores as for radiators and gas-engine cylinders, where the core must be able to be cleaned out with exceptional ease. The cost of linseed-oil core-making can be lessened by filling the voids in the sand with finer sand, which allows a core equally good to be made with less oil. Quartz rock ground extremely fine is very useful here. The use of the finer sand in this manner is not so objectionable with linseed oil as it would be with paste binders like flour. The oil is drawn to the contact edges and corners of the sand grains by capillarity, thus leaving the inter-granular spaces open for the passing of gases. Paste binders have a greater tendency to dry in the space between the sand grains and thus make the cores more difficult to vent. Also, because the oil tends to collect at the contact points, stronger cores can be made with less binder than with those substances that do not collect in this way, as the flour and starch pastes.

It should be remembered that linseed oil dries by oxidation, as explained on page 295, and that an abundant supply of oxygen, such as is furnished by a proper circulation of air in the core oven, is of great advantage in drying the cores.

Because of its high price, linseed oil is liable to be found adulterated with other oils, as rosin oil, cottonseed, corn, soya bean, fish oil, etc.

Molasses mixed with water makes a very good core, but because it absorbs moisture it cannot be left in the mould over 12 hrs. Also, its water solution is subject to fermentation and this lessens its adhesive qualities.

Sulfite Pitch.—In the sulfite process of preparing wood pulp for the manufacture of paper, the pulp is boiled under pressure with calcium or magnesium acid sulfite. The waste liquor is known as "sulfite liquor" and contains a considerable amount of water soluble material obtained from the wood. This liquor is evaporated to the consistency of a thick syrup, having a density of about 35°Bé. and is sold under the name of "glutrin" for a core binder, for which purpose it seems to serve quite admirably. There is, however, one possible objection. When the "glutrin" is used with sharp sand it is largely carried to the surface of the core and deposited there by the evaporating water, thus producing a core with a very hard surface but non-coherent interior. It is said that the binder may be more uniformly retained if a little clay is mixed with the sand.

References

Ries: The Laboratory Examination of Moulding Sands, *Fdy.*, July, 1906.
Longmuir: Fireclays and Moulding Sands, *Eng. Mag.*, March, 1906.
Field: Moulding Sand and Its Composition, *Fdy.*, May, 1906.
Ries and Rosen: Foundry Sands, Mich. Brd. Geol. Sur., 1907.
Ries: Relative Value of Physical and Chemical Examination of Moulding Sands, *Fdy.*, July, 1908.
Meade: Foundry Use of Pulverized Coal, *Fdy.*, June, 1909.
Shaw: The Comparative Examination of Moulding Sands, *Castings*, March, 1910.
Shaw: The Properties of Moulding Sands for Green Sand Work, *Fdy.*, April, 1910.
————Rational Analysis of Sand, *Fdy.*, September, 1911.
Condit: Moulding Sand Tests, *Trans.* Am. Fdy. Assn., **21** (1912).
Johnson: Testing Moulding Sands at Wentworth Institute, *Trans.* Am. Fdy. Assn., **22** (1914).
Evans: Foundry Flour, with Simple Methods of Testing, *Fdy.*, January, 1914.
Moldenke: Moulding Sand Tests, *Trans. Am. Fdy. Assn.*, 23, 690 (1916).

CHAPTER IX

BUILDING STONES

Although practically all rocks are suitable for building construction from the point of view of their mechanical strength, there are other requirements which they may not be able to fill, and which may cause the stone to be rejected as a material unfit for building purposes. Among the important properties that must be taken into account are the durability of the stone, its tendency toward discoloration, and its resistance to the effects of fires. In the present chapter it is proposed to discuss building stones with special reference to these three factors.

Durability.—Baker,[1] in commenting upon this subject, says:

"Rock is supposed to be the type of all that is unchanging and lasting; but the truth is that, unless the stone is suited to the conditions in which it is placed, there are few substances more liable to decay and utter failure. The cities of northern Europe are full of failures of stones of important structures. The most costly building erected in modern times, perhaps the most costly edifice reared since the Great Pyramid —the Parliament House in London—was built of stone taken on the recommendation of a committee representing the best scientific and technical skill of Great Britain. The stone selected was submitted to various tests, but the corroding influence of a London atmosphere was overlooked. The great structure was built and now it seems questionable whether it can be made to endure as long as a timber building would stand, so great is the effect of the atmosphere upon the stone. This is only one of the numerous instances that might be cited in which a neglect to consider the climatic conditions of a particular locality in selecting a building material has proved disastrous."

Merril[2] in referring to instances of a similar sort makes the following statement:

"The Executive Mansion and portions of the Capitol buildings in Washington are of a sandstone so poor in enduring qualities that it has

[1] Treatise on Masonry Construction, p. 4.
[2] Stones for Building and Decoration, p. 419.

been found necessary to paint them periodically in order to keep them in a condition in any way presentable."

The destructive agents at work upon stone may be classified under two heads, mechanical and chemical.

Of the *mechanical agents*, frost is probably most worthy of note. The susceptibility to the action of frost is dependent upon the porosity and friability of the stone.

If the stone is to be used in a location where friction is great, as for stair treads, resistance to abrasion must be considered as a factor in the choice of the stone. There is a wide variation in the ability of stones to withstand abrasive action.

Of the class of *chemical agents* the acid-forming gases in the air are the most destructive. The carbonate stones, namely, the limestones, dolomites, and marbles, as well as the calcareous sandstones, that is, sandstones in which the quartz grains are cemented together with calcium carbonate, are often badly affected by atmospheres containing much carbon dioxide and sulfur dioxide. These gases are especially prevalent where smelters, furnaces or factories exist. The carbonates are converted into soluble compounds and carried away by the rains. Stones consisting principally of silica are practically unaffected by such acid action. However, it should be observed in this connection that chemical composition is not always determinative as regards the weathering of stones. Structure and density are also strongly influential. Thus, a dense, fine-grained carbonate rock may weather well; while a coarse-grained, porous stone, even though composed almost entirely of silica, may weather very poorly.

Life of Stones.—Ries[1] says that the life of a building stone may be considered as the length of time it will stand exposure to the weather without showing signs of disintegration or decay. To indicate the life of various stones, he quotes the following table compiled by Dr. A. A. Julien of New York City:

Kind of stone	Life in years
Coarse brownstone	5–15
Fine laminated brownstone	20–25
Compact brownstone	100–200

[1] Building Stones and Clay Products, p. 86.

14

Bluestone (sandstone), untried, perhaps centuries.
Ohio sandstone (best siliceous variety) perhaps
from one to many centuries. .

Coarse fossiliferous limestone	20–40
Fine oölitic (French) limestone	30–40
Marble, coarse dolomitic	40
Marble, fine dolomitic	60–80
Marble, fine	50–100
Granite	75–200

Gneiss, 50 years to many centuries.

Discoloration.—Certain stones after having been placed in the building show changes in color. The pink granites often change to a much lighter color because of the weathering of their feldspar. Also, green stones sometimes have a tendency to fade, but these changes do not necessarily indicate a weakening of the stone.

Sandstones, granites, and others containing iron compounds are often changed in color due to the oxidation of the iron. Iron pyrite, which occurs quite commonly, is especially susceptible to oxidation. The stone may in this way be changed from a light color to a brownish yellow or a deep brown. The pyrite, FeS_2, is oxidized to the hydrated ferric oxide (iron rust), and if the pyrite is present in large grains or lumps, the resultant rust may be carried over the surface of the stone in exceedingly unsightly blotches or streaks. Under the influence of water containing carbon dioxide, the rust is converted into the soluble bicarbonate of iron, which enters the pores of the stone and is practically impossible to remove.

Fire Resistance.—Judging from experience there is no stone that can withstand the extreme heat developed by the burning of large buildings. Granites seem to be the least resistant and sandstones most. Granites are particularly likely to crack and spall if they are suddenly cooled from a highly heated state, as by the application of water. Limestones show good resistance at temperatures below 600°C., but at about this point the stone begins to dissociate into quicklime and carbon dioxide.

Kinds of Rocks.—Classified according to origin, rocks are divisible into two major groups, igneous and sedimentary.

Igneous rocks are those that have been produced by the cooling of fused material. Of course, it is supposed that the total solid portion of the earth was so formed originally and hence igneous, but it is doubtful if any of this original formation still exists. The igneous rocks here discussed are those produced later by the cooling of fused material forced up through outer layers of rock. In most cases this fused material cooled before it reached the surface and is now exposed only by the removal of overlying strata. Igneous rocks formed in this way are called *plutonic*. Those that cooled on the surface of the earth are called *volcanic*.

Sedimentary rocks are those that resulted from the decomposition of other strata. The decomposed material was carried by water usually and deposited as a bed of sand, clay, or limestone. These loosely deposited layers upon being hardened formed rock.

Metamorphism of Rocks.—Frequently the structure, and in some cases the chemical composition, of both igneous and sedimentary rocks have been changed by heat, pressure and other agencies. Sometimes the alteration has been so great that it is difficult to determine whether the original rock was of igneous or sedimentary origin. Rocks that have been altered in this way are known as *metamorphic rocks*.

IGNEOUS ROCKS

Structure.—Depending upon whether or not they have been metamorphosed, igneous rocks are divided into two classes. Those that solidified quietly from a state of fusion and were not later subjected to severe external stresses, are made up of crystals irregularly arranged and are called *massive* igneous rocks. On the other hand, when rock of the same composition was subjected to pressure, either during or after cooling, a banded structure was often developed. In some cases, the constituent minerals are found in more or less definite alternating bands, while in others the mixed crystals are merely arranged with their long axes in the same direction. But in either case the rock is called a *gneiss*.

Chemical Composition.—The igneous rocks range from 35 to 80 per cent. silica with alumina as the second most abundant constituent. The following is an average by F. W. Clarke of the analyses of 830 American igneous rocks:[1]

	Per cent.
Silica	59.71
Alumina	16.01
Ferric oxide	2.63
Ferrous oxide	3.52
Lime	4.90
Magnesia	4.36
Potash	2.80
Soda	3.55
Water	1.52

The term *acid igneous* is applied to those igneous rocks high in silica and relatively low in lime, magnesia, iron oxide, etc. *Basic igneous* signifies that the rock is high in lime, magnesia, etc., and low in silica.

Commercial Classification.—The scientific classification of igneous rocks is highly detailed but it is too complex to be introduced here. The following simple classification is used by quarrymen and builders:[2]

1. Granites
2. Traps
3. Serpentines.

Granites.—The granites are crystalline, plutonic, igneous rocks, composed largely of quartz and feldspar, with usually some mica and occasionally hornblende. Because of the excess of silica, granites are designated as acid rocks.

Quartz is native silicon dioxide, or silica, SiO_2. Its crystals, which occur in the hexagonal system, vary in luster and color depending upon the impurities contained. Quartz is the hardest of all the common minerals. It has a specific gravity of 2.6. It is infusible in the blowpipe flame and resists all acids except hydrofluoric. It is very resistant to weathering.

Feldspar is essentially a silicate of alumina, combined with the oxide of some other metal. Its color is light, usually whitish, gray, pink or light green. The common or potash feldspar, known as orthoclase, is pink. It has the composition indicated by the formula, $K_2O \cdot Al_2O_3 \cdot 6SiO_2$. Plagioclase is a lime-soda

[1] Quoted from ECKEL: Building Stones and Clays, p. 23.
[2] ECKEL: *Loc. cit.*, p. 30.

feldspar and some of this is usually found in granites. Feldspar weathers rather readily, being converted into clay.

Mica in granite may be of two kinds, either the white called muscovite, or the black known as biotite. The former is a silicate of alumina and potash and the latter a silicate of iron, magnesia, alumina, and potash. Mica occurs in granite in the form of smooth, shining scales, which cleave readily into thin, elastic leaves. It weathers rather readily but unless present in the granite in large quantity it is rot considered injurious.

Hornblende is a silicate of iron, lime, magnesia, and alumina. Its color is usually dark green to greenish black. It occurs in compact crystals and is more resistant to the weather than mica.

Good granites are usually either gray or reddish in color, these being composed of the minerals that are more resistant to weathering. The black or greenish granites are usually less durable for the opposite reason. Whitish and yellowish granites should be examined carefully to see that the feldspar and mica are fresh, since such colors are often produced by the weathering of these constituents.

Trap Rock.—The word trap, as here used, is derived from the Swedish *trapp*, meaning stairs, havirg reference to the occurrence of the rock in the quarry. Trap rock consists mainly of feldspar of the soda or lime (not potash) variety, and hornblende. The rock is basic in composition, the content of silica being notably low. It is fine-grained, compact, and dense, and may occur either in the massive or gneiss formation. Because of the way the rock is usually cracked, it is difficult to obtain large blocks of specified dimensions for building purposes. Because of the low silica content and the presence of considerable iron compounds, the color is usually dark—from dark gray and green to black.

Like all basic rocks, trap rock is liable to weather considerably, producing a chalky or clay-like product.

The strength and density of trap rock exceeds that of granite, and hence from this standpoint alone it is very suitable for use as a building stone. But because of the great toughness of the blocks it is difficult to dress them, and beside they have a tendency to split into irregular masses so that they are seldom used as a dimension stone for building. However, because of these very properties; that is, their density, strength, and toughness, they are exceptionally suitable for paving blocks, railway ballast, and concrete aggregate.

Serpentines are generally igneous in origin, but not directly so, having been produced by the hydration of basic silicates, and so may be classed as metamorphic rocks. Pure serpentine is a hydrated silicate of magnesia, $3MgO \cdot 2SiO_2 \cdot 2H_2O$, containing about 13 per cent. of combined water. It has a soapy feel and is soft enough to be easily cut with a knife. It is not much used as a building stone since in nearly all serpentine quarries, irregular joints occur which make it extremely difficult to obtain large-sized blocks. It is quite suitable for interior use, being commonly green or greenish yellow (although it may be other colors if impure) and is susceptible to a fine polish, unless, as is sometimes the case, it should contain hard crystals of pyrite, chromite or magnetite. For exterior use it is defective since it weathers irregularly, cracks and fades.

Soapstone is closely allied to serpentine, being also a hydrated magnesian silicate, $3MgO \cdot 4SiO_2 \cdot H_2O$. It differs from serpentine in that it is both less basic and less hydrous, containing only about 5 per cent. of combined water. It is used frequently for sinks, wash tubs, laboratory table tops, etc., being often sold under the trade name of *"alberine stone."*

SEDIMENTARY ROCKS

Sedimentary rocks are those that have been formed of material usually carried by and deposited from water in the form of clays, sands, etc. Having been formed as sediments, these rocks possess a characteristic bedded or layer formation, the layers varying in composition, color, etc., depending upon the conditions prevailing at the time of deposition. In using these rocks for building purposes it is necessary to take account of this layer formation, since they split more readily along the bedding planes than in other directions, and because of this, they are less suitable for massive structures than the igneous rocks. They should never be set in the wall on edge, since thin layers may split or scale off where exposed to the weather. Sedimentary rocks, like those of igneous origin, may be found greatly modified by metamorphism.

The sedimentary rocks considered in this discussion will be taken up in the following order: slates, sandstones, limestones, and marbles.

Slates.—With but few exceptions, roofing slates are the hardened, and more or less metamorphosed siliceous clays that collected as fine silt on ancient sea bottoms.[1] In a few cases they are derived from rocks of igneous origin. By long-continued pressure a plane of cleavage has been formed. This cleavage plane may be parallel to the original bedding plane, but more usually it diverges at some angle from it, not infrequently at a right angle. Eckel says:[2]

"It has been proven experimentally that a distinct *slaty cleavage* can be produced in any fine-grained homogeneous mass by long-continued, heavy pressure, so that we may fairly assume that the cleavage of roofing slate was produced in this way."

The original bedding planes are discernible in the form of bands on the cleavage surface and are known as ribbons. Slates differ from clay rocks and shales in that the clays and shales break into more irregular masses, the easiest breaking occurring in planes that run either parallel to the bedding, or parallel to joint systems.

Composition.—Since slates are derived from clayey rocks, they consist essentially of the silicates of alumina, lime, magnesia, potash, and soda, and contain 3 or 4 per cent. of combined water. In some varieties, mica scales are quite abundant, adding considerably to the elasticity and toughness of the slate.

Color.—The commonest color of slate is some shade of gray or bluish-gray, and then follow in order, the black, green and purple slates. In most cases these colors are dependent upon the amount and condition of two substances—organic matter and the iron oxides. If the slate contains considerable finely divided carbonaceous matter, it will probably show a glossy black color; if high in ferric iron, it will probably be a red or purple slate; if high in ferrous iron, it will probably be green in color, unless this is obscured by organic matter, in which case the color will likely be black. The glossy black slates are, as a rule, finer and more

[1] MERRIL: Stones for Building and Decoration, p. 75.
[2] *Loc. cit.*, p. 96.

even-grained, softer and considerably smoother than slates of
other colors. The gray slates, on the other hand, are coarser in
grain and much harder, while the reds and greens are intermediate.

Discoloration of Slates.—The discoloration of slates is due to
the presence of the very unstable ferrous carbonate in a finely
divided form, which gradually oxidizes to the brown, hydrated
iron oxide. The original ferrous carbonate is green, but this
color is not necessarily imparted to the slate since the green may
be masked by graphitic or carbonaceous matter, in which case
the slate will be black. Hence some black slates may discolor
or fade and others may not. This difference is exhibited by two
grayish-black slates from Pennsylvania. The slates from the
Peachbottom district have excellent weathering qualities and are
permanent in color. Some slates from Lehigh and Northampton
Counties have a tendency to fade, due to the oxidation of the
ferrous iron content. Since the amount of this ferrous iron often
varies in the different beds of the same quarry, it may happen that
patches of slate on the same roof may fade while others may not,
thus presenting a very undesirable blotched appearance.

Sandstones.—These are made up essentially of grains of quartz
bound together by some cementing material. The cementing
material may be silica, iron oxide, calcium carbonate or clay, and
upon this basis the sandstones are divided in the same order into
siliceous, ferruginous, calcareous, and argillaceous stones, although it not infrequently happens that more than one kind of
cementing material may be present in the same stone, and beside
.the quartz grains and cementing material, other minerals are often
found, as mica, pyrite, etc. Upon the nature and amount of
the cementing material depends in a large measure the color,
strength, and durability of the stone.

Colors.—The red and reddish brown colors are due to the presence of ferric oxide, the yellow and yellowish brown to the presence of the same oxide in the hydrated condition. The faint
bluish or greenish tints are due to iron sulfide, iron carbonate,
or more rarely to iron silicate in finely divided form. Carbonaceous material may cause a gray or black appearance.

Durability.—The durability of sandstones varies considerably.
Stones with clay seams are liable to split with repeated thawing
and freezing. Those containing mica scales along the bedding

planes are liable to flake off, especially if the stone is set in the wall on edge.

Of all varieties, the siliceous sandstone is most resistant to the action of the weather. But if silica alone is present, they are

FIG. 48.—Weathered sandstone. Guard wall of bridge after 18 years, exposure.

frequently so hard that it is difficult to dress them. A good example of the siliceous type is the Berea sandstone of Ohio.

The ferruginous stones stand second in respect to durability. The "brownstones" of the Eastern States are generally sand-

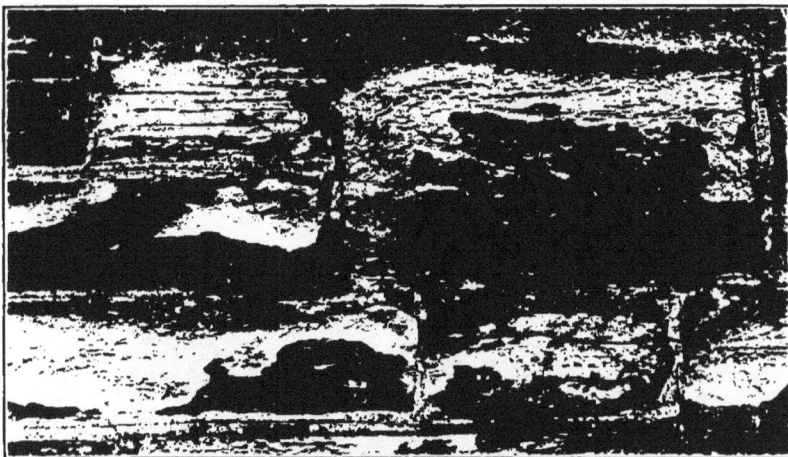

FIG. 49.—Weathered sandstone.

FIG. 50.—Weathered sandstone. A corner of the same wall shown in Fig. 49. The stone has been softened to such an extent by weathering that a knife blade may be readily thrust into it.

stones of this sort. They are easy to dress but are not nearly so durable as the siliceous stones. In many cases they have shown very early failures, although this may have been due to the presence of calcium carbonate in the cementing material.

The calcareous sandstones are generally light-colored or gray and, although strong, are soft and easily worked. They resist the action of the atmosphere very poorly, since the calcium carbonate is leached out by the action of the carbon dioxide in the presence of water.

The argillaceous sandstones are the least durable of all. They absorb water readily and thus render the stone more susceptible to the action of frost. Being comparatively soft they are easily dressed.

Some sandstones contain little if any cementing material, but owe the strength they possess merely to the pressure to which they were subjected during formation. Certain varieties of such stones are used it the manufacture of grindstones and whetstones.[1] Since they do not contain much cementing material, they crumble slowly away, always presenting fresh sharp surfaces to be acted upon. In some cases, they possess sufficient cohesion to be used as building stones.

"Quarry Water."[2]—All sandstones, when freshly quarried, contain a certain amount of water. This causes them to be soft and easily worked, but on the other hand, renders them very susceptible to injury if allowed to freeze when freshly quarried. Some quarries cannot be worked during the winter on this account. As the "quarry water" evaporates, the stone gains considerably in hardness and becomes more difficult to work. This has been explained on the basis of the assumption that in the freshly quarried stone a certain amount of the material exists in the dissolved state, which as the water is evaporated, is deposited, thus binding the grains more firmly together.

Limestones.—Limestones are stratified rocks consisting essentially of calcium carbonate, $CaCO_3$. They have been produced largely from deep-sea deposits of the calcareous remains of animal organisms. In some cases the shell formation of the individual organism is distinguishable, the different particles being cemented

[1] MERRIL: *Loc. cit.*, p. 125.
[2] MERRIL: p. 126.

together by calcium carbonate deposited from solution. Pure calcium carbonate, as a mineral, is known as *calcite*. But limestone is not pure calcite. It contains usually iron oxide, silica, clay, and carbonaceous matter in varying quantities. With a sufficient quantity of silica, the limestone becomes a calcareous sandstone, with a sufficient quantity of clay, a calcareous shale. Magnesium carbonate is also usually present, replacing the calcium carbonate. The amount of this replacement may be considered to range from none at all to 100 per cent.; in the latter case, the carbonate being entirely that of magnesium, the mineral is known as *magnesite*. When the two carbonates are present in molecular weight proportions, $CaCO_3 \cdot MgCO_3$, the rock is called *dolomite*. Formations of this sort contain 54.35 per cent. of calcium carbonate and 45.65 per cent. of magnesium carbonate. The term *limestone* is usually restricted to those rocks that contain less magnesium carbonate than the dolomite, while those that contain more are known as impure magnesites.

Limestones may be white, gray, blue or black, depending upon the impurities present. Their hardness is very variable. Some are so soft that they may be easily cut with a saw or knife, while others are a great deal harder. Both limestones and dolomites are moderately durable, but are not to be compared with good sandstones and granites in this respect. They are dissolved by the carbonic acid of natural waters, although dolomite does not yield so readily to this solvent action as the limestone.

Marbles.—In accordance with strict geologic definition, marbles are those highly crystalline formations that have been developed from limestones or dolomites by heat and pressure, being, therefore, metamorphic rocks. The original limestone or dolomite, which was of ordinary type as far as composition or structure was concerned, was converted into a semifluid mass by the intense heat and pressure and then assumed the crystalline state. The pressure in this case was sufficient to prevent the evolution of the carbon dioxide which would have occurred at this temperature under ordinary conditions.

According to the commercial conception, the term marble has a much broader meaning than the preceding definition shows. It includes any limestone, crystalline or not, that is capable of taking a high polish, and which when polished will show pleasing color effects. Eckel says[1] that according to the commercial classification there are three varieties of marble:

"1. Highly crystalline marbles, showing distinct crystalline structure and fracture. These are usually white, though gray, black or other markings may be present scattered over a white ground.

"2. Subcrystalline or fossiliferous marbles, in which crystalline structure is rarely very noticeable, the value depending rather on color effect than upon texture.

"3. Onyx marbles, translucent rocks, showing color banding, due to the fact that they were formed layer after layer by chemical deposition from cave or spring waters."

The chemical composition of the marbles differs in scarcely any respect from that of the ordinary limestone. It is liable to contain the same impurities as the limestone, two or three of which are objectionable. Mica, if abundant and present in bands or patches, interferes with the polishing of the stone, and also since it weathers readily it is liable to leave the surface pitted. Of course, this weathering is not liable to occur if the marble is used for interior work. Iron pyrite is quite objectionable. It oxidizes readily with the formation of iron rust and sulfur dioxide. The latter by further oxidation may be converted into sulfuric acid, which, of course, exerts a solvent action on the marble.

Color of Marbles.—Marbles are exceedingly variable in color, some being snow white and others varying from gray to black. Also they may be shades of red, pink, yellow, green or brown, these shades being due chiefly to iron oxide. The gray and black are due to carbonaceous matter. The coloring matter may be uniformly distributed or present in patches, producing a mottled appearance.

Some ornamental marbles show that which is termed a *brecciated structure*. These are made up of angular pieces of crushed rock, the spaces between being filled with mineral matter afterward deposited.

Durability.—The weather resisting properties of the marbles is about equal to that of the limestones, but those having the brecciated structure and those containing considerable mica have a considerably lessened durability.

[1] Building Stones and Clays, p. 166.

References

MERRIL: Stones for Building and Decoration, New York, 1903.
RIES: Economic Geology of the United States, New York, 1908.
BAKER: A Treatise on Masonry Construction, New York, 1909.
ECKEL: Building Stones and Clays, New York, 1912.
RIES: Building Stones and Clay Products, New York, 1912.
MILLS: Materials of Construction, New York, 1915.

CHAPTER X

LIME AND GYPSUM PRODUCTS

Lime is the oxide of calcium, CaO. Although in non-technical language the term lime is somewhat indiscriminately applied to a variety of calcium compounds, it properly refers to this one compound only. Frequently the terms "quicklime" and "caustic lime" are used to more definitely indicate this oxide. On a commercial scale, lime is obtained by calcining or "burning" limestone, $CaCO_3$. At the temperatures employed the calcium carbonate dissociates into lime and carbon dioxide, which gas escapes. If the stone be pure calcium carbonate the loss in weight resulting from the separation of the gas is equal to 44 per cent. of the weight of the original stone. Commercial limes are not pure calcium oxide because limestones are not pure calcium carbonate.

Limestones.—Limestones belong to the class of sedimentary rocks, which are those that have resulted from the decomposition of other formations, the material produced in this way being carried by and deposited from water. Limestones have been produced largely by deep-sea deposits of the calcareous remains of animal organisms. In some cases the individual organisms are plainly visible, being merely cemented together by calcium carbonate deposited from solution. A pure limestone corresponds to the white mineral calcite, $CaCO_3$, but usually there are present varying amounts of magnesium carbonate, silica, iron oxide, clay and carbonaceous matter. Frequently it happens that the calcium carbonate and magnesium carbonate are present in molecular weight proportions, $CaCO_3 \cdot MgCO_3$, this being equivalent to 54.35 per cent. of the former and 45.65 per cent. of the latter. Such rocks are called *dolomite*. Limestones are found that contain magnesium carbonate in any amount from 0.0 per cent. to 45.65 per cent., but usually the amount is near either the upper or lower limit, the intermediate grades being comparatively rare.

Manufacture of Lime.—Lime is produced by calcining limestone in kilns. The temperatures employed may vary from 900 to 1,200°C., the reaction produced being essentially:

$$CaCO_3 + heat \rightarrow CaO + CO_2$$

There are various types of kilns in use, but that which is generally employed is some form of shaft kiln. Rotary kilns, like those used in the Portland cement industry, are used to some extent, but for these the stone must be crushed. The shaft kilns are chiefly of the continuous type, the lime being drawn periodically from the bottom, while the charge is continuously introduced at the top. An older type is the intermittent kiln in which a charge is burned, allowed to cool and then drawn. This is, of course, less economical because of the heat wasted by cooling the kiln each time. Also the shaft kilns differ in respect to the manner of heat application. In some the fuel is mixed with the charge, first a layer of coal or coke and then a layer of stone, these layers being repeated until the kiln is filled. Fire is started at the bottom and works its way up. In this type the lime always contains the ashes of the fuel. In another type there is a separate firebox where the fuel is burned and the hot products of combustion only pass into the kiln. Lime produced in this type is, of course, not contaminated by the ashes. There are various modifications of these general types.

Physical Properties of Lime.—Lime is a non-crystalline substance, and if pure, is snow white. Often it may have a grayish, yellowish or brownish color, due to the oxides of iron or manganese contained in the clay or other impurities in the stone.

Lime produced by burning a pure limestone has a specific gravity ranging from about 3.08 to 3.30. The density depends somewhat upon the kiln temperature. It approaches the higher limit as the kiln or calcining temperature increases. Of course, it is always more porous than the rock before it was calcined. Lime, as packed, weighs about 60 lbs. per cubic foot.

Kinds of Lime.—As has been indicated elsewhere, aside from the 5 to 10 per cent. of impurities, such as silica, alumina and iron oxide, commercial limes may contain from a trace to 44 per cent. of magnesia. On the basis of the quantity of magnesia present, limes are divided into classes. Those that contain less than 5 per cent. of magnesia are known as pure or high-calcium limes, while those that contain over 5 per cent. of magnesia are known as magnesian limes. However, the magnesian limes on the market usually contain above 30 per cent. of magnesia.[1] Because the properties of these two classes vary considerably, they will be discussed separately.

High-calcium limes contain generally from 90 to 95 per cent. calcium oxide, the remainder consisting of magnesia, alumina, iron oxide, silica, and a little carbon dioxide and water. High-calcium limes are generally desired for preparing building mortar because they slake readily, and also because they swell considerably during slaking and thus produce a larger volume of paste.

A lime that contains less than 5 per cent. of impurities such as silica, alumina, iron oxide, etc., is often called a *"fat"* or *"rich"* *lime.* These slake exceptionally rapidly and produce more plastic mortars, being described as more "smooth" under the trowel than limes containing greater amounts of impurities, which are known as *"lean"* or *"poor"* *limes.*

Magnesian Limes.—By definition,[2] these are limes that contain above 5 per cent. of magnesia, but usually the commercial magnesian limes contain above 30 per cent. The magnesian limes slake less rapidly than the high-calcium limes, develop less heat, and expand less in slaking. In mortars they do not possess as great early strength but develop greater ultimate strength than the high-calcium limes.[3] They are often considered superior to the high-calcium limes for finishing, but opinions vary as to this and both kinds are used.

Slaking of Lime.—The slaking of lime is a process of hydration, the calcium oxide entering into chemical combination with water to form calcium hydroxide, as:

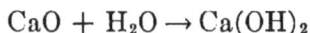

$$CaO + H_2O \rightarrow Ca(OH)_2$$

This reaction is accompanied by the evolution of a great amount of heat, so that some of the water added may be converted into steam, this being especially noticeable in the slaking of "fat" limes. It is important to consider this heat evolution with

[1] ECKEL: Cements, Limes and Plasters, p. 98.

[2] ECKEL: pp. 97 and 98.

[3] *Municipal Engineering,* **28,** 4–7 (January, 1905).
15

reference to the *fire danger*. According to the work of the Royal Saxon Mechanical-Technical Experiment Station,[1] it is easily possible for wood to be ignited by the heat evolved in slaking even "poor" or "lean" limes, for such limes, although poor in calcium oxide, may reach a temperature of 270° to 300°C., at which point wood chars in air. High-grade or "fat" limes are still more dangerous, because in slaking, they can produce a temperature of 400° or over. It is not safe to ship or store lime in contact with wood, since water may accidentally come into contact with it.

In slaking lime, the weight of water that must unite with it in order to convert it into the hydrated form is 32.1 per cent. of the weight of the lime, assuming the lime to be absolutely pure calcium oxide. But, because the lime practically always contains inert impurities, somewhat less than this amount is theoretically required. On the other hand, some water is lost as steam, the amount varying with conditions. For mortar work, about twice the theoretical quantity of water is employed, the slaked product being then in the form of a stiff paste. For best results, it is important that the proper amount of water be added at once, not in portions, and that it be distributed uniformly and quickly. A great excess must be avoided. If so much is used that the hydrated product is in the form of a thinly fluid paste, its binding power is apparently much lessened in the mortar made from it. This may be due to the fact that mortars made from such hydrated limes contain more water than is customary.

If the lime is wetted with enough water to start the hydration but not enough to complete it, the unhydrated portion becomes inert,[2] being then described as "burnt." This condition may be brought about accidentally when partially slaked lumps of lime are pushed out of the water by the swelling of the mass beneath. Just what constitutes this "burning" seems not to be understood. If an insufficient amount of water is added in the beginning and then a further addition is made during the process, the slaked lime will be granular and lumpy. When properly slaked, the steam evolved by the heat of the reaction bursts the lumps and aids in the slaking. But if water is added during the

[1] *Jour. Ind. Eng. Chem.*, **7**, 542 (June, 1915).
[2] U. S. Bureau of Standards, *Technologic Paper* 16, p. 77.

process the evolution of steam is checked. Further, the added water causes a pasty coating of slaked lime to be formed on the surface of the lumps, which renders the entrance of water to the interior of the lumps difficult, if not impossible.[1]

The bulk of the lime increases from 2.5, to 3.5 times during slaking, the more rapid the slaking the greater being the volume increase. Impure calcium limes, magnesian and dolimitic limes slake only about half as fast as the high-calcium limes and expand far less.

It has been observed, also, that the degree of plasticity is much increased by rapid slaking, this being due to the fact that rapid slaking produces a *colloidal hydrate*, whereas slow slaking tends to produce the crystalline condition. The colloidal hydrate is produced very rapidly, but crystals require much more time to form. When the hydrated product is crystalline the volume is less, the mass is granular and gritty and the sand-carrying capacity is much reduced. The slaking reaction may be hastened by using warm water or by covering the slaking lime with a tarpaulin.

Magnesian limes are slow in slaking, and as has just been explained, slow slaking operates to form a less plastic product. On this account magnesian limes can be mixed with less sand for use. Quite aside from its effect in lowering plasticity, magnesia is said to have a very marked effect in increasing the ease and smoothness with which the paste may be spread with a trowel. Magnesian limes will slake readily only if especial care is taken to burn them at low temperatures.

The *porosity* of the lime, which results largely from the evolution of the carbon dioxide gas from the stone, is of importance in respect to the rate of slaking since it allows the water to enter more readily and the slaking to proceed more rapidly. All limes are not equally porous. High burning temperatures make denser limes. This is, at least partly, due to the fact that high temperatures cause some of the lime to unite chemically with the impurities present, thus producing silicates and aluminates of lime. These compounds are much more readily fusible than the lime and as they soften, close the pores of the stone. This fusion is especially noticeable on the outside of the lumps, where

---

[1] BAKER: A Treatise on Masonry Construction, p. 105.

a more or less vitreous coating may form, particularly if the temperature exceeds 1,200°C. to any extent. Limes in which the calcium oxide has reacted with the impurities are said to be "over-burned." This overburning is, of course, more likely to occur if the impurities are high. It is very difficult to overburn a pure lime because calcium oxide is a very refractory substance.

If mortar is made of lime that for any reason has been incompletely slaked, it is very likely to be defective, because the lime is almost certain to slake eventually and then blisters or more serious faults will show in the wall or other structure where the mortar is used.

Common Method of Slaking Lime.—In the ordinary practice of slaking, the lumps of lime are placed, usually in a water-tight box, in a layer from 6 to 8 in. deep. A quantity equal to two, or two and a half times the actual volume of the lime (which is an amount a little greater than two-thirds the weight) is then poured on. This is about twice the amount of water theoretically required, but the excess tends to insure thorough slaking. If the slaking is properly carried out, the slaked product will be in the form of a stiff paste. One barrel of lime (230 lbs.) will make about 8 cu. ft. of such paste, which is frequently known as "lime putty." The paste may be kept almost indefinitely if it is kept from drying out. It is sometimes stored in casks or in the wide shallow boxes in which it is slaked. It cannot harden under water. The lime should be slaked at least one day prior to its use, in order that it may be more certain that the slaking is complete.

Fine Lime and Lump Lime.—If the lime has been made from broken rock it will generally be in lump form, the occurrence of powder or dust in this case indicating that the lime is old and has been poorly stored, which has allowed it to slake partially by absorption of moisture from the air. Also, when thus slaked due to long exposure, it takes up carbon dioxide from the atmosphere and becomes more or less inert depending upon the extent to which this reaction has proceeded. Such lime is known as "*air-slaked*" *lime*. When completely air-slaked it is useless for the purposes to which lime is usually put.

However, limes that are made from shells, soft chalk, marl, highly crystalline marbles, shelly limestones or limestones that contain much water or organic matter will usually come from the kiln in a more or less finely divided state. Fineness is no indication of air slaking in such cases. Also, the product of the rotary kiln is fine. Moreover, some of the manufacturers grind their product because it will keep better. It is practically impossible to preserve lump lime for any extended period, especially if it is in loose bulk, since air has such ready access to all parts of it. If it is contained in casks or barrels it may be preserved for some time, if stored in a dry place and kept covered with tarred paper, canvas or a tarpaulin. In the case of ground lime, the top layer slakes and this forms a coating that serves to protect the lime underneath.

When properly burned and fresh from the kiln, lime should be free from water and should not contain more than 0.5 per cent. of carbon dioxide.

Hydrated Lime.—This is a somewhat recent commercial product prepared to avoid the unsatisfactory results that attend the old method of slaking lime in the mortar box by unskilled labor. It is essentially calcium hydroxide; that is, it is quicklime that has been slaked, and is usually prepared by the manufacturer of the lime. The process consists of first crushing the lime, if it is in lumps, to pieces not greater than 1 in. in diameter, and then this crushed lime is agitated in one way or another with the proper amount of water. After slaking, the now hydrated lime is sieved and the coarse particles discarded, or in some cases ground and mixed with the hydrated lime, which is a fine, dry, white powder. Its specific gravity varies from about 2.15 to 2.24. It may be used for practically any purpose in place of lump lime.

If properly made and stored, hydrated lime will contain less than 1 per cent. of carbon dioxide, this being the equivalent of 2.27 per cent. of carbonated lime. The determination of the carbon dioxide may be used as a test for the quality of the hydrated lime.

Emley[1] says that the chief advantages to the consumer of hydrated lime are as follows: It can be handled more easily because it is in powder form. It will keep better than lump lime for the reasons explained in the preceding discussion of fine

[1] U. S. Bureau of Standards, *Technologic Paper* 16, p. 18.

lime and lump lime, although there is no evidence to show that
it will keep better than quicklime of an equal degree of fineness.
It is necessary only to soak it with water to prepare it for use.
This saves time and labor and eliminates any danger of loss of
lime due to unskilled slaking. The hydrated lime should contain
less refuse than lump lime because the inert material, such as
produced by over- or underburning, can be screened out after
hydrating. On the other hand, hydrated lime contains from 15
to 25 per cent. water on which the consumer must pay the
freight.

Also, according to Mills[1] there is another objection, in that the
paste made of hydrated lime is very noticeably non-plastic, and
low in sand-carrying capacity. He says that some hydrates are
so lacking in colloidal properties that they are absolutely gritty,
and on this account are not generally suitable for plastering or
finishing.

"Alca" Lime.[2]—The term Alca is derived by combining the
symbols for aluminum and calcium and is used as a trade name to
apply to a hydrated lime that has been specially prepared in
order that it may be more plastic and suitable for wall plaster.
It is made by incorporating with the hydrated lime about 15
per cent. of calcium aluminate derived from a special blast-furnace slag. The product prepared in this way has been found to
have increased sand-carrying capacity, to be more easily applied
with a trowel, and to set more rapidly than the ordinary limes.
It is thought that the initial setting is dependent upon the hydration of the calcium aluminate, as described in the discussion of
the setting of Portland cement. In using the Alca lime, Mills
says that best results are obtained if the sand and lime are
first thoroughly mixed dry and then wet with about 16 per cent.
of water and allowed to stand for not less than 1 hr. before being
used. Or, if it can be allowed to stand for a longer time, over
night for example, it will be improved in plasticity and working
qualities.

Hydraulic Limes.—As was stated in the discussion of over-
burned "lean" limes, if some of the calcium oxide reacts with
the silica and alumina during burning, then the lime will have

[1] Materials of Construction, p. 55.
[2] MILLS: p. 56.

more or less well-defined hydraulic properties. The reaction
mentioned produces compounds of the same sort as those that
occur in Portland cement.

In the manufacture of hydraulic limes proper, limestones are
used that contain from about 10 to 17 per cent. of impurities,
and sometimes the limestone may be of the magnesian variety.
Higher burning temperatures are used than in the production
of ordinary lime in order that the lime may react with the silica
and alumina. Hydraulic limes are not manufactured in the
United States but considerable quantities are imported, especially
from France.

Hydraulic limes slake in a manner similar to ordinary lime
only much more slowly. Some of them have only slight hydraulic properties, while others may harden quite satisfactorily
under water.

Grappier cements are obtained by grinding the unhydrated
cores that are left in the slaking of hydraulic limes. Like the
hydraulic limes they are not manufactured to any great extent
in the United States but are imported from Europe.

The use of hydraulic limes and grappier cements depend upon
the fact that they are very white and do not stain the stone, as
for example in marble work, as does Portland cement. This
non-staining property is due to their low content of iron oxide
and soluble salts.

Lime-sand Mortar.—The lime paste alone is not suitable for
use as a mortar. If the paste is allowed to dry it shrinks greatly
and cracks. In practice, sand is always added to it, the proportions being about 2½ or 3 of sand to 1 of paste. The addition
of this inert material almost eliminates shrinking and prevents
the cracking. Also, the addition of sand makes the mass more
porous, so that the penetration of the carbon dioxide, necessary
for the hardening of the mortar, is facilitated. It is of some
importance that the sand and paste should be mixed in about
the proportions indicated, since with these amounts the paste
will just fill the voids in the sand. If much more than this
amount of sand is used the mortar will be weak because it will be
too porous. If less is used, the resultant excess of paste will
cause a shrinkage upon drying. However, if sands of different
degrees of coarseness are mixed, the voids between the particles

will be lessened and then a smaller proportion of paste may be used without the mortar being rendered too porous.

Hardening of Lime-sand Mortar.—Compared to the hardening of Portland cement, the hardening of lime-sand mortar is a very simple process. The initial "set" is due to the evaporation of the water. If used with porous bricks it may set very quickly because much of the water is taken up by the brick.

But not until after the mortar has become practically dry does the real hardening begin. This is due to the absorption of carbon dioxide from the air, the following reaction occurring:

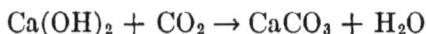

$$Ca(OH)_2 + CO_2 \rightarrow CaCO_3 + H_2O$$

The calcium carbonate so produced crystallizes about the grains of sand and holds them together. The depth of the carbonate formed is from 0.10 to 0.12 in. the first year, but it is slower after that because the surface layer becomes more impervious to the passage of the necessary carbon dioxide gas as the crystals of the carbonate form in the pores. The inner portions of thick walls show the presence of calcium hydroxide often after a century or two. On this account lime-sand mortar is not suitable for use in extremely thick walls. After about 25 years, however, the hardening has been completed in ordinary walls.

Because it cannot set when air is excluded, lime-sand mortar cannot be used for structures under water, nor in soil that is constantly wet. And because it never acquires any great amount of strength it should not be used where great strength is required.

GYPSUM PRODUCTS

Gypsum.—The mineral gypsum, from which various kinds of commercial plasters are prepared, is the hydrated calcium sulfate, $CaSO_4 \cdot 2H_2O$. It frequently occurs in the massive form known as *rock gypsum*, or it may be found as a granular variety known as *gypsum sand*. Sometimes it occurs relatively pure, but generally it contains a few per cent. of impurities as silica, alumina, iron oxide and the carbonates of calcium and magnesium. An earthy variety known as *gypsum earth or gypsite* may contain 50 per cent. or more of such impurities, especially silica and the carbonates mentioned. Pure gypsum is white, and when crystalline is more or less translucent. It is so soft that it can be readily scratched with the thumbnail. *Alabaster* is a pure variety of gypsum used to some extent for statuary.

Preparation of Gypsum Products.—Briefly stated, the various plasters, such as plaster of Paris, wall plasters, etc., made from gypsum, are prepared by heating gypsum until its combined water is given off. It is important to note the temperatures to which the gypsum is heated. If temperatures below 200°C. are employed, only three-fourths of the combined water is removed, as:

$$CaSO_4 \cdot 2H_2O + heat \rightarrow CaSO_4 \cdot \tfrac{1}{2}H_2O + 1\tfrac{1}{2}H_2O$$

In this manner plaster of Paris is produced. When this plaster is mixed with water, it readily unites again with an amount equal to that given up and reverts to a hydrated form that is chemically equivalent to gypsum. As a result of this reaction with water, the plaster "sets" and hardens.

If the calcining temperature much exceeds 200°C., practically all of the combined water is driven off, and the product is then known as "hard-burned" or "dead-burned" plaster. Although the plaster is called "dead-burned" and sometimes even "anhydrous," it is generally understood that it is not absolutely so, there being slight traces of hydrated crystals still remaining, which serve as nuclei about which other crystals of hydrated sulfate may form. All plasters that result from calcination at these elevated temperatures are very slow in hydrating, the rate being dependent upon the intensity and prolongation of the heating during manufacture, and upon the degree of fineness of the gypsum. It is possible, in fact, quite easy to render the plaster incapable of hydrating at all, in which case no doubt all crystalline structure has been destroyed. But for practical purposes the gypsum is calcined under such conditions that the setting and hardening are not prohibited altogether, but only greatly prolonged, the hydration then requiring hours or days, instead of minutes as with plaster of Paris. When finally formed, the hydrated product resulting from these slowly setting plasters is both harder and stronger than that produced by the ordinary plaster of Paris.

Classification of Plasters.—On the basis of the degree of dehydration, the gypsum products may be divided into two major classes, and these, in turn, may be subdivided into two classes on the basis of whether or not the product is pure or impure. The following classification is essentially as given by Eckel:[1]

(A) Produced by the *incomplete* dehydration of gypsum, the temperature employed not exceeding 200°C.

1. Produced by calcining a *pure* gypsum, no foreign material being added either before or after calcining. *Plaster of Paris.*

2. Produced by calcining a gypsum containing certain natural *impurities*, or by adding to a calcined pure gypsum, certain materials that serve to retard the setting of the product. *Cement Plaster.*

(B) Produced by the *complete* dehydration of gypsum, the calcining being carried out at temperatures exceeding 200°C.

1. Produced by calcining a *pure* gypsum. *Flooring Plaster.*

2. Produced by calcining at red heat or above, gypsum to which certain substances (usually alum or borax) have been added. *Hard Finish Plaster.*

In commercial classification, the term *calcined plaster* is used to designate a burned plaster to which nothing has been added, and may include both plaster of Paris and cement plaster, it being the latter if the original gypsum used was impure.

Plaster of Paris.—As has been indicated in the preceding classification, plaster of Paris is made by calcining a pure gypsum so that three-fourths only of the combined water is driven off, producing thereby $CaSO_4 \cdot \frac{1}{2} H_2O$. This hydrates and "sets" very rapidly—in from 5 to 15 min.—in accordance with the following equation:

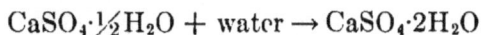

$$CaSO_4 \cdot \tfrac{1}{2} H_2O + \text{water} \rightarrow CaSO_4 \cdot 2H_2O$$

Stucco is practically the same as plaster of Paris, since it is made from fairly pure gypsum, but usually it is not so finely ground as plaster of Paris.[2]

Cement plasters are made in a manner similar to plaster of Paris, but set more slowly because of the presence of retarding agents. These retarding agents may be either impurities that

[1] Mineral Resources of the United States, 1905, p. 1112.

[2] The term stucco is used also in a somewhat different sense to indicate a mortar finish on exterior walls. The material used for such work consists essentially of a cement-sand mortar to which usually has been added a certain amount of lime and fiber.

were present naturally in the gypsum or may consist of small amounts (less than 1 per cent.) of added substances, as glue, sawdust, etc. Cement plasters set in from 1 to 2 hrs. and may be used for structural work.

Wall plasters are practically the same as the preceding, but in addition to the retarding agent they contain also from $1\frac{1}{2}$ to 3 lbs. per ton of hair or about 100 lbs. per ton of wood fiber. Since the advent of metal lath, the use of calcined gypsum in wall plasters has met with some objection. Calcium sulfate is a compound that results from the reaction of a strong acid with a relatively weak base, and in the presence of water dissolves slightly and ionizes to produce an acid reaction, thereby acting as a corrosion accelerator when in contact with the lath. On this account there has been a tendency to eliminate gypsum to a certain extent for wall plasters and substitute Portland cement or calcium aluminate, which also set rather rapidly by hydration, and beside have an alkaline reaction which apparently inhibits the corrosion of the metal.

Unless the plaster is made from the earthy variety of gypsum, and therefore, already contains clay, it is necessary to add clay or hydrated lime to increase the plasticity and sand-carrying capacity. For this purpose about 15 per cent. of hydrated lime is usually added, which, beside increasing the working qualities of the mass, corrects its tendency to bring about corrosion.

Flooring plasters differ from the preceding forms in that they are calcined at such temperatures that they are practically anhydrous. They set very slowly and require very fine grinding, but ultimately they develop great strength and hardness. Eckel[1] says that these plasters are manufactured and used on a fairly large scale in certain European countries, but have not been used in the United States.

Hard Finish Plasters.—Like the flooring plasters, these are practically an anhydrous product, but in addition they have been treated with certain salts. After a preliminary burning at a red heat, the calcined product is dipped into a solution of alum or other salt, as borax, and then burned again. This salt treatment aids in increasing the hardness. After the final burning, the product is very finely ground. It is necessary to use a very

[1] Cements, Limes and Plasters, p. 75.

pure gypsum, since even a trace of iron oxide would detract from the necessary whiteness. Representatives of these plasters are known as "Keene's Cement" and "Parian Cement." They are manufactured in the United States and a considerable amount is imported.

References

A

ECKEL: Cements, Limes, and Plasters, New York, 1905.
BAKER: Treatise on Masonry Construction, New York, 1909.
MILLS: Materials of Construction, New York, 1915.
ROGERS and AUBERT: Industrial Chemistry, New York, 1915.
DANCASTER: An Elementary Treatise on Limes and Cements, London, 1916.

B

BARTLETT: Manufacture of Plaster of Paris, *Eng. Ming. Jour.*, **82**, 1063 (1906).
GRIMSHAW: Industrial Application of Gypsum, *Sci. Am. Sup.*, **61**, 25,476 (1906).
PERRING: Gypsum as Fireproofing Material, *Eng. Rec.*, **62**, 649 (1910).
SPACKMAN: Aluminates as Addition to Wall Plaster, *Proc. Am. Soc. Test. Matls.*, **9**, 315 (1910).
MEADE: Manufacture and Properties of Hydrated Lime, *Eng. News*, **65**, 554 (1911).
NOBEL: The Prevention of Corrosion in Metal Laths, *Cement*, **12**, 48 (1911).
EMLEY: Manufacture of Lime, U. S. Bureau of Standards, *Technologic Paper* 16 (1913).
SPACKMAN: Effect of Hydrated Lime on Portland Cement Mortars, *Eng. Rec.*, **69**, 25 (1914).
EMLEY and YOUNG: Strength of Lime Mortars, *Proc. Am. Soc. Test. Mtls.*, **14**, 338 (1914).
MARANI: Gypsum as a Fireproofing Material, *Jour. Cleveland Eng. Soc.*, **7**, 233 (1914).
MACGREGOR ET AL.: Report of Committee C-7 on Standard Specifications for Lime, *Proc. Am. Soc. Test Matls.*, **15**, Pt. 1, 167 (1915).

CHAPTER XI

PORTLAND CEMENT

Portland cement is a finely ground mixture of calcium compounds capable of setting and hardening by chemical combination with water. The main essential ingredients are lime, silica, and alumina, which are combined to form the silicates and aluminates of lime.

Historical.—Some knowledge of cements similar to Portland, at least in respect to the fact that they possessed the property of setting under water, was possessed by the Romans and recorded in their literature, but the wide usefulness of such materials seems not to have been appreciated for a great many centuries. Perhaps the first great impetus given to the development of the hydraulic cements was that contributed by Smeaton, who in 1756 was assigned the task of erecting a new lighthouse on Eddystone Rock after the old wooden structure had been destroyed. By experimenting he found that the best mortar for work under water consisted of that made from lime, produced by burning limestone that contained considerable clayey matter. About 50 years later, Vicat, a French chemist, extended this knowledge a step by producing a similar product by burning finely pulverized chalk and clay after he had mixed them in the form of a paste. But the first patent taken out for the making of a Portland cement was that issued in 1824 to Aspdin, a Yorkshire bricklayer, who heated finely pulverized chalk with the clayey mud of the river Medway. Because of a certain resemblance, which was in reality slight,[1] between his product when set and a well-known limestone quarried near Portland on the southern coast of England and extensively used for building under the name of Portland stone, he named his product "Portland Cement." This name has been almost universally adopted. Portland cement was first manufactured in the United States in about the year 1874.

[1] ECKEL: Cements, Limes and Plasters, p. 294.

Materials.—Portland cement is made by fusing together two materials, one rich in lime, as limestone, marl or chalk, and one rich in silica and alumina, as clay, shale, slate or blast-furnace slag, the former being known generically as calcareous and the latter as argillaceous materials. In the Lehigh district of Pennsylvania is found a certain clayey limestone that contains the calcareous and argillaceous materials in almost the right proportion. This is technically known as "cement rock." If it contains too much lime, clay is added; or if too much clay, lime is added. A large proportion of the cement produced in the United States is made from this. Also, a considerable quantity is made from a mixture of pure limestone and clay or shale. . In some of the Central States cement is made by combining clay with marl, an amorphous calcium carbonate containing much water and organic matter. The Illinois Steel Co. produces Portland cement from a mixture of blast-furnace slag and limestone.

Manufacture.—Most of the cement manufactured in the United States is made by the so-called dry process, in which the materials are mixed in the form of a dry powder, although in the plants using marl, because this itself contains much water, the materials are mixed in the form of a thin sludge or "slurry."

In the dry process, the materials are rather coarsely crushed and dried in inclined rotating cylinders, after which they are held in storage bins until analyzed. From these bins the materials are drawn and mixed in the proportions indicated by the analysis. The mixture is now ground in suitable mills to such a degree of fineness that at least 90 per cent. of it will pass through a 100-mesh sieve. The grinding also insures an intimate mixing of the materials. The finely ground mixture is now conveyed into a rotary kiln for calcining. A rotary kiln is a sheet-steel, brick-lined cylinder, which may vary from 70 to 150 ft. in length and from 6 to 8 ft. in diameter. It is inclined at an angle of about 15° to the horizontal and is caused to rotate at the rate of one turn or a half-turn per minute. The material is fed in at the upper end and gradually works through the kiln. The temperature within the kiln may range from 1,450° to 1,600°C., the heat being generated usually by burning powdered coal, which is blown in at the lower end. Also oil and gas are sometimes used. In passing through the kiln the mixture is heated to the point of incipient fusion and leaves the lower end in the form of small lumps known as "clinker." After cooling, which may be accomplished by allowing it to lie in piles or by passing it through mechanical cooling devices, the clinker is hard, glassy and of a blackish color. When cooled it is ground so that at least 92 per cent. will pass a 100-mesh sieve.

Since ground clinker alone would set too rapidly, it is necessary to add a retarding agent. This may consist of gypsum, plaster of Paris, or anhydrous calcium sulfate, but because it is cheaper, gypsum is the form usually employed. The quantity added amounts to 2 or 3 per cent. of the weight of the clinker, and since the grinding process brings about thorough mixing, the addition is usually made before the clinker is ground. After grinding, the cement is prepared for shipment.

Composition.—As has been said before, Portland cements consist chiefly of lime, silica and alumina. As given by Meade[1] the cements of good quality usually fall within the following limits of composition:

	Limits, per cent.	Average, per cent.
Lime	60–64.5	62.0
Silica	20–24.0	22.0
Alumina	5– 9.0	7.5
Magnesia	1– 4.0	2.5
Iron oxide	2– 4.0	2.5
Sulfur trioxide	1– 1.75	1.5

Since iron oxide occurs in nearly all clays and shales, it is found in all ordinary Portland cements. Cements that have no iron in their composition are usually white.

Experience has shown that the ratio between the percentage of lime and the combined percentages of silica, alumina and iron oxide should be about 2 on an average, or should not fall below 1.9 or go above 2.1, as:

$$\frac{CaO}{SiO_2 + Al_2O_3 + Fe_2O_3} = 1.9 \text{ to } 2.1$$

Also, the ratio between the percentages of silica and alumina should be as follows:

$$\frac{SiO_2}{Al_2O_3} = 2.5 \text{ to } 4$$

[1] ROGERS and AUBERT: Industrial Chemistry, p. 269

In manufacturing the cement, attention is paid to these ratios. If too much lime is present, the cement will expand and crack during setting, being then described as "unsound." If too little lime is present, the cement may "set" too quickly. Also, cements containing high percentages of alumina set rapidly. High-silica cements harden very slowly and have, therefore, a low strength at first because of this. Magnesia is generally considered harmful if the amount exceeds 4 per cent. Its effect is not exactly understood, but it is thought to produce disintegration by the expansion resulting from the formation of magnesium hydroxide, $Mg(OH)_2$. Also cements should not contain more than 1.75 per cent. of SO_3, this being the equivalent of 2.97 per cent. of calcium sulfate.

Constituents.—The oxides shown in the analysis and discussed in the preceding paragraphs do not, of course, exist in the cement as such, but in the form of compounds. The lime acts as a basic, and the silica and alumina as acid anhydrides, therefore calcium silicates and aluminates are formed. At least two silicates and two aluminates exist in the Portland cement clinker. These are dicalcium silicate, $(CaO)_2 \cdot SiO_2$, tricalcium silicate, $(CaO)_3 \cdot SiO_2$, dicalcium aluminate, $(CaO)_2 \cdot Al_2O_3$, and tricalcium, aluminate, $(CaO)_3 \cdot Al_2O_3$. The dicalcium silicate, called also calcium orthosilicate, is the chief constituent.[1]

The Setting and Hardening of Portland Cement.—The changes that take place during the setting and hardening of Portland cement are only imperfectly known. Much experimental work has been done and many explanations have been offered to account for the facts observed, but some of these explanations conflict at times in very important points. Of course, it is understood that the setting and hardening depend upon the hydration of the cement compounds, and the composition of the hydrated products and the manner in which they are formed have been quite generally agreed upon.

Hydration of the Portland-cement Compounds.—The explanation of this process is based on the early work of Le Chatelier,[2] who observed that when a cement is gaged with water, its

---

[1] U. S. Bureau of Standards, *Technologic Paper* 43, p. 63.

[2] Experimental Study on the Constitution of Hydraulic Mortars, 1887.

tricalcium aluminate is hydrated in a manner very similar to plaster of Paris, or in accordance with the following equation:

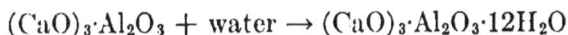

$$(CaO)_3 \cdot Al_2O_3 + \text{water} \rightarrow (CaO)_3 \cdot Al_2O_3 \cdot 12H_2O$$

Since the tricalcium aluminate is more readily active than the other compounds in the cement, it is generally considered that upon its hydration and hardening is dependent the initial setting of the cement.

The tricalcium silicate reacts next, splitting up into amorphous, hydrated mono-calcium silicate and hydrated lime, as:

$$(CaO)_3 \cdot SiO_2 + \text{water} \rightarrow CaO \cdot SiO_2 \cdot 2\frac{1}{2}H_2O + 2Ca(OH)_2$$

To the reaction represented by this equation is ascribed the hardening of the cement that takes place during about the first week. The amount of combined water taken up by these compounds seems to depend somewhat upon conditions, since other investigators who have examined the hydration products have reported various numbers of molecules. Still this may be due to the differing degrees of drying to which the hydrated substances were subjected in preparing them for examination.[1]

The Colloidal Theory of Cement Hardening.[2]—The colloidal theory is but one of the theories that have been advanced to explain the setting and hardening of cements, but since colloidal chemistry has become more familiar, it has been quite generally accepted. It is essentially different from the others in that it insists that the hardness, strength and durability of the cement depend upon the fact that the products of hydration are formed in the non-crystalline, colloidal state, and Michaelis, the chief exponent of this theory, considers the later crystallization which takes place, a detriment ultimately, whereas the exponents of the other theories count the development of crystals as of fundamental importance in producing hardness and strength.

Before proceeding with the colloidal theory, it will be necessary to consider a few points concerning colloids in general. For

[1] See KLEIN and PHILLIPS: The Hydration of Portland Cement, U. S. Bureau of Standards, *Technologic Paper* 43, pp. 17 and 51.

[2] For a more complete discussion of this theory, see MICHAELIS: The. Hardening Process of Hydraulic Cements, 1907. For general colloidal chemistry, see TAYLOR: Chemistry of Colloids.

this purpose the student is referred to page 361 of this text, where a brief discussion is given.

In discussing the colloidal theory of cement *hardening* it will be assumed that the processes of *hydration* are essentially as shown in the equations given by LeChatelier. Further, according to this theory, the hydrated substances are produced in the colloidal condition. And it does not seem unreasonable to assume that such is the case. At least, the microscope shows that the freshly hydrated aluminates and silicates are definitely noncrystalline, and that these amorphous substances do not commence to crystallize until 10 to 20 days have passed—long after the cement has become quite hard. Then, proceeding upon the assumption that the hydrated products are colloidal, the hardening can be explained in the following manner.

Since the hydration must begin at the surface, the grains become covered with a coating of colloidal gel, which possesses adhesive properties, and causes the grains to stick together. Because the gel is not very permeable to water, the centers of the grains may not become hydrated. They can become hydrated only as water can slowly diffuse through the colloidal coating.

At this stage, the cement is not hard. The thin coating of colloid that surrounds the grains contains absorbed water; that is, it is wet, soft, and jelly-like. The hardening of the cement depends upon the drying of this gel, just as in the familiar case of the glue jelly, which becomes very hard and strong as it dries out. In order that the cement gel may dry out, it is not necessary for the water to evaporate, although some water is usually lost in this way, when the cement hardens in air. Its *free* water may be taken up by the unhydrated centers of the grains and converted into *combined* water, in which condition it can no longer manifest itself as water, just as water of crystallization does not cause crystals containing it to appear wet, even though they may be more than half water by weight. It is not difficult to see that the cement gel can dry out in this way even when under water. Water cannot pass between the grains (this does not refer to the pore spaces) since the colloidal coatings cohere, and it cannot readily pass through the colloid itself because it is not sufficiently permeable. Then, the centers of the grains, by continued hydration, use up the free water in the gel faster than it can be supplied from the outside, and the gel becomes dry and hard. Moreover, the cement gel is an irreversible colloid, and cannot be softened again by water absorption after it has become hard.

In accounting for the fact that the cement in hydrating does not become increased in volume, like a glue jelly for example, Michaelis[1] says that as the grains of the pulverized cement begin to hydrate they acquire the colloidal state and immediately stick together and this interferes with their individual expansion. Further he has demonstrated[2] that if the cement is continuously agitated in water so that the particles are kept sepa-

Fig. 51.—Showing increase in volume due to the swelling of the colloidal gel in Portland cement.

rated, they do swell and a flocculent, voluminous mass is produced. In the case cited by him, the swelled colloid formed by the agitation of equal parts by weight of Portland cement and slag cement with 40 times as much water, was allowed to harden under water for 5 years, and although it had contracted somewhat during hardening, it still possessed a volume from 15 to 20 times that of the original cement. In a similar manner, the author has obtained the increase shown in Fig. 51. Portland cement was gaged with sufficient water to make a paste of medium consistency. One portion was at once pressed into a small porcelain crucible of 6 c.c. capacity, and allowed to harden, producing the small block shown in the center of the illustration. Two other

[1] *Loc. cit.*, p. 14.
[2] *Loc. cit.*, p. 21.

equal portions were continuously agitated for 6 hrs. with 200 c.c. of distilled water and were then filtered into Gooch crucibles. The filtering was carried out with suction, in order that the material might be compacted as much as possible. In this way the larger blocks were formed. When hardened and dried the bulk was equal to about 24 c.c., which was four times the volume of the original paste.

Michaelis says further that a dough may be prepared by agitating cement in water, that may be handled like paper, slabs being formed of it by pressure, which when hardened resemble slate.

Rate of Hydration and Development of Strength.—The following summary of the hydration of Portland-cement compounds is given by Klein and Phillips of the U. S. Bureau of Standards.[1] A short time after the cement is gaged with water, hydrated tricalcium aluminate is produced in amorphous form, this later crystallizing. At the same time sulfo-aluminate crystals, $3CaO \cdot Al_2O_3 \cdot CaSO_4 \cdot xH_2O$, are formed, these being produced by the reaction of the aluminate with the gypsum added to retard the setting. Also, at this time the low-burned and finely ground lime is hydrated.

"The next compound to react is the tricalcium silicate. Its hydration may begin within 24 hrs. and is generally completed within 7 days. Between 7 and 28 days the amorphous aluminate commences to crystallize and the beta orthosilicate (dicalcium silicate) begins to hydrate. Although the latter is the chief constituent of American Portland cements it is the least reactive compound. The early strength (24 hrs.) of cements is probably due to the hydration of free lime and the aluminates. The increase of strength between 24 hrs. and 7 days, depends upon the hydration of the tricalcium silicate, although the further hydration of aluminates may contribute somewhat. The increase between 7 and 28 days is due to the hydration of the beta-calcium orthosilicate (dicalcium silicate), but here are encountered opposing forces in the hydration of any high-burned free lime present and in the crystallization of the aluminate. It is to this hydration that the falling off in strength between 7 and 28 days of very high-burned, high-limed cements is due, whereas the decrease shown by the high-alumina cements is due to the crystallization of the aluminate."

[1] Hydration of Portland Cements, *Technologic Paper* 43, p. 63.

Retardation of Set by Gypsum.—The exact manner in which gypsum retards the initial set of cement is not known. For reasons that are pointed out by Klein and Phillips;[1] the usual explanation that the retardation is due to the formation of the sulfo-aluminate crystals, $3CaO \cdot Al_2O_3 \cdot 3CaSO_4 \cdot xH_2O$, having a selective action for water is insufficient. From their observations it appears that the retarding effect of the gypsum is due to its action as an electrolyte on a colloid solution; that is, in this case, preventing the hydrated aluminate from precipitating from its sol to form a gel.[2] They found that with 2 per cent. plaster of Paris, the hydrated substance could coagulate and separate as a solid least readily, while with other concentrations of plaster, the separation of the solid occurs in a shorter time, and the "set" takes place more readily, therefore. Because the gypsum retards the coagulation, the hydration of the tricalcium aluminate becomes more nearly complete since the impervious coating about the cement grains is not formed so readily.

Retardation of Set Due to Aeration.—The time required for the initial set of the cement is lengthened by the exposure to the air. The aluminate grains absorb a small amount of water from the moisture of the air and become coated with a thin layer of more or less impermeable colloid. Then, when the cement is gaged with water, this layer retards the passage of the water to the unhydrated center of the grains, so that a longer time is required for hydration and coagulation to take place. On this account an old cement is likely to set more slowly than a fresh one.

Effect of Regaging the Cement.—The following explanation of the effect of regaging is quoted from Klein and Phillips:[3]

"The effect of regaging is merely to hydrate a greater amount of aluminate than would be done by one mixing, but a very important factor is the amount of gypsum present. From this work it is very evident that the early strengths in cement are due not to the crystallization of the aluminates but to the formation of a dense, hard, amorphous mass of hydrated aluminate. Then it follows that the greater the amount of hydrated aluminate formed in 24 hrs., the greater the early

[1] *Loc. cit.*, pp. 29 and 61.
[2] See p. 362.
[3] *Loc cit.*, p. 54.

strength, and for this reason the high-alumina cements always show high early strengths. It has already been noted that aluminates without gypsum set up almost instantaneously with water, but that a very small amount is hydrated. In other words, the grains hydrate along the edges and adhere, so that the addition of large amounts of water has little effect since it can but slowly permeate through the amorphous material which is formed. The addition of gypsum, however, allows a greater amount of water to penetrate toward the center of the grains before there is coagulation with the formation of the impermeable wall through which the water penetrates very slowly.

"Manifestly, therefore, the slower setting the cement the less it will be benefited by remixing, as the effect of the latter is to mechanically remove the hydrated material from the surface of the grains, thus exposing a fresh surface to be hydrated. Conversely, a quick-setting cement will be benefited by remixing or continuous mixing, until a point is reached at which the cohesion between the hydrated particles is reduced. A cement takes up water very slowly, and if 20 per cent. of the water be used in mixing, this is not combined even after the cement has hardened, but part of it is combined and part is mechanically held between the grains, where it slowly penetrates from the outside to the center of each grain. It is evident, therefore, that an excess of water is necessary for complete hydration. However, the greater the excess, the less the cohesion between the particles of the hydrated material and the lower the resulting strength tests. Therefore, additions of gypsum to the cement and the remixing the cement play the same rôle in allowing the access of water to the aluminate grains."

Effect of Fineness on Setting and Hardening.—Generally the American specifications for Portland cements require that the degree of fineness be sufficient that 92 per cent. will pass a 100-mesh, and 75 per cent. a 200-mesh sieve. And it is very important that the cement be fine. It has been shown[1] that in a cement of normal fineness, only about half the particles have hydrated when the cement has hardened. The reaction with water is always very incomplete because of the impervious coating formed about the grains, and the larger the grains, of course, the greater will be the amount of the material unhydrated in the center. Desch[2] says that if a cement which has once hardened be reground to the specified fineness and then regaged with water, it will develop a very considerable degree of strength due to the

[1] DESCH: Chemistry and Testing of Cement, p. 117.
[2] P. 119.

hydration of the previously unchanged material. And this is well known. The unhydrated centers of the grains have no function other than could be more cheaply fulfilled by sand. Thus it is evident that the finer cements allow the use of more sand in the mixture.

Other Factors Affecting the Setting Rate.—A thin mixture with water sets more slowly than a stiff mixture. The greater amount of water does not retard the hydration, but does lessen the setting because it lessens the cohesion between the hydrated particles.

Benson[1] says that the temperature of the water used has also a modifying effect; that cold water retards while warm water hastens the setting and also increases the ultimate hardness. Hence it would appear that cements which set in warm weather would develop greater hardness than during cold weather.

The Effect of Freezing During Setting.—If freezing occurs during the setting of the cement, the colloidal mass is to some degree prevented from consolidating. During the time prior to the completion of the set, a certain amount of free water exists in the interstices between the hydrating grains of cement. Under normal conditions, this water diffuses through the colloidal coating of the grains and is used up by continued hydration. Then as the grains continue swelling they stick together and the mass consolidates. But since freezing retards diffusion and hydration, these processes may be delayed to such an extent that the water will evaporate from the mortar and not leave enough for sufficient hydration when the temperature rises. This evaporation of water may occur, even though the temperature remains below the freezing point. The effect of this drying is most noticeable on exposed surfaces, and is very similar to that produced by using a cement mortar with a dry porous brick.

Further detriment results from the expansion that attends the freezing of the uncombined water. The expansion forces the grains apart, and although the hydration should continue after thawing, the consolidation would be imperfect and the structure lacking in strength.

The Cause of "Hair Cracks."—When a cement sets and hardens in air, there is a slight contraction due to the drying out of the

[1] Industrial Chemistry, p. 281.

colloidal gel.　This contraction is, of course, lessened by the sand
or other inert substances.　But in mixtures that are too rich,
that is, contain too little inert material, this contraction is
manifested by the development of fine hair cracks.　It is noticeable especially in rich facing mortars.　The remedy is, of course,
to use a mortar containing less cement, the whiteness being
secured, if this is especially desired, by the use of marble dust.

Hair cracks may be produced also when the mixture is too
wet, since the volume is then partly dependent upon water, and
shrinkage occurs when this dries out.

Action of Destructive Agents.—*Heat.*—Cement and concrete
begin to disintegrate under heat when a temperature of about
300°C. is reached, because then the combined water is expelled.
But in a concrete that is made of proper aggregate and suitably
proportioned, the conductivity is so low that this disintegration
is likely to occur in only a thin surface layer.

In reinforced concrete the coefficient of expansion of the concrete is practically the same as that of the steel,[1] and the value
of reinforced concrete in fire resistance is due largely to this fact.
But the *heat conductivity* of the steel is much greater than that
of the concrete.　Consequently, if the steel is covered with only
a thin layer of concrete at any point, in case of fire, it will become
heated more rapidly, expand at a greater rate and so set up
internal stresses that may be disastrous.

Frost.—If cracks or "voids" due to improper proportioning
exist in the cement or concrete, frost may prove destructive.
The enormous expansive force manifested by the freezing of the
water contained in these cracks and voids causes the mass to
disintegrate.

Carbon Dioxide.—A water solution of carbon dioxide exerts a
destructive action on the cement because the soluble bicarbonate
of calcium is formed from the constituents of the cement.　Marsh
waters are active in this way.

Action of Sea Water.—If not well made, concrete structures are
subject to destruction by the action of sea water.　However, the
action seems to be due more to mechanical than to chemical
forces.　It generally occurs when a porous concrete structure is
alternately exposed to the water and then to the air by the tides.

[1] DESCH: *Loc. cit.*, p. 201.

Crystallization of the dissolved salts in the pores is brought about by evaporation of the water and the expansion resulting from this crystallization causes disintegration. Porous stone, brick and other substances are affected in the same way. Therefore, if it is required to resist sea water, it is very essential that the concrete be dense.

In respect to the chemical action of sea water the report of Bates, Phillips and Wig of the U. S. Bureau of Standards[1] says that although in laboratory tests it is found that various sulfate and chloride solutions have a chemical action on cements, in service this action is much retarded if not entirely suspended in most cases, due probably to the carbonization of the lime of the cement near the surface, or to the formation of a protective coating by deposits of salts. The report further states that properly made Portland-cement concrete, when totally immersed, is apparently not subject to decomposition by the chemical action of sea water, and that the metal reinforcement is not subject to corrosion if imbedded to a depth of 2 in. or more from the surface of well-made concrete.[2]

Testing of Cements.—Cements are tested chiefly by mechanical means, the tests usually applied being the test for soundness (constancy of volume), tensile strength (breaking test), fineness, time of setting, and specific gravity. The first two of these are the most important.

The test for soundness is applied to determine whether or not the cement will cause mortars or concretes made of it to disintegrate in any way. The lack of soundness may be due to free or slakable lime, or to coarse grinding. Free lime, in slaking, expands and may cause crumbling. If the quantity present is not too great, it may be rendered inert by air slaking, which may be accomplished by storing the cement for 2 or 3 weeks. Thoroughly slaked lime, such as is sometimes added to Portland-cement mixtures for waterproofing, or for the purpose of increasing its plasticity or smoothness under the trowel does not have this undesirable effect on soundness. Coarseness of grinding tends

---

[1] Action of Salts in Alkali Water and Sea Water on Cements, *Technologic Paper* 12, 1912, p. 100.

[2] For a general discussion of the protection afforded the reinforcement by concrete see "Corrosion," p. 144.

to lessen soundness because the hydration of the larger particles is deferred, and when it does occur a disintegrating action may be manifested.

The degree of fineness is determined by sieving. As has been said before, fineness is essential, because only the finer of the particles are actively hydraulic, the coarser ones fulfilling merely the function of sand. Therefore, the finer the cement, the greater is the proportion of sand that can be mixed with it.

The determination of the setting rate is important, since in some cements, the initial set may occur before sufficient time has elapsed to allow the mortar or concrete to be placed. On the other hand, after it has been placed, hardening must not be too long deferred, since the longer the time required for the development of hardness and strength, the greater is the chance that it may be injured by accidents.

Storing of Cement.—In order that cement may be kept in a condition fit for use it must be properly stored. It should be kept in an atmosphere as dry as possible, and on a raised floor. Moreover, it should be kept covered with canvas or roofing paper. It absorbs moisture from the air quite readily, and hence should be shielded from moisture-laden currents of air as much as possible. When it has become wet it "sets" and then cannot be used. Lumps are sometimes formed by pressure during storing, and these must not be mistaken for lumps caused by dampness. Those caused by pressure are readily pulverized.

CONCRETE

Sand.—To be suitable for the making of concrete, sand should be coarse-grained, and reasonably clean—free from clay, loam, silt, and vegetable matter. It is not advisable to allow the use of sand containing more than 5 per cent. of such material, and when more than 10 per cent. is present the sand should be rejected or else it should be washed. When the sand is not clean, no amount of cement will make strong concrete.

An idea of the amount of clay, loam, etc., present may be obtained in the following manner.[1] Put 4 in. of sand in a pint preserving jar, then fill to within 1 in. of the top with clean water,

[1] U. S. Department of Agriculture, *F. Bull.* 461, p. 7.

and close the lid tightly. Shake vigorously for 10 min., stand upright and allow to settle. If more than ½ in. of clay or loam shows as a layer on the top of the sand, the sand should be rejected or washed. The line of separation between the clay and sand is clearly indicated by the difference in color and fineness.

It is of considerable importance that the sand be not too fine. Tests have shown[1] that an exceedingly fine sand required seven times the amount of cement required by a coarser sand retained on a 20-mesh screen, without increasing the strength of the concrete. Also when the sand particles are about of uniform size, the strength of the concrete made from it is not nearly so great as when the particles are of different sizes. When the size of the particles vary, the voids, that is, the open spaces between the particles, are much less and the resultant concrete is more sound.

Stone and Gravel.—The best rocks for concrete are in general the traps, granites, and some varieties of sand- and limestone. The shale, slate, and very soft limestone and sandstone should not be used, for the strength of the concrete, of course, can never be greater than the strength of the material used in it.

Water.—The water for mixing should be clean and free from all strong alkalies and acids.

The Theory of Concrete.—The theory of the concrete mixture is that the stone and sand are so proportioned that the sand will just fill the voids in the stone, while the cement will fill the voids in the sand. In this manner a solid mass is obtained. This is the ideal mixture, but it is rarely, if ever, exactly reached in practice. Due to varying sized particles, the voids in both the stone and sand vary. Hence, in order to be absolutely safe a little more sand and cement than will just fill the voids are used. Generally the mixture contains half as much sand as stone or gravel. For example, the following proportions make a very good concrete for walks and floors, although mixtures containing much less cement are often used: screened gravel or crushed rock, 5 parts; sand, 2½ parts; Portland cement, 1 part. Proportions vary greatly, of course, in accordance with the character of the work being done. They may range from 1, 2 and 4 to 1, 3½ and 7 or even more.

[1] U. S. Department of Agriculture, *F. Bull.* 461, p. 7.

WATERPROOFING CONCRETE

The cement concrete, as it is ordinarily made, is more or less porous and permeable to moisture. In a great many of its uses it would be a great deal more efficient if it were not so, as for example, in constructing basement walls and floors, vaults, reservoirs, drains, etc., as well as in all of the reinforced work, the permanency of which would be less doubted if it were waterproof, since the corrosion of the reinforcing rods would certainly be prevented if the moisture were excluded.

A great deal can be done toward making concrete waterproof by merely having the ingredients of the concrete *rightly proportioned*, so that the voids are eliminated. It is said that it can be made entirely waterproof in this manner.

However, in the methods that are commonly employed, some special waterproofing material is used, this being either applied as an external coating, or mixed with concrete before it is placed in the forms. Searle[1] makes a distinction between dampproofing and waterproofing, and says that the term *dampproofing* should be used to denote the methods and appliances employed to keep out the dampness from the superstructures of buildings, and the term *waterproofing* for the methods used for treating work subject to hydrostatic pressure, and for vessels intended to contain or retain water.

Dampproofing.—The materials employed in dampproofing are divided into three classes: transparent coatings, opaque cement coatings, and bituminous coatings.

Transparent coatings consist of those materials that do not change the appearance of the surface treated. Perhaps the oldest coating of this class is that produced by alternate applications of solutions of soap and alum or other aluminum salt. As these materials react within the pores, an insoluble aluminum soap is formed, which in addition to being insoluble has a distinct water-repelling property. Because it is necessary to repeat the process a number of times to secure a sufficient amount of the insoluble soap, this method is not economical and is not much used at present.

Paraffin wax is another example of the transparent coatings,

[1] Cement, Concrete and Bricks, p. 197.

and it may be applied either as melted wax after the exposed surface has been carefully heated or it may be applied as a solution in petroleum naphtha.

Also, solutions of water glass or sodium silicate are employed. The silicate is decomposed by the action of carbonic acid, and gelatinous silica is set free in the pores. There is an objectionable feature accompanying this process, in that it is difficult to cause the silicate to penetrate sufficiently.

The aluminum, magnesium, and zinc silicofluorides or fluosilicates, known sometimes as fluates, are also made use of.

Opaque Coatings.—Paints are included under this head, but ordinarily paints cannot be safely applied to cement coatings, since the calcium hydroxide that is set free during the setting of the cement, reacts with the oil and destroys the durability of the film. To prevent this destructive action on the oil, the cement surface is first treated with a solution of zinc sulfate which reacts with the calcium hydroxide to produce calcium sulfate and zinc hydroxide, both of which are quite stable, and do not react with the oil. Sometimes a casein paint is used as a first coat before the application of the oil paint. Or the casein paint may be applied alone followed with a treatment with formaldehyde to render it waterproof and resistant to moulds, etc. Caesin is obtained from milk by precipitating with dilute acids.

If there is no objection to the color, tar makes a very serviceable coating. The tar is heated to expel moisture, is treated with lime to neutralize its acidity, and then is applied hot, or may be applied as a paint after thinning it with petroleum naphtha or benzole.

Special bitumens[1] are not used as surface coatings, but are applied to the interior of concrete walls, a coating of plaster being applied directly over them, with which they form a very good bond.

Waterproofing.—The methods of waterproofing are divided by Searle into two classes: "integral" or rigid method, in which a waterproofing compound is incorporated with the concrete mass; and "membrane" or bituminous shield method, in which the concrete is insulated from contact with water by the use of a continuous, waterproof, bituminous layer.

[1] SEARLE: *Loc. cit.*, p. 200.

The "Integral" Method.—The materials employed here include hydrated lime and calcium soaps. The soap may be previously prepared and added in the dry state, in which case it is very difficult to mix uniformly, or it may be formed in the concrete by using a water solution of sodium soap in place of water only in mixing.

According to Page,[1] concrete may be made thoroughly waterproof by mixing with it a quantity of heavy mineral oil, equal to from 5 to 10 per cent. of the weight of the cement used. He says:

"The admixture of oil is not detrimental to the tensile strength of mortar composed of 1 part cement and 3 parts of sand when the oil added does not exceed 10 per cent. of the weight of the cement used. The compressive strength of mortar and of concrete suffers slightly with the addition of oil, although when 10 per cent. of oil is added the decrease in strength is not serious. Concrete mixed with oil requires a period of time about 50 per cent. longer to set hard than does plain concrete, but the increase of strength is nearly as rapid in the oil mixed material as in the plain concrete."

However, Desch[2] says that from general experience it has been found that the oil decreases the strength about 30 per cent.

The "Membrane" Method.—In this method, no effort is made to treat the concrete, the object being merely to insulate it from contact with water by the use of a waterproof shield. Coatings of felt and burlap saturated and coated with bitumen are extensively used. This method has the advantage over the preceding, in that it in no wise affects the strength of the concrete.

References

A

ECKEL: Cements, Limes, and Plasters, New York, 1905.
MICHAELIS: The Hardening Process of Hydraulic Cements, Chicago, 1907.
WEST: Modern Manufacture of Portland Cement, London, 1910.
LEWIS and CHANDLER: Popular Hand Book for Cement and Concrete
    Users, 1911.
MEADE: Portland Cement, Easton, 1911.
DESCH: Chemistry and Testing of Cement, London, 1911.

[1] U. S. Office Public Roads, *Bull.* 46.
[2] *Loc. cit.*, p. 213.

BUTLER: Portland Cement; Its Manufacture, Testing, and Use, London, 1913.

SEARLE: Cement, Concrete, and Bricks, London, 1913.

TAYLOR: Practical Cement Testing, New York.

RICHARDS and NORTH: A Manual of Cement Testing, New York.

ROGERS and AUBERT: Industrial Chemistry, New York, 1915.

DANCASTER: An Elementary Treatise on Limes and Cements, London, 1916.

B

POULSEN: Long-time Tests on Concrete in Sea Water, *Eng. News*, **64**, 3 (1910).

MOYER: Use of Mineral Oils Mixed with Concrete, *Eng. Rec.*, **61**, 10 (1910).

KEMPSTER: Portland Cement from Blast-furnace Slag, *Jour. Ind. Eng. Chem.*, **3**, 33 (1911).

PAGE: Oil Mixed Concrete, U. S. Office Public Roads, *Bull.* 46.

HAGAR: Manufacture of Portland Cement from Slag, *Eng. News*, **66**, 314 (1911).

CARPENTER: Adulteration of Portland Cement, *Eng. News*, **67**, 94 (1912).

BATES, PHILLIPS, and WIG: Action of the Salts in Alkali Water and Sea Water on Cements, U. S. Bureau of Standards, *Technologic Paper* 12 (1912).

LARNED: A Study of Sand for Use in Cement Mortar and Concrete, *Jour. Assoc. Eng. Soc.*, **48**, 189 (1912).

KLEIN and PHILLIPS: Hydration of Portland Cement, U. S. Bureau of Standards, *Technologic Paper* 43 (1914).

SCHUMANN: Oil and Portland Cement, *Concrete*, **8**, 21 (1916).

CHAPTER XII

CLAY AND CLAY PRODUCTS

Source and Composition of Clay.—Pure clay is a hydrated alumina silicate, $Al_2O_3 \cdot 2SiO_2 \cdot 2H_2O$, technically known as kaolin, and is pure white. It is produced by the weathering of feldspars, a group of mineral substances, consisting of the silicates of alumina, potash, soda, and lime. Potash feldspar, known as orthoclase, is $K_2O \cdot Al_2O_3 \cdot 6SiO_2$. By the action that occurs during weathering of orthoclase, the potash, together with two-thirds of the silica is dissolved out, water is combined with the residue, and kaolin is formed. Pure kaolin is produced only by the weathering of pure feldspar, but this occurs in nature very rarely.

The common clays are formed by the weathering of igneous rocks, shales and clayey limestones, the composition of the clay being dependent upon the character of the rock from which it was formed. Iron oxide is present in most clays and small quantities of lime, magnesia, and the alkalies are of frequent occurrence. Also, since the rock may contain constituents that are not susceptible to the weathering action, these may be found mechanically mixed with the clay. For example, in the weathering of granite, which is a notable source of common clay, the feldspar constituent is weathered as described; but granites contain also a great deal of quartz, and since quartz is practically unaffected by the weathering process, the clay that results will contain it, usually in a finely divided state. In this way a sandy clay is formed.

In the production of clay, chemical action is not always of fundamental importance as in the weathering of feldspar. Sometimes it amounts to very little, as in the weathering of shales, where frequently the action is merely a physical disintegration. In such cases, the composition of the clay differs very little from that of the original rock.

Clays may or may not be found where first formed. Kaolin

is very light and is easily washed away and deposited in some other place. Clays that have been carried in this manner are known as *transported* or *sedimentary clays*, while those that remain in place where formed are called *residual clays*. In transit, sedimentary clays may become contaminated with considerable amounts of other materials, and thus be very impure; but the substance is called clay as long as sufficient kaolin is present to form a plastic mass when properly mixed with water.

Properties of Clay.—Among the noteworthy properties of clay are its plasticity and imperviousness to water when moist, and the ease with which it can be converted into a stone-like mass when heated, or as it is technically known, "fired" or "burnt." Burnt clay is probably the most enduring of all materials, as is evidenced by the state of preservation of specimens found in the ruins of remotest antiquity. Clay is unique in respect to the property it has of being readily fashioned into any desired form, which form may then without difficulty be made permanent.

Clays for Bricks and Other Clay Products.—Although all clays can be moulded when properly moistened, and can be converted into a hardened form by heat, they are not all suitable for making bricks. For this purpose, the clay must quite accurately retain its shape during burning, and not warp. When burned it must possess suitable hardness, porosity and color. It must be as free as possible from pebbles, especially of limestone; since these become calcined during burning and after long periods, hydrate in the wall and produce defects by the resultant expansion.

The clay must not be too plastic. Highly plastic clays often show an excessive shrinkage, both during the drying of the paste and during the subsequent burning. Pure kaolin becomes so cracked and distorted due to shrinkage that it cannot be used without admixture of sand, previously burned clay, or similar material. The shrinkage of brick clay is controlled in the same manner. The object is not to prevent shrinkage altogether. A certain amount—about a total of 8 or 9 per cent.—is considered desirable, since it aids in producing a more compact product.

Highly plastic clays are also hard to work. This difficulty is corrected by the addition of the same material that is used to offset shrinkage. But too much of this material must not be

17

used, for although the clay works easily, the resultant brick will
be soft and porous.

Fluxing Constituents.—The soda, potash, lime, magnesia, and
ferrous oxide present in the clay are known as the fluxing constituents, since they combine with the silica during burning to
produce fusible compounds. If they are notably absent, the
burnt clay is open and porous; if present in sufficient amount,
the granules of clay become fused together so that a more or less
vitrified product results.

Color of Clay and Clay Products.—Iron oxide is the coloring
agent of both the burned and unburned clay. The oxide is
generally ferric, but it is probable that it does not occur in the
free state in the raw clay, but rather as an almost colorless
hydrated silicate, which decomposes upon heating into free
ferric oxide, silica, and water. All clays, including those of a
greenish or grayish color, will burn red if burned in an oxidizing
flame. If burned in a reducing flame, ferrous silicate will be
formed, which may impart a purplish color to the burned product.
If present in sufficient amount, it will color the burned clay black.
When iron pyrite is present, the sulfur content is oxidized leaving
ferrous oxide, which readily combines with the silica, producing
the black color as just described. If the pyrite is present in
large-sized granules, a hole is left where the granule existed. The
sulfur escapes as sulfur dioxide, and the ferrous silicate is absorbed
by the brick, a black-lined opening being thus produced. Openings of this sort are commonly met with in hard-burned building brick. If considerable pyrite is present, the brick may be
rendered too porous for use. Carbonaceous matter in clay acts
as a reducing agent and converts ferric iron into the ferrous condition. In this way ferrous silicate may be formed from ferric
oxide during burning.

Clays free from iron oxide will likely burn white, since other
oxides capable of producing color are not commonly present.
Buff and cream bricks are made by the burning of clays containing very little iron oxide. However, if a great quantity of lime
is present, the coloring effect of the iron oxide is overcome, even
though it may be present in considerable quantity. Hence, cream
or buff bricks may be produced from clay either low in iron oxide
or high in lime.

Brick-making Methods.—The suitably prepared clay may be moulded into shape by any one of three methods. These methods are designated as the soft-mud, stiff-mud, and dry-press processes.

In the *soft-mud process* the clay is mixed with water to the consistency of a soft paste and pressed into wooden moulds. In order that the brick may be delivered from the mould the more easily, the mould is usually sanded each time before being filled. It is not difficult to recognize soft-mud bricks since they will show five sanded surfaces with the sixth comparatively rough, being made so by scraping off the excess clay from the top of the mould. Unless re-pressed, soft-mud bricks lack sharp corners and straight edges. Also, their interior is likely to show more pebbly particles than bricks made by other processes.

In the *stiff-mud process* the clay is mixed with just sufficient water that it is plastic, but still quite stiff, and is then forced through a rectangular die in the form of a bar, upon a cutting table. This bar is cut into brick lengths by means of wires. Since the wire makes a somewhat tearing cut, the cut surfaces are always recognizable on a stiff-mud brick. Another characteristic of these bricks that renders recognition quite certain depends upon the manner of mixing the clay. This is accomplished by means of a rotary shaft supplied with blades, the clay being forced through the die by means of a tapering screw. The twisting action resulting from this treatment produces usually a laminated structure within the brick. A stiff-mud brick will have four smooth surfaces and two roughly cut surfaces. Upon the cut surfaces the laminations are sometimes visible.

In the *dry-press process* the clay is allowed to dry in the air until its moisture content is reduced to at least 12 or 15 per cent. The clay is then granulated and screened. After this, it is conveyed to the pressing machine where it is pressed in steel moulds. The process produces bricks with sharp edges and smooth surfaces, it being commonly used for front or face brick. If the dry-press bricks are not well vitrified in burning they are liable to show a granular structure; the grains of clay cannot coalesce nearly so well as when the clay is mixed in a more or less plastic state.

Re-pressing.—The soft-mud and stiff-mud bricks are sometimes re-pressed shortly after moulding. This straightens the edges and removes the surface roughness. Usually, also, during the re-pressing some design is stamped upon the surface of the bricks. Re-pressing makes the bricks denser and stronger and more resistant to weathering.

Burning.—Most of the common bricks are fired at about 1,050°C., while the pressed bricks are heated to a temperature of about 1,280°C.

Rejected Brick.—Because of overfiring or perhaps for some other reason, many bricks come from the kiln discolored, roughened, or twisted. These must be cast aside and are known as "culls." The brickmaker is rarely able to obtain more than 85 per cent. of good bricks from the kiln, and often the percentage is not so great.

Comparison of Bricks Made by Different Processes.[1]—Soft-mud bricks are generally more homogeneous in structure than the bricks made by other processes, and they are less liable to be affected by frost action if properly burned. However, they lack smooth surfaces and sharp edges, unless re-pressed.

Stiff-mud bricks also usually possess a good structure, but laminations may be pronounced and separations may occur between them.

The dry-pressed bricks present the most finely finished surfaces, but because of the granular structure, which results from the manner in which they are made, if they are not well burned, they are liable to be softer, with less cohesion between the particles than bricks made by other processes.

Efflorescence on Bricks.—Efflorescence is a coating—usually white—which appears on stored bricks or upon bricks laid in the wall. It consists usually of the soluble sulfates of magnesium, sodium and potassium. Sometimes it is due to calcium sulfate, and less commonly to the sulfates of iron and aluminum. These salts may have been present naturally in the clay or water used to make the bricks, or may have been produced by the oxidation of minerals during burning. Also sometimes they are produced by the reactions of weathering, as by the action of atmospheric sulfur dioxide upon certain silicates existing in the brick. Or in a similar manner, pyrite left in the brick may bring about salt

[1] For a further discussion of bricks, especially for requisite qualities and methods of testing, see RIES: Building Stone and Clay Products, pp. 284–319.

formations, by oxidizing to produce sulfuric acid. In rare cases, efflorescence may be due to mortar.

But, however they may be formed, the incrustation is produced by the leaching out of these salts from the brick by water, the salts being deposited upon the surface of the brick as the water evaporates. The incrustation, or efflorescence, is most pronounced near the eaves and downspouts, under window sills, etc., where the water soaked into the bricks either through leakage from the eaves and spout or through being caught and drained from the sills.

Aside from being unsightly, efflorescence exerts a disintegrating effect upon the bricks. The salts producing it are somewhat hygroscopic, and even when not actually wetted by rains, they dissolve by taking up moisture from the atmosphere and saturate the bricks with their solutions. Then when dry weather arrives they bring about disintegration of the bricks by the expansive force of their crystallization. It may have been observed that during damp or wet weather the efflorescence often disappears, returning, however, with the coming of dry weather.

The remedy for efflorescence lies chiefly with the maker of the bricks. Water should be used for mixing with the clay that is free from salts. Hard burning is also of great value. This volatilizes the alkalies and increases the density of the bricks so that less water can find entrance for leaching. Treatment with barium salts is also effective, since the soluble sulfates are converted into the insoluble barium sulfate.

It is not an uncommon practice to remove efflorescence by the use of washes containing nitric or hydrochloric acids, but this is very objectionable. Although the incrustation is for a time removed, the real evil is only aggravated. By the action of the acid, the salts present are converted into more soluble forms, which means an increase in the quantity dissolved, and a consequent increase in the amount of incrustation left by the evaporated water. The only really successful way is to cut off the source of water supply to the wall. If the wall can be thoroughly dried, good results may be secured by removing the efflorescence with a scratch brush or similar device, and then applying a waterproof coating such as described for waterproofing cement and concrete, which see.

Terra Cotta.—The term *terra cotta* means baked earth, and this name is applied to a variety of forms of burned clay used for structural purposes. The forms are moulded usually from low-grade fireclays and are hard-burned but not vitrified. In this discussion, the terra-cotta products will be considered under three heads: (1) decorative terra cotta; (2) building blocks; and (3) fireproofing.

Decorative terra cotta is either moulded in plaster casts or is modeled by skilled hand work. After drying, there is sprayed upon the surface a coating consisting of clay, quartz, feldspar, and other ingredients, ground in water to the consistency of a thin cream. This is known as the "slip layer." In addition to this layer the ware may subsequently be glazed. The slip layer may be colored and the ware decorated to show various designs.

Building blocks are made usually from clays and shales or a mixture of these with low-grade fireclays. They are generally hard-burned but not vitrified. No surface coating is applied, since they are used strictly for structural purposes and are then covered by other material as mortar or plaster. They are made in a variety of forms; for example, those known as hollow blocks, furring blocks, hollow bricks, book tiles, etc. They are widely used because of their light weight, and their non-conductivity of sound and heat.

Fireproofing is a general name applied to the hollow ware used for flat arches in floors, for partitions and furring in fire-proof buildings. It serves to protect the metallic members from the intense heat in case of fire, and so prevents warping and other injury. It serves also as a non-conductor of heat at all times. **Terra cotta lumber** is a name given to that sort of fireproofing that has been made soft and porous by the addition of straw, sawdust, etc., to the clay mixture. The straw and sawdust burn out in the kiln leaving the product in such condition that nails or screws may be readily driven into it.

Roofing Tile.[1]—The procedure in making roofing tile is very similar to that followed in the manufacture of pressed bricks. The pressed forms are carefully dried and burned. It is said that

[1] For manufacture and properties of roofing tile see *Bull.* 11, 4th Series, Ohio Geological Survey.

a softer-burned tile is a poorer conductor of heat and hence provides a cooler roof, but when all factors are considered there is much difference of opinion as to whether a porous or vitrified tile is best. Vitrified tiles have a greater frost resistance and are probably more enduring on this account.

Wall tiles are pressed tiles and may be made from either dry or plastic clay. Those made from dry clay are known as "dust-body" or "dry-press" tiles. However, the clay used in this process is not dust-dry, since after it has been dried and pulverized it is steamed to render it slightly moist. Most of the wall tiles are "dry-pressed." After pressing, the tiles are set in fireclay boxes to protect them from direct contact with the flame and are then burned. When burned, the face of the tiles is dipped in a glazing preparation, consisting usually of the silicates of lead, potassium, calcium, and aluminum, ground with water to the consistency of a cream. If a colored glaze is desired, the coloring matter is introduced in the form of metallic oxides. By suitable employment of these, almost any color or shade desired may be obtained. After dipping and drying, the tiles are again fired until the glaze is fused, producing in this way a lustrous surface. Matte or dull glazes also are much used, and these, too, may be colored. The dulled appearance is produced by applying a thick coating of the surface material.

Floor tiles are made by the dry-press process, and since they are rarely glazed, only one firing is necessary. Floor tiles may be divided into two kinds—plain and encaustic tiles.

The **plain tiles** are made of one clay throughout, and are burned sufficiently hard to be vitrified or semi-vitrified. Vitrification results from surface fusion of the component grains. Examples of the plain tiles are the *mosaics*.

The **encaustic tiles** have a facing of one kind of clay and a backing of another. The surface may have a design of several colors. It is first put down in the moulding box by the aid of a brass cell frame. The various-colored clays are sifted into this frame, and then the frame is lifted out. When this is done the mould is filled with a clay backing. Encaustic tiles cannot be burned to vitrification, since the colors are largely dependent upon iron compounds in the clay and such clays burn to buff or red when vitrified.

Sewer pipes are formed from plastic clay in a special form of press. Odd-shaped sections, such as traps, sockets, etc., are usually made by hand in plaster moulds. After careful drying the ware is fired. When the kiln has reached a temperature of about 1,200°C., common salt, NaCl, is thrown into the fires. This volatilizes and as it comes into contact with the earthenware, reacts with the clay and produces a readily fusible sodium aluminum silicate which covers the surface of the articles as a glaze. This glaze is known as the "salt glaze," and is the usual form applied to sewer pipe in the United States. The clay of which the pipe is made must be chosen with this end in view, since all clays will not take a salt glaze. A glaze is sometimes secured by coating the article with an easily fusible "slip clay" consisting of a mixture of litharge, PbO, and clay, or finely ground feldspar, and subjecting to a white heat. But this glaze is in no way superior to the salt glaze, and besides is not so cheap.

Drainage tiles are made of red-burning clay by the stiff-mud process, the product issuing upon the cutting table as a hollow cylinder, which is divided into lengths by wires. After drying, the tile are fired, but are not vitrified or glazed since in this case porosity is desired.

SANITARY WARE

Sanitary ware may be divided into two main classes: (1) that produced by the application of a glaze to a clay-product body; and (2) that produced by enamelling articles made of cast iron.

In both classes it is vitally essential that the coating applied have the same coefficient of expansion as the material composing the body. This, of course, is presumably taken care of by the manufacturer, but sometimes only within certain ranges of temperature. Hence, cracking of the enamel may result from contact with boiling water, when atmospheric changes of temperature would have no effect.

None of the ordinary glazes or enamels of sanitary ware are acidproof, since they were not made with this end in view. The composition is too high in basic oxides, especially of the heavy metals, and too low in silica. However, glazes and enamels can be made that are proof against at least the weaker acids, by suitably adjusting the composition.

That class of ware produced by the application of a glaze to a clay-product body is of two kinds, these being known to the trade as vitreous ware and "solid porcelain."

Vitreous ware is made from a mixture of pure white-burning clay, whiting, $CaCO_3$, feldspar and silica. The forms are made by hand in moulds, and then are carefully dried and fired. After the first firing, the ware is said to be in the "biscuit" condition. It is now dipped in a mixture made up of a variety of substances, such as feldspar, kaolin, boric acid, silica, and certain metallic oxides, as those of zinc, tin, and lead, all finely ground and in suspension in water. The mixture is so composed that it will fuse at a temperature lower than that at which the body of the ware will fuse. During the second firing the glaze fuses and fills the pores. Water closets, tanks, lavatories, and drinking fountains are made in this manner.

"Solid Porcelain."—Although this ware is called solid porcelain, it must not be confused with the true porcelains of which laboratory ware, as evaporating dishes, and fine table ware are examples, since it does not resemble these in composition of body, glaze or properties. The body of the solid porcelain ware is made of fresh fireclay, mixed with a certain amount of previously burned fireclay, finely ground, which is used to lessen the shrinkage. To this body is applied a layer that burns to a white vitreous lining. And to this in turn, is applied a glaze very similar to that employed for the vitreous ware, except that the metallic oxides are omitted. The white lining and the glaze are applied to the unburned ware and the whole is fired in one firing. Great care is necessary in applying the glazes that the coefficient of expansion for the glaze is the same as that for the body, otherwise cracking of the glaze results. Of the so-called solid porcelain, bath tubs, urinals, sinks, lavatories, etc., are made.

Enamelled Cast-iron Ware.—The iron employed for the manufacture of the metallic body of this ware must be of special composition. Grünwald[1] says that a cast iron of the following composition is generally used:

[1] Theory and Practice of Enamelling on Iron and Steel.

	Per cent.
Carbon	3.5
Silicon	2.0
Phosphorus	1.4-1.8
Manganese	0.5-0.7

No doubt the very high phosphorus content is essential to the production of the thin sections of the hollow-ware castings, since it confers much increased fluidity upon the molten iron; also, it lessens shrinkage. The casting must be carefully pickled to insure a clean surface.

The enamel applied to the iron article is in some respects a glass, but it is not an ordinary glass. It consists fundamentally of the following: silica, derived from quartz, flint or sand; alumina obtained by the use of feldspar and clay; and lime from fluorspar or calcite. In addition to these substances, cryolite, Na_3AlF_6, is introduced to render the enamel non-transparent. Borax is used to introduce the boric anhydride, B_2O_3, which among other desirable features makes the enamel more ductile and elastic. Soda ash and pearl ash are used as fluxes. Sodium and potassium nitrate are employed in light-colored enamels and manganese dioxide in dark-colored enamels as oxidizing agents. A suitable mixture of these substances is made and melted, and when properly molten is allowed to flow into cold running water. By this treatment the mass is cracked and shredded into extremely small particles, and is now known as *"frit."* The frit is mixed with a small quantity of white plastic clay, and the whole is ground with about 50 per cent. of water in a ball mill for 30 hrs. The product is a fluid having about the consistency of a rich cream.

This cream-like product is applied to the iron in various ways, as by dipping, brushing, or spraying with compressed air. After application it is dried and then fired. The enamel is applied in two or more coats—a ground coat and then one or more cover coats. It is interesting to note that the enamel for the ground coat is almost without exception made up containing a certain amount of cobaltic oxide, which seems almost unique in respect to its great power of causing the enamel to stick to the iron. The reason is not well understood. Because of the use of cobaltic oxide, the ground coat is usually colored blue; but color is no object in the ground coat.

In the cover coat or glaze, the whiteness is secured by the use of about 3 per cent. of tin oxide, SnO_2. Although this is rather expensive, an entirely satisfactory substitute has not been found. Enamels may be colored as desired by the use of proper oxides. Bath tubs, sinks, etc., are made of enamelled cast iron.

References

A

RIES: Clays; Their Occurrence, Properties, and Uses, New York, 1906.
GRÜNWALD: Theory and Practice of Enamelling on Iron and Steel, translated by HODGSON, London, 1909.
BOURRY: Treatise on the Ceramic Industries, translated by Searle, London, 1911.
RIES: Building Stones and Clay Products, New York, 1912.
GRÜNWALD: Technology of Iron Enamelling and Tinning, translated by HODGSON, London, 1912.
HECHT and CRAMER: The Collected Writings of Herman Seger.
SEARLE: Cement, Concrete, and Bricks, London, 1913.
SEARLE: Bricks and Artificial Stones of Non-plastic Material, London, 1915.

B

HOWARD: The Strength and Absorption of Brick, *Eng. News,* **61,** 273 (1909).
ASHLEY: The Colloidal Matter of Clay, U. S. Geol. Sur. *Bull.* 388 (1909).
GREAVES-WALKER: Efflorescence on Brick Work, *Jour. Ind. Eng. Chem.,* **2,** 373 (1910).
UNDERHILL: Enamelled Cast-iron Sanitary Ware, *Fdy.,* February, 1910.
MARSTON and ANDERSON: Standard Tests for Drain Tile and Sewer Pipe, *Eng. News,* **65,** 258 (1911).
NASH: Porcelain Enamelled Iron and Its Relation to Sanitation, *Eng. Rev.,* June, 1912.
MAHR: Studies in Absorption of Water by Building Brick, *Jour. Ind. Eng. Chem.,* **6,** 800 (1914).
VOGT: The Manufacture of Sewer Pipe, *Clayworker,* **62,** 273 (1914).
WILLIAMS: How to Prevent Efflorescence in Finished Brick Work, *Brick and Clay Rec.,* **47,** 341 (1915).

CHAPTER XIII

PAINTS, VARNISHES, STAINS AND FILLERS

Paint Materials

Paints are preparations designed to be applied as surface coatings for the purpose of decoration or protection or both, and consist generally of mineral pigments ground in oil. The oil used must be of such character that upon exposure to the air in a thin layer it will oxidize, either spontaneously or by the aid of catalysts, and form a tough, rubbery film. However, films produced in this way from oil alone are more or less porous and permeable to water, and lack wearing qualities; hence mineral pigments are ground with the oil. It is the function of the pigments to close the pores, and supply hardness and strength to the film. Although linseed oil is the most widely used paint oil, there are several others that may be employed, especially for certain particular purposes. The number of pigments that may be used is very great, and each possesses certain characteristic properties upon which, in a large measure, depends the value of the paint in which it is employed. Because the properties of the paints are so largely determined by the constituents from which they are made, in the present chapter considerable attention will be devoted to raw materials.

Paint Pigments[1]

Definition of Terms.—In describing the various physical and chemical properties of paint pigments the following terms are commonly used.

Whiteness.—All white pigments when ground in oil do not produce white paints. For example, barytes, which is white,

---

[1] The information given in this chapter on "Paints and Paint Pigments" has been derived largely from the *Bulletins* issued during 1909 and 1910, by the Scientific Section of the Paint Manufacturers Association of the United States.

becomes transparent when ground in oil. Lithopone, when ground in oil and painted out, produces a whiter surface than any other pigment but it darkens upon exposure. Zinc oxide is also very white when ground in oil and because its whiteness is more permanent, it is used as a standard for whiteness.

Opacity.—This term is used to designate the degree to which a pigment is able to hide the underlying surface.

Stable and Chemically Active Pigments.—Stable pigments are such as barytes, China clay, silica, etc., which neither react with the oil nor with atmospheric gases. Chemically active pigments are such as white lead, which reacts with the atmospheric gases and also either reacts with the oil or causes its disintegration in some other manner so that chalking results.

Chalking.—This term is used to designate the development of a condition of the paint coat which allows the pigment to be removed after the manner of chalk dust by rubbing with a cloth or with the hands. It is due to the disintegration of the oil film.

Checking.—This has reference to the development on the paint surface of fine hair-like lines, usually interlaced and forked. It is the result of a lack of elasticity and is also sometimes an attendant of chalking.

"Tooth."—This term applies to the manner in which the paint works under the brush. It implies a certain drag or resistance to spreading. The proper degree of tooth allows sufficient spreading without too much greasiness or excessive flow.

Spreading Values.—Some pigments cause the paint to spread out to an exceedingly thin surface, too thin to be durable, while others spread with difficulty.

Antisettlers and Reinforcing Pigments or Paint Coat Strengtheners.—Some pigments when ground in oil settle to the bottom and form a hardened mass on standing. When admixed with pigments of this sort, asbestine and China clay have a certain ability to prevent the settling. Asbestine occurs in long needle-shaped crytsals and China clay in flake form, and they form a web that tends to hold up the other pigments. Also, in a paint coating they act in a manner similar to rods or webs in concrete, hair in plaster, or woven wire in fireproofed glass. They increase the abrasion resistance of the film.

Inhibiting, Accelerating, and Indeterminate Pigments.—These terms are used in reference to pigments applied to iron structures. Inhibiting signifies that the pigment has a tendency to retard rusting; accelerating, that it has a tendency to hasten the rusting; and indeterminate that it is neutral or has no pronounced tendency either way.

THE WHITE PIGMENTS

Basic Carbonate White Lead (*Corroded White Lead, Dutch Process White Lead*), $(PbCO_3)_2 \cdot Pb(OH)_2$.—This pigment is made by supporting lead grids in clay pots in the bottoms of which is acetic acid about as strong as vinegar. The pots are stacked in tiers in tan bark, which ferments, producing heat and carbon dioxide. The heat vaporizes the acetic acid, and the vapors attack the lead, producing basic lead acetate. The carbon dioxide finds its way into the loosely covered pots and converts the basic acetate into the basic carbonate. The process is very slow, requiring from 100 to 130 days. When complete, the white lead is broken up, ground in water and floated to separate the "blue" or uncorroded lead and then is dried in copper pans. It is sold either dry or ground in oil.

It is about 75 per cent. carbonate and 25 per cent. hydroxide, or 85 per cent. lead oxide and 15 per cent. carbon dioxide and water by weight. Of all the pigments it has the greatest opacity or hiding power and it is fairly white. It consists of particles of variable size, and for this reason paints made of it are relatively free from unfilled spaces known as voids. Also, its particles are fairly coarse. About 10 per cent. of them are quite coarse and even its finer particles are coarse in comparison to the particles of basic sulfate white lead or the still finer particles of zinc oxide. The gradation in the size of its particles explains its low oil-absorbing power, its opacity and its denseness or compactness.

Its specific gravity is 6.8, and it grinds in 9 per cent. of oil to a stiff paste, 100 lbs. of which can be prepared for use by mixing with 4 to 6 gal. of oil, a quart of turpentine, and a pint of drier. When applied to iron and steel structures it has good inhibitive qualities. The greatest defect chargeable to white lead as a pigment is that paints made of it are very much subject to chalking. There is some difference of opinion concerning the exact cause of this. One explanation is, that because of the basic nature of the pigment it reacts with the linseed oil, saponifying it to a certain degree, thus bringing about a disintegration of the linoxyn film; but zinc oxide reacts in a similar manner and does not produce chalking. Another explanation is that the chalking is due to the reaction of the pigment with the carbon dioxide of the air, whereby the lead hydroxide is converted into

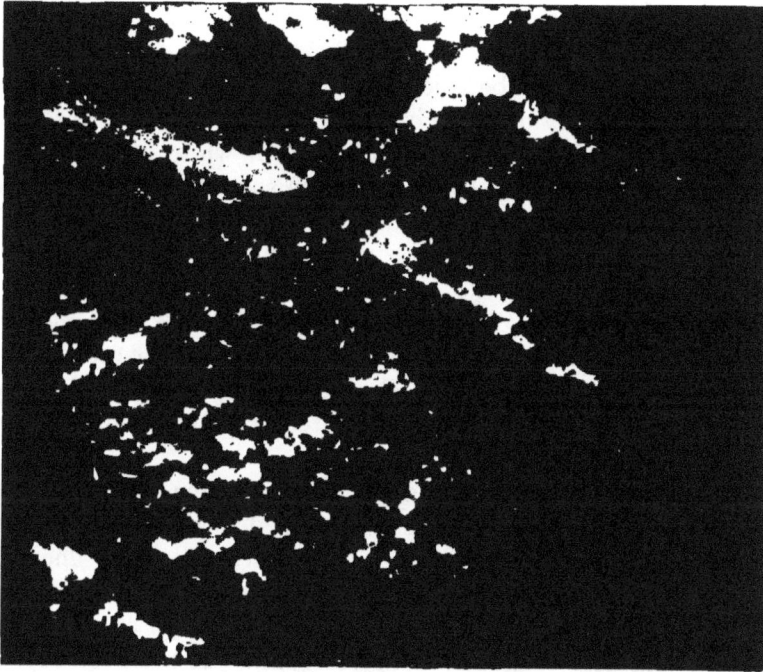

Fig. 52.—Type of paint decay known as chalking. Magnified view. (Gardner.)

the carbonate. Also, lead carbonate forms the soluble bicarbonate of lead in the presence of water containing carbon dioxide and this is perhaps a factor in the chalking. But whatever is the cause, the chalking is very characteristic of white lead. An attendant evil is that after the chalking has started, the paint holds very tenaciously the dust that settles upon it, so that it soon becomes unsightly due to this cause. Paint in this condition cannot be cleaned by sponging with water.

Paint containing the basic carbonate white lead is also subject to rapid checking. And because of its basic reaction this pigment damages delicate colors with which it may be tinted. Unless reinforced it settles badly, thus producing non-uniformity on a painted surface. It is much damaged by atmospheric gases, turning dark in the presence of the very prevalent hydrogen sulphide, due to the formation of the black lead sulphide. For this reason it should not be used with a pigment containing sulphur in any form. It is somewhat deficient in spreading power.

In general it may be said that white lead is weak in those respects in which zinc oxide is strong, and *vice versa*.

Testing of White Lead.—The purity of white lead can in a measure be determined by its complete solubility in nitric acid. Barytes, sand, etc., will remain undissolved. If ground with oil, the oil must previously be removed with several washings of benzine.

Zinc Oxide (*Zinc White, Chinese White*), ZnO.—The American process zinc oxide is obtained by the roasting of zinc ores, as the sulfide, carbonate, etc. The French process oxide is obtained by sublimation and oxidation of spelter or metallic zinc. There are various grades of each.

Because of its chemical stability and extreme whiteness it is an excellent pigment to use as a base for delicate colors. It has good opacity and spreads well. Its particles are very fine and it has no tendency to chalk and hence can be mixed to advantage with the basic carbonate white lead, the particles of which are coarser and which has a tendency to chalk. It has good drying powers of itself, hence with it little added drier is needed. If very rapid drying is desired, litharge is the drier that should be used. Because of its continued oxidizing effect on the oil it produces a very hard paint, but for the same reason one that is very inelastic. Although it reacts with the hydrogen sulfide of the atmosphere it does not darken on this account because the zinc sulfide is almost white.

It has a specific gravity of 5.2 and grinds in 16 per cent. of oil to a stiff paste, 100 lbs. of which can be thinned for use with 64 lbs. of additional oil. The reason why so much more oil is required by zinc oxide than by white lead depends partly upon the fact that a certain weight of zinc oxide is much more voluminous than an equal weight of the basic carbonate white lead. Also it is partly due to the fact that the particles of zinc oxide are of approximately equal size and so do not make as dense a mass as do the graded particles of white lead, thereby leaving more space to be filled with oil.

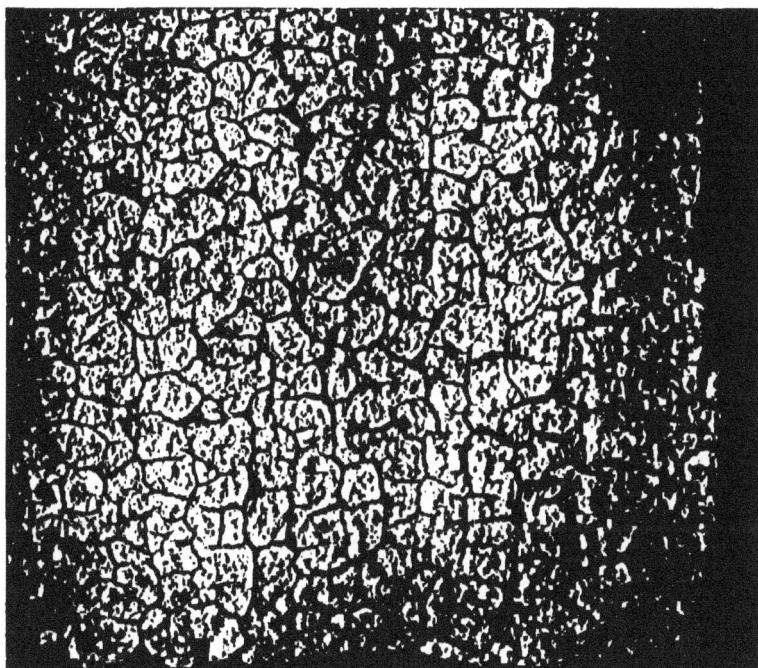

Fig. 53.—Type of paint decay known as checking. Magnified view. (Gardner.)

Zinc oxide is excellent for use in enamels, because of the gloss and whiteness it confers upon the paint. It has good inhibitive qualities.

Its greatest defect is that when used alone it has a tendency to cause the paint to check, crack and peel badly, because of the hardness and inelasticity of the film. Also it reacts with the carbon dioxide of the air, increasing about twice in so doing. The inelasticity coupled with the increase of bulk causes the paint to peel or flake off. This makes it impossible to produce a

18

smooth surface in repainting without removing the old paint. When used alone, the paint possesses excessive flowing qualities and spreads out too thin. These defects can be eliminated by proper reinforcement, as for example with basic carbonate white lead, calcium carbonate, etc.

Adulterations may be detected in samples of supposedly pure zinc oxide by determining its solubility when gently warmed with dilute sulfuric acid. If pure, it will dissolve to a clear solution without effervescence. Effervescence would point toward the presence of calcium carbonate or whiting. Of course, lead carbonate will effervesce also. Insoluble residues indicate the presence of silica, China clay or some compound of calcium, barium or lead, present either as the sulfate or converted into the sulfate by the action of the acid.

Sublimed White Lead (*Basic Sulfate White Lead*), $(PbSO_4)_2 \cdot Pb(OH)_2$.—This is obtained by oxidizing with air the fume produced by roasting and volatilizing galena ore, PbS.

It is a fine, white, amorphous pigment, composed of 75 per cent. lead sulfate, 20 per cent. lead oxide and 5 per cent. zinc oxide. It is chemically stable and does not blacken by action of sulphur compounds. Also, because of its stability it has no action on delicately tinted paints, and so is useful as a base for colors. It is very dense, and more permanent than the basic carbonate; and it surpasses the basic carbonate for painting on wood, having most of its good properties and not its defects. It has a specific gravity of 6.2 and grinds to a stiff paste in about 10 per cent. of oil. For 100 lbs. of this paste 38 lbs. of oil are needed to thin it for application. Because of its fineness it has a tendency to exhibit excessive flowing and spreading qualities.

Lithopone, $BaSO_4 \cdot ZnS$.—This pigment is made by the double precipitation of the insoluble barium sulfate and zinc sulfide by mixing solutions of the soluble salts, zinc sulfate and barium sulfide, the reaction being:

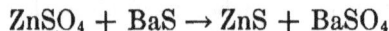

$$ZnSO_4 + BaS \rightarrow ZnS + BaSO_4$$

The precipitate contains between 70 and 75 per cent. of barium sulfate and the remainder consists of zinc sulfide. The mixed precipitates are not, however, suitable for use as a pigment without further treatment, but when the mass is heated to dull redness, suddenly plunged into water, ground in the pulp state, thoroughly washed and dried, its characteristics are totally changed and it makes a very effective and durable pigment for paint purposes.[1]

Lithopone is composed of extremely fine particles and is more dense than zinc oxide. It is very durable, is not affected by gases and it spreads well. It is probably the whitest pigment known and is much used for making enamels. However, it is very peculiarly susceptible to the chemical rays of the sun, so that in the presence of sunlight and moisture it darkens. In the absence of moisture or the direct rays of the sun it does not darken and hence is excellent for inside use. For outside use it is necessary that it be mixed with some pigment as zinc oxide or calcium carbonate to overcome the darkening. White lead pigments or driers containing lead compounds must not be used with lithopone because of the formation of the black lead sulfide.

Reinforcing Pigments.—In relatively small amounts, the following pigments are of great value in overcoming the defects of other pigments, but they must not be used in excess.

Asbestine (*Magnesium Silicate*).—This pigment is prepared by grinding the native mineral, asbestos, which is very fibrous in nature, and even upon grinding finely, retains a rod-like form. Hence it is an indispensable paint-coat strengthener, and in mixed paints will hold up the heavier pigments, preventing settling, It is very stable and retards chalking of paints applied outside, but it lacks opacity in oil and does not spread well. It has a specific gravity of 2.7 and grinds with 32 per cent. of oil to a stiff paste.

The *talcose* variety of magnesium silicate, instead of being fibrous, is plate like in structure. It has the same properties and is put to the same use as asbestine.

China Clay (*Hydrated Aluminum Silicate, Kaolin, Pipe Clay*). —This is produced by the weathering of feldspar. It is a very permanent, white, amorphous powder. It is very fine, is a good spreader, prevents settling, and retards chalking of white lead. Its particles are not all of the same size, hence it is high in density. It has a specific gravity of 2.6 and grinds to a paste in 28 per cent. of oil.

[1] TOCH in ROGERS and AUBERT's Industrial Chemistry, p. 330.

Because of its fineness, it lacks tooth and if used in excess is liable to induce "slicking," thus brushing out too thin. It lacks opacity or hiding power.

Silica (*Silex, Quartz Silica, Infusorial Earth, Decomposed Silica*), SiO_2.—This is obtained by grinding white sand. The particles are not of uniform size, and have sharp angular edges and corners, which accounts for the excellent tooth it possesses. It is very stable and resistant, and when mixed with white lead and zinc oxide produces an excellent wearing paint. It has good spreadng qualities, but because of its crystalline structure and consequent transparency in oil it lacks hiding power. However, because of this transparency, it makes an excellent filler for natural hard woods.

Whiting (*Calcium Carbonate, Paris White, White Mineral Primer*), $CaCO_3$.—This pigment is obtained by grinding chalk. In small quantities it is very useful in neutralizing any free organic acid of the linseed oil, also acids in certain prepared iron oxides. It has moderate opacity, thickens the paint, and increases its density. Also it does not settle much, and is a good spreader. Its specific gravity is about 2.8, and it grinds in 20 per cent. of oil to a stiff paste. It possesses but little opacity, and if used in large quantity, it thickens the paint too much.

Calcium Carbonate (Precipitated Form), $CaCO_3$.—This pigment is prepared by treating lime water with carbon dioxide. It possesses more opacity, but otherwise is similar to whiting. It is much used in making putty. It is not a suitable pigment for use on iron structures, since it contains some soluble impurities, gathered up during its manufacture, which stimulate corrosion. It has a considerable tendency to chalk.

Gypsum (*Terra Alba, Hydrated Calcium Sulfate*), $CaSO_4 \cdot 2H_2O$. —This pigment is prepared by grinding the native mineral. It has excellent opacity when ground in water, and is, therefore, much used as a base for distemper colors. It is chemically stable and possesses good tooth. For painting upon wood, it is very serviceable as a reinforcing pigment to overcome some of the defects of the fundamental pigments, such as the excessive spreading of zinc oxide, or the chalking of white lead. It is not suitable as a reinforcing pigment in paints to be applied to iron, since it is somewhat soluble in water, and dissociates into calcium and sulfate ions, which latter serve to accelerate corrosion. It lacks opacity in oil.

Barytes (*Natural Barium Sulfate*), $BaSO_4$.—Barytes occurs as a native mineral, and is prepared for use as a pigment by grinding and washing in hydrochloric acid to free it from iron compounds, that its color may be white. It possesses good tooth, and prevents excessive spreading of the paints in which it is used. It is an excellent base on which to precipitate colors in the formation of "lake pigments."

Blanc fixe is precipitated barium sulfate. It has practically the same properties and uses as barytes.

Influence on Corrosion.—Of the reinforcing pigments, gypsum, barytes and blanc fixe have a stimulating effect upon corrosion when applied to iron, and asbestine, talcose, China clay, silica and whiting are indeterminate in their action.

THE COLORED PIGMENTS

Concerning the Use of Colored Pigments.—In preparing a colored paint, the *color* of the pigment is not the only point to be considered in choosing it. In addition to having the required tint or shade it must not be chemically active with any other material used in the paint. For example, ultramarine must not be mixed with white lead. Also the color of the pigment must be "fast" to light. That light induces chemical changes is true not only of direct sunlight but in a less degree also of diffused daylight.

At times paint is required to be applied direct to walls rather recently finished with lime or cement plaster, as in the case of decorative or fresco painting. In this event the pigment must be such that it is not affected by alkalies. Prussian blue is an example of a pigment thus affected. The alkalinity of freshly plastered walls can be corrected by washing them with a solution of zinc sulfate. The substances produced by the reaction of the zinc salt with the calcium hydroxide of the plaster are zinc hydroxide and calcium sulfate both of which are insoluble. Whether or not a pigment is affected by alkalies may be determined by shaking it with lime water and allowing it to stand for several hours.

If the painted surface is to be varnished, the mineral colors should be used in the paint if possible, since the "lake" colors are often dissolved by the varnish solvents.

THE BLUE PIGMENTS

Prussian Blue (*Paris Blue, Chinese Blue, etc.*), $Fe_4(FeC_6N_6)_3$. —This pigment is manufactured in a variety of slightly different ways and the depth of color varies somewhat with the method. It is usually made by treating a freshly prepared water solution of ferrous sulfate with a solution of potassium ferrocyanide. A white or greenish-white precipitate of the ferrous ferrocyanide is formed, which is oxidized with some oxidizing agent to the ferric compound. The oxidizing is usually done with bleaching powder.

The coloring powers of this pigment are very great. It is the best of the blues in this respect. One-half ounce will tint 20 lbs. of zinc white a perceptible blue. It is quite permanent and resists the action of atmospheric agencies and light. It absorbs much linseed oil which it tends to preserve, presenting a glossy surface even after long exposure. Its particles vary much in size and are of an amorphous structure.

It cannot be used with whiting, basic carbonate white lead or other basic pigments since it reacts with these, assuming a reddish tinge.

Ultramarine blue is made by fusing together silica, China clay, sodium carbonate, and sulfur in crucibles. The materials react chemically, but it is not known in exactly what manner. In the initial stages of fusion, the product is green, but with further heating a very bright blue is developed. The final product is ground very fine and then is bolted.

Formerly this pigment was gotten from the earth as a mineral, being obtained from lapis lazuli, found in Persia, China, and Siberia.

There are two varieties of the manufactured ultramarine: (1) sulfate ultramarine (so named because in its production sodium sulfate is substituted for a part of the sodium carbonate), of a pale greenish-blue color; and (2) soda ultramarine, of a violet-blue color.

Ultramarine is not discolored by light, and is unaffected by alkalies. Sulfate ultramarine is the palest blue pigment made.

Ultramarine is not affected by heat, hence can well be used where resistance to heat is essential.

Although ultramarine is an excellent pigment for painting on wood it is not a good pigment for painting on iron, because its sulfur content is liable to hasten corrosion. It cannot be used with pigments containing lead, such as white lead or chrome yellow, because of the darkening that results from the lead sulfide formed. Its blue color is discharged by acids and acid vapors.

THE GREEN PIGMENTS

The green pigments may be divided into two classes: (1) those having an inherent or natural green color; and (2) those prepared by an intimate mixture of blue and yellow. Guignet's green and emerald green are examples of the former class, and chrome green of the latter.

Chromium Oxide (*Permanent Green, Guignet's Green*), Cr_2O_3.— This pigment is made by fusing 1 part of potassium dichromate with 3 parts of boric acid and then boiling with water, washing and filtering.

It is of a brilliant dark green shade possessing unequaled permanence. It is not affected by atmospheric agencies and light, nor if pure, by acids or alkalies. Very frequently it is mixed with chrome yellow, and this cannot withstand acids. It is very stable in respect to high temperatures, being resistant up to 800°F., above which, however, it turns brown. When mixed with the chrome yellow it is resistant to only 300°F.

Pure chrome green can then be well used in painting motor parts. Because of its permanence it is used for signal work on railroads. It does not affect, nor is it affected chemically by other pigments. It has good body and covering power. In respect to corrosion it has an inhibitive action. The one objection to this green is its rather high cost.

Emerald Green (*Paris Green, Schweinfurt Green, Copper Aceto-arsenite*), $Cu(C_2H_3O_2)_2 \cdot 3Cu(AsO_2)_2$.—In preparing this pigment, a sodium arsenite is first produced by treating white arsenic with washing soda. Into this is run a hot solution of copper sulfate which produces the olive-green copper arsenite. There is then added dilute acetic acid or vinegar and the mixture is allowed to stand. From it crystallizes the emerald green.

It is an exceedingly beautiful, bright green pigment. In oil it has moderate opacity and covering power. It dries well, is not affected by light and has good spreading qualities.

It is exceedingly poisonous. It is darkened by hydrogen sulfide of the· atmosphere, but is not affected by other gases. It cannot be mixed with cadmium yellow, ultramarine, and other pigments containing sulfur, due to the formation of the black copper sulfide. It is soluble in acids and alkalies, including ammonia.

Chrome Green (*Brunswick Green, Prussian Green, Green Vermilion*).—This pigment consists of an intimate mixture of Prussian blue and lead dichromate. It can be made by the dry method by grinding together Prussian blue and chrome yellow, but it is more often made by precipitating together these two colors upon barytes.

Chrome green is produced in two tones, blue and yellow. The blue tone is the one last described, and contains from 75 to 85 per cent. of some white pigment. The yellow tone contains no white pigment, and hence is much the stronger in coloring value.

Chrome green has good covering power, better than any other green pigment, and is comparatively low-priced. It has a specific gravity of 4.4 and grinds in 25 per cent. of oil to a stiff paste.

On iron it inhibits corrosion when pure, but it not infrequently contains water-soluble impurities, which detract from this property. Also lead chromate itself is somewhat soluble in water. It cannot be used with pigments containing sulfur, because the sulfur would react with the chrome yellow and darken the green. It is likewise affected by hydrogen sulfide. It should not be used with the basic carbonate white lead or with whiting, since the basicity of these pigments affects both the blue and yellow components, turning the green into a red. It cannot be used with acids, which dissolve the chrome yellow, thus causing the green to turn blue.

THE YELLOW PIGMENTS

Chrome Yellow (*Neutral Chromate of Lead*), $PbCrO_4$.— (a) *Light Yellow Chrome* (Sulfur or Canary Chrome).—This is obtained by precipitating lead chromate from a solution of a lead salt by the use of potassium dichromate. By the addition of sulfuric acid, or some soluble sulfate, there is precipitated with this, lead sulfate also, which serves to make the pigment light in color. Hence, the sulfur chrome yellow is really a mixture of lead dichromate and lead sulfate, and may contain as much as 60 per cent. of the latter. Also it is often adulterated by the addition of gypsum, China clay, and barytes, of which materials it is able to hold a great deal without the shade being sensibly affected.

(b) *Medium yellow chrome* is pure neutral chromate of lead and should contain no white pigment.

The chrome yellows are very brilliant and have the greatest covering power of all the yellow pigments. The particles are fine, the color is unaffected by light, and they have an exceedingly high tinting power.

Although theoretically these pigments are good inhibitors, their practical value in this line is not high, because they usually contain water-soluble impurities, which is generally true of all precipitated pigments.

Because of their lead content, they are darkened by hydrogen sulfide, and by ultramarine, cadmium yellow, and other pigments containing sulfur. They are turned orange by basic pigments, as white lead, because the basic chromate of lead is formed. They have a tendency to cause the paints to dry slowly.

Zinc Chrome (*Citron Yellow*), $ZnCrO_4$.—This is prepared by precipitating the pigment from zinc salt solutions by potassium dichromate, or by boiling zinc oxide with sulfuric acid and potassium dichromate.

Zinc chrome has a delicate, bright, sulfur-yellow color, and has good body and covering power, but it is not equal in this respect to the lead chromes. It is not affected by light or atmospheric gases. Hydrogen sulfide does not affect it nor is it affected by other pigments.

For metals it has a higher inhibitive value than any other pigment. The Scientific Section of the Paint Manufacturers Association of the United States recommends a 2 per cent. addition of this pigment to all paints to be applied to iron or steel surfaces. However, it is somewhat soluble in water and is dissolved in weak acids, both organic and mineral. Heat decomposes it, causing it to turn to a gray-violet color due to the formation of the zinc and chromium oxides. It is slow in drying, requiring considerable added drier.

Cadmium Yellow (*Cadmium Sulfide*), CdS.—This is prepared by treating the soluble cadmium sulfate with either sodium or hydrogen sulfide.

It is a brilliant, beautiful yellow, ranging from pale to deep orange. It has good covering power and is not affected by light and gases. It cannot be mixed with white lead or emerald green because of its sulfur content. In the presence of moisture it has a tendency to oxidize to cadmium sulfate and hence fades. It is rather high-priced.

Yellow ochers are pigments found native in many localities and consist of either clay or calcium carbonate colored with about 20 per cent. of hydrated ferric oxide. They are prepared for use by grinding and levigating. This consists of stirring the pigment in a gentle current of water which carries the fine particles into a settling tank, leaving the coarser material behind. The fine material is used as the pigment and after drying is ready for use. Sometimes the ochers are calcined (strongly heated) and the color is thereby changed to a reddish tone, due to the formation of the dehydrated ferric oxide.

The ochers are usually of a golden-yellow color. They have a moderate opacity usually, although this varies much according to the source of the pigment, some being quite transparent in oil. They have a moderate spreading power. Other pigments are not affected by them nor are they affected by any other pigments. Also, they are not affected by gases.

Sienna is also an earthy pigment, belonging to the ocher class, and is of similar composition, but it contains more iron oxide and considerable brown manganic oxide, and hence is darker in color. The name sienna is taken from the name of the town in Italy where this pigment was first obtained. The pigment upon being burnt or calcined becomes, according to its composition, brown or reddish orange to dark red.

It has properties similar to the ochers. It is almost transparent in oil and, therefore, is used almost exclusively as a glaze or tinting color.

THE RED PIGMENTS

Red Lead (*Minium*), Pb_3O_4.—This is prepared by heating metallic lead in furnaces in the presence of air, whereby litharge, PbO, is first formed. The litharge is then further oxidized into red lead by heating to about 375°C. in a muffle furnace.

This pigment has a bright red color and is permanent to light. Its particles vary in size, hence it makes a very compact coat. It has very strong drying properties, rapidly oxidizing the oil with which it is mixed. On this account it is usually mixed with the oil at the time of application, 30 lbs. of the red lead to a gallon of oil being the proportions most commonly used. This is a very small amount of oil as compared with the amounts taken up by other pigments. Because of the small amount of oil and also its strong oxidizing power, paints made of red lead have a marked tendency to dry to very hard and brittle films and not infrequently scale off. No added drier is needed with red lead paints. It is largely used mixed with more inert pigments. It cannot be mixed with sulfur-bearing pigments and it is turned brown by the action of atmospheric hydrogen sulfide. It has long been used in paints for the protection of iron and steel and is considered very valuable for this purpose.

Venetian red consists of a mixture of ferric oxide, Fe_2O_3, and calcium sulfate, $CaSO_4$. It is prepared by heating a mixture of ferrous sulfate, $FeSO_4$, and quicklime, CaO. The amount of calcium sulfate varies, but may make up as much as half of the mixture. The color of the pigment depends somewhat upon the temperature of the furnace, the lower temperatures producing the brighter colors.

Venetian red is a very permanent red pigment. It does not affect other pigments nor is it affected by them. It forms excellent protective paints for wood, being especially good as a priming coat, but is not a good paint for metal. The calcium sulfate, which it contains, dissolves and ionizes slightly in water, and the resulting sulfate ions are active in stimulating corrosion. Also it may contain some sulfur trioxide not properly neutralized by lime during the calcination. It does not exert a drying action on the oil as does the red lead, and the longer it stands in the full amount of oil it is to receive, the better paint films it seems to produce.

Natural Iron Oxides (*Indian Red, Colcothar*).—These iron oxide pigments are obtained by grinding red hematite iron ores, levigating with water, and then drying. They vary much in composition, being from 88 to 95 per cent. iron oxide and the remainder clay and silica.

Their properties are very similar to those of the Venetian reds, but for metals the natural oxides are much better, since they contain no ionizing calcium sulfate nor any free acid.

Chrome Red (*American Vermilion, Basic Lead Chromate*), $Pb_2(OH)_2CrO_4$.—This is the most highly basic of the lead chromate pigments. It is obtained by boiling white lead with a solution of potassium dichromate, and then treating the mass with a small quantity of sulfuric acid to brighten the shade.

Its particles are granular and coarse. Its general properties resemble those of chrome yellow, which see. For painting on metal it has very excellent inhibiting powers, being among the best of the pigments for this purpose.

THE BROWN PIGMENTS

Prince's Metallic Brown (*Prince's Mineral*).—This is an oxide of iron pigment mined as an ore. It occurs both as the hydrated oxide and as the carbonate, these forms being roasted at a red heat and then ground. The pigment contains a high percentage of alumina and silica.

It has a good brown color and is very permanent. Its particles are generally fine but of varying sizes. It forms compact paint coats that are very good for both wood and metal.

Umber is also a natural pigment consisting of a clay base containing iron oxide. It belongs to the same general class as the ochers but differs from them in that it contains considerable amounts of the brown manganic oxide. The raw umber is of a dull grayish-brown color, but upon burning (calcining) it turns to a reddish brown, which color is very permanent. Its properties in general very much resemble the properties of the ochers.

Vandyke brown consists of a mixture of red iron oxide and yellow ocher, darkened with lamp-black or some bituminous compound. It is very permanent upon exposure, works well in oil, but it seems to cause the paint to dry badly.

The Black Pigments

Lamp-black is made by burning heavy refuse oils in a limited supply of air. It is in reality a soot, averaging about 98 per cent. pure carbon. There are various grades. The soot from the flame is received into different chambers arranged in a series. The black that collects in the first chamber contains a certain amount of oily matter which has passed off unburned. On this account it has a brownish or grayish tint. That in the farther chambers is freer from the oily deposit. Lamp-black in general possesses a grayer tone than many of the other carbon pigments so that it may often be recognized by this fact. Lettering made with lamp-black will appear light on a background painted with ivory black or drop black.

Its particles are very fine and uniform. It is very permanent and seems to preserve the oil with which it is mixed, thereby producing very permanent paints. It outlasts lead and zinc paints for wood. It has a specific gravity of 1.82 and grinds in 75 per cent. of oil to a stiff paste.

It retards drying. This may be due to the effect of the hydrocarbon oil it contains. Toch says that less than 1 per cent. of this unburned oil will retard the drying to such an extent as to render the pigment unfit for use.

Carbon Black (*Gas-black*).—This also contains about 98 to 99 per cent. pure carbon and is obtained by burning hydrocarbon gases, as natural gas or acetylene gas, in a limited supply of air. It very closely resembles lamp-black in most of its properties.

Drop Black (*Vine Black, Ivory Black, Frankfort Black*).— This is made by heating a variety of materials, such as waste cuttings from ivory, vine shoots, cork cuttings, cocoanut shell, etc., in retorts or crucibles, the residue containing the greater part of the carbon of the original material. The name "drop black" is applied to this substance because the product after grinding is mixed with a little glue water and made into pear-shaped drops for sale. Drop black always contains mineral matter resulting from the ash of the burned material, particularly potassium salts, which are somewhat hygroscopic, and so it should be washed and dried carefully. It is very black, lamp-black appearing gray in comparison.

Graphite is an allotropic form of carbon formerly obtained only native as a mineral, but more recently it has been prepared artificially, notably by the Acheson process (see page 58). The native variety occurs both in the crystalline and amorphous form. Both the natural and artificial varieties contain impurities, such as silica, iron oxide and alumina. The natural form contains the greater amount of these and, therefore, the less carbon, there being in it about 60 per cent. of the latter, while the artificial variety contains about 90 per cent. carbon.

Graphite is known to be a pigment that causes very slow drying of the oil with which it is used. The long life of graphite paints may be due to this fact. For not only will the formation of the linoxyn be delayed but also the continued oxidation of the linoxyn after it has formed will be slow. It is this continued or progressive oxidation of the linoxyn of many paint coatings, notably those containing pigments possessed of marked drying properties, that causes the linoxyn to soon become very inelastic and subject to early failure. Since the oxidation is delayed in the graphite paints, the film remains elastic for a longer time and, therefore, is much less readily disintegrated.

The Lake Pigments[1]

The lake pigments are made by precipitating some organic coloring matter, usually a coal-tar dye, upon some suitable mineral base, as barytes. The base here is similar in its action to the cloth in ordinary dyeing, being merely a carrier of the color. In addition to barytes, many other white pigments are used as bases. The colors used should be insoluble in water, oil, and the ordinary solvents used in paints and varnishes. But it often happens that they are not thus insoluble and then the

---

[1] The term "lake" has its origin in the Italian word "lacca," which was the term used to designate the scum that gathered on the top of the vats when dyeing with lac dye, which was much used by the old Italian dyers. Tin and aluminum oxides were used to fix the color on the cloth, and the scum consisted of these oxides colored. Since the present lake pigments consist of minerals colored by means of dyes, the English modification of the Italian name has been applied to them.

coloring matter of the pigment dissolves in the vehicle, and in repainting may even diffuse through the superimposed layer of paint. If the top coat is of a different color, this is very objectionable. This diffusion of color is known among painters as "bleeding." It may be overcome by taking advantage of the well-known property of carbon in absorbing coloring matters. Charcoal, lamp-black, bone-black and similar substances possess this property to such a degree that they are used to clarify colored solutions. In repainting over a paint containing a lake pigment, any tendency toward "bleeding" may be overcome by applying first a coat of black paint containing pigments of this sort.

Lake pigments are made up in all colors and are much used in painting. They are usually fairly permanent in respect to light and atmosphere and very beautiful effects can be obtained by their use.

PAINT AND VARNISH OILS

The oils used as the vehicle or medium with which the pigment in paints or the resinous material in varnishes is mixed are those that are known as drying oils. They are so named because they have the property of being converted into a more or less hardened film when exposed to the air. These oils make up a rather limited group, or subdivision included under the general head of fatty oils, and so before proceeding to a discussion of the properties of the drying oils, it will be advantageous to consider the fatty oils in general.

The Fatty Oils.—Under this head are included both the liquid and solid fats. There is no sharp line of demarcation between the two. All of the liquid fats become solid at low temperatures, although some of them remain fluid at temperatures considerably below the freezing point of water. On the other hand, even the hardest fats become fluid at about 50°C. Still those that are fluid below 20°C. are usually called oils, and those that are solid above this temperature are known as fats.

The *specific gravity* of the oils and fats ranges between the limits of 0.910 and 0.975 at 60°F.

The fatty oils when pure are *odorless, tasteless,* and *colorless,* that which is considered characteristic in these respects being due to the presence of certain foreign substances.

Solubility.—The fatty oils are almost completely insoluble in acetone and cold alcohol, but the solubility increases in boiling alcohol. They dissolve very readily in ether, chloroform, carbon tetrachloride, carbon bisulfide, benzole, paraffine oils, and petroleum naphtha. Castor oil is, in respect to its solubility, a distinct exception to the other fatty oils. It is quite soluble in cold alcohol, while it is only slightly soluble in paraffine oils and petroleum naphtha.

Fatty oils do not evaporate upon exposure, and they cannot be distilled even under reduced atmospheric pressure. They cannot be vaporized by heat without destruction; that is, in the true sense they cannot be boiled. If the products driven off by heat are collected and condensed, the original oil is not obtained, but in its stead there will be true gases, and waxy, resinous, or tar-like substances. Because of this characteristic the fatty oils are often called *fixed oils*. The decomposition begins at about 250°C., but they can be heated to just below this point with practically no change.

Composition.—The fatty oils do not consist of single compounds, but are made up of mixtures of substances known as triglycerides. These glycerides are products resulting from a union of the trivalent glyceryl radicle, C_3H_5, of the alcohol glycerine, $C_3H_5(OH)_3$, with three monovalent fatty acid radicles of such acids as stearic, oleic, palmitic and others. Although it is not likely that reactions of this sort occur in the living organism producing the oil, the following equation between glycerine and **palmitic acid** will serve to illustrate their composition:

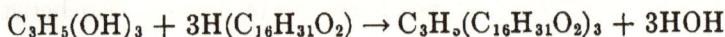

$$C_3H_5(OH)_3 + 3H(C_{16}H_{31}O_2) \rightarrow C_3H_5(C_{16}H_{31}O_2)_3 + 3HOH$$

The products are triglyceryl palmitate, known also as palmatin, and water. Other acids, whose triglycerides occur commonly in these oils, are:

Stearic acid, $H(C_{18}H_{35}O_2)$ or $C_{18}H_{36}O_2$, belonging to the $C_nH_{2n}O_2$ series.

Oleic acid, $H(C_{18}H_{33}O_2)$ or $C_{18}H_{34}O_2$, belonging to the $C_nH_{2n-2}O_2$ series.

Linoleic acid, $H(C_{18}H_{31}O_2)$ or $C_{18}H_{32}O_2$, belonging to the $C_nH_{2n-4}O_2$ series.

Linolenic and isolinolenic acids, both of which have the same formula, $H(C_{18}H_{29}O_2)$ or $C_{18}H_{30}O_2$, belonging to the $C_nH_{2n-6}O_2$ series.

The corresponding triglycerides of these acids are stearin, olein, linolein, linolenin, and isolinolenin. Palmatin and stearin are white crystalline or semi-crystalline solids, melting at 61° and 72°C. respectively. These two occur in the greatest proportion in the solid fats, the others mentioned being fluid at atmospheric temperatures. Olein is the principal constituent of the liquid fats, especially of olive oil. Olein, linolein, linolenin, and iso-linolenin are unsaturated compounds; that is, they have certain double bonds, and are able to absorb more or less oxygen. Olein has two, linolein four, and linolenin and isolinolenin have six double bonds, and these molecules can absorb one, two and three atoms of oxygen respectively.

Drying and Non-drying Oils.—It is to the presence of the oxygen absorbing triglycerides that paint oils owe their drying ability. When exposed to the air in thin layers they oxidize to form dry, hard, and generally transparent films, which are insoluble in water, alcohol, and to a very great extent in ether. If the oil is in a rather finely divided state, as in the meshes of oily cotton waste or rags, so that a larger surface is exposed for atmospheric action, the heat developed by the resultant vigorous oxidation is sufficient to raise the temperature of the substance to its kindling point and spontaneous combustion will occur.

The non-drying oils when exposed to the air do not harden but become disagreeable in smell and acid in taste, and are then said to be rancid. When long exposed they may oxidize and decompose sufficiently to become viscid but they never harden to a solid mass even though they be exposed to the atmosphere for years. Examples of this sort are whale blubber oil, lard oil, tallow oil, olive oil and others. Certain oils have properties midway between the drying and non-drying oils and are therefore known as semi-drying oils, but the lines of division between the classes are not distinct.

Cause of Rancidity.—It is not known just what the change is that causes rancidity, but Lewkowitsch says it is something more than mere acidity, although a certain amount of free acid is no

19

doubt always produced first. But even fresh vegetable oils usually contain considerable free acid and are not rancid.

Absolutely dry air, if light is excluded, has no action on any of the fatty oils. But it is practically impossible to completely exclude moisture in storing the commercial oils, so in any case small amounts of free acid will be produced. The formation of free acid is due to the fact that the oil becomes to a slight extent *hydrolized;* that is, the direct reversal of the action shown by the equation in the preceding discussion of composition takes place. This hydrolysis reaction is probably brought about, according to Lewkowitsch, by the presence in the oil of certain soluble ferments derived from the oil-producing seeds. It is also helped along by albuminous and gelatinous substances present in unrefined oils. The fatty acid and perhaps the glycerine also, which are produced, then further decompose and the oil becomes rancid.

The fatty oils are quite readily hydrolyzed by heating with water to 200°C. or more. The reaction occurs at much lower temperatures in the presence of acids—even as low as 100°C. in the presence of hydrochloric acid.

Saponification.—When the fatty oils are treated with alkalies, glycerine is liberated and soaps are formed, as is shown by the action of caustic soda on stearin:

$$3NaOH + C_3H_5(O \cdot OC_{18}H_{35})_3 = 3Na(O \cdot OC_{18}H_{35}) + C_3H_5(OH)_3$$

The soap produced in this case is a sodium stearate.

Examination of Oils and Fats.[1]—When oils and fats are under examination to decide upon their suitability for special purposes, certain physical and chemical properties are determined. These properties vary in amount in the different oils, and so aside from showing what the oil is good for, their determination serves to identify the oil. The numbers which show the measure or degree of these properties are referred to as "values" or "numbers" or often as "constants." This latter term must not be construed to mean that the numbers are in reality constant for any certain oil, because the numbers will vary for different pure specimens of the same kind of oil, although the variation will be within well-defined limits.

[1] See Lewkowitsch: Chemical Technology and Analysis of Oils, Fats and Waxes.

The values or numbers commonly determined are specific gravity, the saponification number, acid number, iodine number, and for paint oils especially, the eliadin test may be applied. Of course, in a complete examination of the oil many other determinations and tests are carried out.

Specific gravity may be determined by means of the hydrometer, the hydrostatic balance, the specific gravity bottle or picnometer, or in other ways. The specific gravity bottle furnishes the most accurate results. By it, the relation between the weights of equal volumes of oil and water at 60°F. is obtained by direct weighing. The weight of the bottle when accurately filled with pure water is first determined, and then its weight when filled with the oil. The hydrometer allows greater speed in making the determination but furnishes less accurate results. For the method of using hydrometers see Chapter XX.

The saponification value shows the number of milligrams of potassium hydroxide required for the complete saponification of 1 gram of the oil sample; that is, for complete neutralization of the *total* fatty acid, whether free or in combination with the glyceryl or other radicles. From this may be calculated the mean molecular weight of the fatty acids present. Also since the hydrocarbon oils, mineral and rosin oils, are unsaponifiable, this serves to show their presence when they are admixed with the fatty oils.

The saponification value is determined by boiling the sample with potassium hydroxide dissolved in alcohol, and then determining the excess of potassium hydroxide by titration with standard hydrochloric acid, using phenolphthalein as an indicator.

As a qualitative test, the following is fairly accurate: Dissolve about ½ gram of potassium hydroxide in about 5 c.c. of 95 per cent. ethyl alcohol, add a few drops of the oil to be tested, and keep at the boiling point for about 5 min. Then add 3 or 4 c.c. of distilled water. If the solution remains clear, fatty oil only is present; but if it becomes turbid, then mineral oil, or other hydrocarbon oil, is indicated. The fatty oil is converted into soap, and this is soluble in water, but hydrocarbon oils are not saponifiable. If the sample is not sufficiently boiled, turbidity may be due to unsaponified fatty oil also. The amount of turbidity will vary according to the mixture, but the test cannot be relied upon to show less than 2 or 3 per cent. of hydrocarbon oil.

The acid value shows the number of milligrams of potassium hydroxide needed to completely neutralize the *free* fatty acid in 1 gram of sample. The difference between the saponification number and the acid number is known as the "ester number." Esters are compounds produced by the reaction of an alcohol with an oxy-acid. Of course, in the fatty oils the esters are the glycerides.

The determination is made by titrating the sample with standard potassium hydroxide.

The iodine value represents the per cent. of iodine absorbed by the sample; that is, the number of centigrams taken up by 1 gram of the oil. The absorption is due mainly to direct addition but may in part consist of substitution of iodine for hydrogen atoms. This value serves to show the proportion of unsaturated fatty acids present, both free and combined. The direct addition of iodine depends upon the fact that in the unsaturated radicals, which are those of oleic, linoleic, and linolenic acids, there exists a number of "double bonds." For each double bonding two iodine atoms are absorbed. The oleic acid absorbs two, the linoleic absorbs four, and the linolenic absorbs six iodine atoms. Since these unsaturated compounds also react quite readily with oxygen, the iodine value is taken as a measure of the ability of an oil to oxidize with the formation of a dry film.

The Elaidin Test.[1]—This is not a quantitative test but still it aids in distinguishing between the non-drying oils and those that are drying or semi-drying. It depends upon the fact that nitric oxide, NO, is able to convert the liquid olein into its solid isomer elaidin; that is, into a substance having the same empirical formula as olein, but a different structural arrangement of the atoms within the molecule. The glycerides of linoleic and linolenic acids are unaffected, hence the drying oils, which contain a greater proportion of these, remain liquid or at most become only pasty. Since lard oil and olive oil consist largely of olein this test is especially applicable to them.

Five grams of oil are mixed with 7 grams of somewhat diluted nitric acid, specific gravity 1.34, and to the mixture is added about

[1] GILL: Oil Analysis, p. 51.

1 gram of copper. The container is placed in cold water, about 15°C., and the mass is well stirred for about a minute. After standing for 2 or 3 hrs. the solidity of the sample is examined.

THE DRYING OILS

Linseed Oil.—For the manufacture of paints and varnishes linseed oil is most important. It is obtained from the seed of the common flax, which is grown quite generally over the central, western and northern portions of the United States, although the bulk of the seed comes from Minnesota and the Dakotas.

FIG. 54.—Presses for the production of linseed oil.

Also, the flax is much cultivated in Canada, Argentine, India and southern and western Russia. The seed is usually the sole object of the cultivation, the fiber of the plant being wasted.

Separation of Oil from Seed.—Linseed oil may be obtained from the seed by any one of three different methods; the hydraulic-press method, the continuous-expeller method, and the extraction method.

The *hydraulic-press method* is the oldest of the three, and much of the oil is still obtained in this way. The seed is crushed to a very fine meal between chilled iron rolls, and the meal is then heated to about 80°C. with constant stirring. When sufficiently

hot, it is moulded into cakes weighing about 20 lbs., and these
are wrapped in cloth and pressed with hydraulic pressure, which
amounts at a maximum to about 2 tons per square inch. The
seed is but once pressed and the resultant cake, which still contains from 4 to 8 per cent. of oil, is used as cattle food. Oil
obtained in this way is known as *hot-pressed oil.*

In order to remove suspended particles of meal the oil is
filtered, but this purification is not sufficient since there is always present in a dissolved state certain amounts of albuminous
or mucilaginous substances and water. To precipitate this as
far as possible the oil is conducted into storage tanks and allowed
to stand from 4 to 6 months. The resultant sediment is known
as "foots." The longer it is stored the better, since it seems to
be characteristic of linseed oil to deposit foots almost indefinitely.
The supernatant clear oil is the raw oil of commerce.

Cold-pressed oil is of superior quality but the quantity produced is relatively small. The oil is lighter in color, more fluid
than the hot-pressed, and contains less sedimentary matter,
probably because it contains less solid fats and foreign substances.
However, about 15 per cent, of oil is left in the cake, and in consequence this method is less in favor among the oil producers.
In the *continuous-expeller method* the crushed seed is forced by a
screw through a cone-shaped grating. The oil trickles through
the grating and the pressed meal comes out at the end. This
process requires less hand labor than the older method, but allows
somewhat more of the oil to remain in the meal.

In the *extraction process*, the oil is dissolved from the meal by
leaching with petroleum naphtha. The naphtha is distilled
from the extract and used over again. It is very difficult to
remove the last traces of naphtha, and the oil obtained by this
method generally has the odor of this solvent.

Properties of Linseed Oil.—The properties of the oil depend
somewhat upon the origin of the seed; for example, oil obtained
from seed grown in the neighborhood of the Baltic Sea has a
considerably higher iodine value than that from Argentine.
The following values have reference to oil obtained from the
North American seed.

At 60°F. the *specific gravity* varies from 0.932 to 0.936. A falling below 0.932 would suggest the presence of fish oil or mineral oil.

Its *saponification value* ranges from 189 to 195. It should contain no more than 1.50 per cent. of unsaponifiable matter. All hydrocarbons, as mineral oils and rosin oil, are unsaponifiable and hence would be shown by this test.

The *acid value* is quite variable, but is usually less than 7. The presence of rosin oil would greatly increase this figure since it is highly acid. However, the acid figure for mineral oil may be zero.

The *iodine figure* ranges from 170 to 187, which is the highest value of all the known fatty oils. Linseed oil does not yield a solid mass in the *elaidin test*.

It *freezes* at −28°C., which is extremely low. Walnut oil is the only oil having a low freezing point. Menhaden oil freezes at about 0°C., rosin oil at −6, cottonseed at −15, and the heavy mineral oils at about −18.

The *adulterants* to be looked for in linseed oil are mineral, rosin, corn, cottonseed, menhaden, hemp, and rape seed oil. The cheapest and perhaps the most frequently used are mineral and rosin oil. These would be shown by the saponification test and low iodine number.

Drying of Linseed Oil.—All of the important uses of linseed oil depend upon its great power to absorb oxygen, whereby it changes from the fluid into the solid state. It possesses this power to a greater degree than any other oil. Of the total glycerides in linseed oil 15 per cent. are solid fats and 85 per cent. are fluid, these latter being endowed with distinct drying powers. Usually drying means evaporation and loss of weight, but as applied to linseed oil it means just the opposite, for when linseed oil dries it may increase in weight as much as 18 per cent. due to its oxygen absorption. In a paint or varnish film, however, the increase is probably not more than 10 to 12 per cent. Oxidized linseed oil is called *linoxyn*, and is a solid, somewhat tough and rubber-like substance.

The drying of linseed oil in the absence of any drier is at first slow, but after a certain stage is reached it dries rapidly. This is probably due to the formation of peroxides during drying which act as catalytic agents and hasten the process. The oil dries more rapidly in the light than in the dark. This is undoubtedly due to the action of the actinic or ultraviolet rays. Also because of the bleaching effect of these rays, white paints that dry in the light are freer from any yellowish tinge than those that dry in the dark.

Linoxyn is practically insoluble in most of the solvents of linseed oil, hence dried paint and varnish films cannot be readily removed by these solvents. Certain solvents for linoxyn are shown later under "Paint and Varnish Removers." Linoxyn is attacked by the alkalis, hence paints and varnishes are injured by contact with them and may be removed by their use if desired.

"Boiled" Oil.[1]—The drying of linseed oil may be hastened by the so-called "boiling" of the raw oil. This boiling consists of heating the oil with driers,[2] sometimes in an open kettle with direct fire, or far more commonly, in a steam-jacketed kettle or with steam coils to about 150°C. During the heating, the driers dissolve to some extent and a certain amount of polymerization[3] of the oil is believed to take place. The driers used consist usually of the linoleates, resinates or borates of lead and manganese. Boiled oils prepared with lead driers contract upon oxidizing while those with manganese expand. Usually the two compounds are used together in quantities sufficient to introduce about 0.5 per cent. of metallic lead and 0.2 per cent. metallic manganese. Raw oil dries in 3 or 4 days when exposed in a thin layer while the boiled oil dries to a hard film in less than 24 hrs., even in as little as 8 hrs. indoors if conditions of temperature and humidity are favorable. This increased rate seems to be due to the fact that the metallic compounds act as catalytic agents. Boiled oil is usually of a dark red color, the color depending upon the amount and kind of drier used, and the length of time and the degree to which it has been heated. Boiled oil costs usually about 1 ct. per gallon more than the raw oil.

A "cold-boiled" or "bung-hole" boiled oil is prepared by

---

[1] It must be understood that the term "boiled" as applied to any of these prepared oils is incorrect, since as was explained on page 288, none of the fatty oils can, in the true sense, be boiled.

[2] For a discussion of driers, see p. 301.

[3] A substance is said to polymerize when it changes into another form having the same elements in the same proportion, but a higher molecular weight, such as would result from a combination of two or more molecules.

mixing with the cold or only slightly heated raw oil comparatively small quantities of concentrated solutions of the metallic compounds or driers. These concentrated solutions consist of metallic resinates dissolved in hot linseed oil and turpentine. Such "boiled" oils are quite inferior products.

Bleached Oil.—As has been stated previously, the glycerides which compose the fatty oils are colorless, but the oils themselves are usually colored to some degree by the presence of certain impurities. The yellow color of linseed oil is objectionable for the making of certain white paints and pale varnishes. This can be removed in a large measure by the oxidation, or some other chemical transformation of the colored substance. The process generally employed consists of the addition of sulfuric acid and the blowing of air into the oil at the same time. A cloudiness develops after which the oil is allowed to stand until settling takes place, or it is filtered. Various other bleaching agents such as oxides and peroxides are employed. Chemically bleached oil is said to wear not so well as the raw oil.

Chinese Wood or Tung Oil.—This is obtained by pressure from the nut of the tung tree native to China and Japan. The color of the oil varies from pale yellow to dark brown and it has a strong characteristic odor, that serves as a ready means of identification.

The following values for tung oil are given by Toch: Specific gravity, 0.935 to 0.941; saponification value, 155 to 190; and iodine value, 156 to 165. Lewkowitsch says that the Chinese variety of tung oil has a specific gravity of 0.941 and that this is the highest of any known oil except castor oil. Tung oil consists chiefly of the glycerides of oleic and elaeomargaric acids, the latter acid radicle being $C_{18}H_{32}O_2$.

Tung oil differs from all the other oils in respect to two rather noteworthy characteristics. When exposed to light in bulk, a flocculent white precipitate is produced, which gradually increases in amount until after some months the oil becomes a solid whitish mass. This change is known as the *"light break."*

Also, when heated to about 250°C. the oil *coagulates* with the formation of a jelly-like mass. After cooling it is not again made fluid by heating nor by the addition of the ordinary solvents.

This gelatinization is believed to be due to polymerization. It is not due to oxygen absorption.

When raw tung oil is brushed out in a thin film, it dries with extreme rapidity in 30 to 50 min. But raw tung oil is not suitable for paints and varnishes, since its dried film is whitish, opaque and wax-like, without elasticity. Also, it has a tendency to wrinkle and does not adhere to the surface on which it forms. However, with proper treatment the oil acquires an ability to dry to a hard, dense, tough and elastic film, said to be more durable than the linseed oil film and more impervious to water. The necessary treatment, according to Toch, consists of heating

Fig. 55.—Wedge press used for extracting Chinese wood oil. Basket of nuts and oil cake in foreground.

the oil to about 180°C. and adding from 12 to 15 per cent. of organic salts of lead and manganese. These organic salts prevent the coagulation. Tung oil is much used in conjunction with linseed oil. In this case the linseed oil is added just subsequent to the organic salt. After sufficient cooling the prepared oil is thinned with turpentine or benzine or a mixture of both. The treated oil dries much more slowly than the raw oil.

A tung-oil film is less spongy and porous than a linseed-oil film, with consequently less tendency to absorb and allow the passage of moisture. On this account it has proven very useful for the manufacture of enamels, waterproof paints and varnishes. It is excellent for out-of-door work. Toch says that by the use of tung oil, paints can be prepared that dry well in damp atmospheres.

The film from the prepared oil is hard and not readily scratched. Also, when it is used in large amount with a small amount of resin, floor varnishes can be produced that do not show heel marks.

According to Ware and Schumann[1] excellent varnishes can be made by the use of tung oil and common rosin or colophony. This is rather remarkable, since rosin lacks hardness and is very brittle. Equally good results are not secured when it is used in this way with other oils. The value of the oil has been said to lie in its ability to form actual combinations with the abietic acid of the rosin. This statement is open to question, but varnishes of superior quality are made from the two.

Soya-bean Oil.—The soya bean is native to Manchuria, but has been rather extensively introduced into the United States, especially the southern portion. There are as many as 280 known varieties, and more may exist. All varieties do not yield oil suitable for paints, much of it being used for making food products and soaps.

From 10 to 15 per cent. of oil is obtained from the soya bean by the method of hot pressure. Excepting that it has perhaps a slightly darker color, it is very similar in all its physical characteristics to linseed oil, so that it is somewhat difficult to distinguish between the two. Soya oil contains about 15 per cent. of saturated fats, largely palmitic. The unsaturated fats are chiefly the glycerides of oleic and linoleic acids, about 56 per cent. of the former and 20 per cent. of the latter. The values are about as follows: specific gravity, 0.925; saponification value, 192; iodine value, 130.

The raw oil produces an elastic film but dries rather slowly, although it can be converted into a very suitable drying oil by the use of the proper driers, which according to Toch, are mixtures of the tungates and resinates of lead and manganese. The tungates are the organic salts produced by the reaction of lead and manganese compounds with tung oil. The resinates are the products of the reaction of the metallic compounds with common rosin. The soya oil is said to dry to a hard resistant film in 24

[1] *Drugs. Oils and Paints*, **30**, 52–54.

hrs. when used with 5 to 7 per cent. of this drier. Also it is very susceptible to the drying action of the tungate of trivalent cobalt, but the cobaltous compound is not active.

The film of soya-bean oil is said not to wear so well as that of linseed oil for exterior use, but is reported to be equally good for interior work. The oil is not used alone but in conjunction with linseed oil.

Fish Oil.—The fish oil used most commonly in paints is that obtained from the menhaden. From May to November, this fish appears along the Atlantic coast, especially off the coast of New Jersey, in enormous quantities, and hundreds of thousands of tons are taken annually. A first quality of oil is obtained by boiling the fish with steam and allowing the oil to rise. A second quality is obtained by pressure. The fish residue is used as fertilizer. The chief objection to the oil is its odor, but this is largely removable by certain processes.

Its specific gravity is about 0.931; saponification value, 190; and iodine value, 150 to 165. It is not suitable for paints in the raw state, but when treated with the tungate and resinate driers becomes a fair drying oil. It is a very good oil to withstand combined heat and exposure to the weather. It does not blister and peel under these conditions as readily as other paint oils, and is, therefore, especially suitable for smoke stacks and boiler fronts.[1] Mixed with linseed oil it is used for waterproof paints. It has been suggested that its content of stearin may account for its waterproofing qualities.

Rosin Oil.—Although this is not a fatty oil and does not possess to any extent the properties that a good paint oil should have, it is nevertheless frequently introduced into paint oils as an adulterant, and on this account mention is made of it here.

When rosin is subjected to destructive distillation, various decomposition products are obtained. Chief among these is rosin oil, there being in the distillate about 85 per cent. of this. The crude oil is dark and viscous, with a marked characteristic odor, and because rosin is carried over mechanically to a small extent during distillation, it is acid in character. The oil is refined and various grades are produced with colors varying from dark red to pale yellow. The refined oil consists almost entirely

[1] Toch: Chemistry and Technology of Mixed Paints, p. 95.

of hydrocarbons and is unsaponifiable. It has a bluish bloom or fluorescence, which is less pronounced in the more highly refined oils. It dissolves readily in ether, chloroform, turpentine and petroleum spirit. It has some drying properties due to oxygen absorption. It is among the cheapest of all the oils and is consequently much used to adulterate other oils. A mixture of linseed and rosin oils is known among painters as "sipes" oil.

DRIERS

Action of Driers.—The drying of the paint and varnish oils is dependent upon their absorption of and combination with oxygen, and most of the paint oils when exposed in a thin film will eventually dry thus with the formation of a hard elastic film, but the time required for this natural drying is generally so long that dust and dirt from the atmosphere collect and are held by the undried film, thus rendering it very unsightly. To lessen the time of drying, certain metallic compounds are introduced into the oil. The method whereby these substances hasten the drying process is not definitely established, still there is but little doubt that their effect is largely catalytic; that is, they act as a medium whereby oxygen from the air is caused to be united with those glycerides in the oil that have the unsaturated bonds, without themselves permanently entering into the product of the reaction. Being catalytic, they are able to work continuously, and on this account only a small quantity of drier is necessary, generally less than 1 per cent. being sufficient.

Although these driers shorten the period of drying, undesirable results may develop from their use. Their action does not cease with the conversion of the oil into a dry film, but goes on indefinitely, so that if enough has been used, the film is finally oxidized to destruction. If a paint or varnish could be allowed to dry without the use of a drier its life would likely be longer.

Materials Used.—The driers consist of various compounds of lead, manganese and sometimes of zinc. The efficiency of the drier seems to depend more upon the amount of *metallic element* present than upon the compound in which it exists. If the different compounds of each metal are used in such quantities that the same amount of the metallic element is introduced into the oil, the drying effects will be about equal.

Classes.—The driers may be divided into two general classes: oil driers and japan driers, the former being used in the dry and the latter in the liquid state.

Oil driers are chiefly the *inorganic compounds* of lead and manganese, and are so named because they are used only with linseed oil. They may be mixed with the cold oil but are not readily assimilated unless the oil is heated as in "boiling."

The following metallic compounds are used as oil driers:

Lead Compounds.—*Litharge* is the monoxide of lead. It is yellowish-orange in color and may occur either in the powdered or crystalline form. The crystalline form is obtained by fusing the powdered litharge and allowing it to cool.

Red lead or minium, Pb_3O_4, is made by heating litharge to just below its melting point in a current of air. Red lead causes a brittleness in paint films, while litharge makes them elastic.

Lead borate, PbB_4O_7, and *lead acetate*, $Pb(C_2H_3O_2)_2 \cdot 3H_2O$, are both white. They are more easily introduced into the oil and darken it to a less extent than the oxides.

Manganese Compounds.—*Manganese dioxide or peroxide*, MnO_2, sometimes called manganese black, has a very vigorous drying action but being black it darkens the oil and this is objectionable. Other manganese salts that are used are the sulfate, acetate and borate, all of which have a pinkish color and, therefore, produce less darkening effect than the black oxide.

Japan driers are liquid driers, consisting of the *organic compounds*, chiefly the resinates and linoleates of lead and manganese, dissolved in benzine or turpentine. Their chief advantage is that they dissolve readily in these solvents and the solution in turn readily mixes with the oils at ordinary temperatures. Also, their drying action is rapid.

Two general methods may be employed for the preparation of both the resinates and linoleates. The metallic oxides may be dissolved in fused or melted rosin, thus producing the resinates. In a similar manner the oxides may be caused to combine with very hot linseed oil, thus producing the corresponding linoleates. Or the soluble sodium resinate or linoleate may be first formed by saponifying rosin or linseed oil in the ordinary way. Then, if to a water solution of the resinate or linoleate, solutions of lead or manganese salts be added, the corresponding lead or manganese resinate or linoleate will precipitate out. This is collected, washed, and dried.

By dissolving these resinates or linoleates, made by either method, in turpentine or benzine the japan driers are produced.

Comparison of Lead and Manganese Driers.—Manganese compounds quickly start the drying action and cause rapid surface drying. Lead compounds cause the drying to proceed more generally through the film and should ordinarily be used in larger quantities than the manganese compounds. The percentages[1] found to be most efficient are 0.5 per cent. of lead and 0.05 per cent. of manganese (that is, sufficient of the drier to furnish this quantity of metallic content), when used alone in linseed oil. Or still better results are obtained by a combination of the two, in which case 0.5 per cent. of lead and 0.02 per cent. of manganese are employed.

SOLVENTS AND DILUENTS

Under this head are included those light, volatile liquids capable of dissolving oils, resins, etc., and which are on this account used largely to thin or "reduce" paints and varnishes. Although several of the liquids that will be mentioned have but little connection with paints, except perhaps as paint removers, they are important in connection with varnishes and stains which are discussed later, hence to avoid returning to this topic they all will be described here. These liquids are all very volatile; that is, evaporate easily upon exposure, and their vapors are, with the exception of chloroform and carbon tetrachloride, very inflammable and explosive when mixed with air. Hence they should be kept in well-stoppered containers, and away from all lights and fires. They are all recognizable by their characteristic odors. Generally they do not mix with water, but alcohol and acetone are exceptions.

Turpentine is obtained from a balsam exuded from certain varieties of pine found chiefly in the Carolinas, Georgia, and Florida. The chief producer is the long-leaf pine (Pinus Palustris) known also as the Southern yellow or hard pine. This balsam or oleo-resin is an aromatic fluid consisting of a mixture

---

[1] Theory of Driers and Their Application, *Bull.* 23, Scientific Section of the Paint Manufacturers Association of the United States, 1910.

of a resin with a volatile liquid, and is commonly called "crude turpentine." In preparing the commercial product the usual practice is to distill the balsam in copper stills with open steam. Sometimes it is distilled with direct fire, but much care is needed since overheating injures the residue. Recently distillation has been carried out in electrically heated retorts in which the heat can be readily controlled. The distillate is the *spirit or oil of turpentine*. In steam distillation it will be mixed with water, which separates on standing. The residue left behind in the retort is *rosin*. The balsam yields from 10 to 25 per cent. of turpentine, which seems to be a product resulting from a reac-

Fig. 56.—Turpentine forest. Showing method of scoring trunks to secure yield of turpentine.

tion occurring during distillation, since if the rosin is dissolved in it, the original balsam is not obtained.

Turpentine is a water-white volatile liquid, consisting chiefly of pinene, $C_{10}H_{16}$. It is used in paints and varnishes to dilute and increase their flowing qualities. It is exceptionally suitable for this purpose, because, although it is volatile, it does not evaporate rapidly, and so keeps the freshly applied film fluid for a sufficient length of time that the marks of the brush, which are unavoidable, will flow out and disappear. Many of the turpentine substitutes, for example, certain of the light petroleum products, that is gasolines and naphthas, are too highly volatile, allowing the film to lose its fluidity before the irregularities have been eliminated. However, petroleum products can be prepared that very closely resemble turpentine in this important property.

Upon continued exposure to the air turpentine undergoes oxidation with the formation of an aldehyde, $C_{10}H_{16}O_3$. The sharp odor of old turpentine is due to this. By continued oxidation it turns yellow, thickens and forms a semi-solid, sticky, resinous mass. By this reaction with oxygen, small amounts of ozone and hydrogen peroxide are produced. (Note the bleaching of the cork stopper of a bottle in which the turpentine has stood for some time.) Turpentine that has thus thickened is no longer desirable as a paint and varnish solvent or diluent. All

Fig. 57.—Turpentine still.

turpentines are not equally susceptible to oxidation, some remaining clear and thinly fluid for a long time, while others thicken within a few weeks. The change is more liable to occur when the turpentine is kept in containers that are only partly filled, or when they are loosely stoppered or opened frequently.

Turpentine is quite expensive. It is consequently liable to adulteration with mineral-oil products, usually benzine and kerosene, which cost considerably less than half as much. Adulteration may be detected by a determination of the following properties:

Specific Gravity.—The specific gravity of pure turpentine lies between 0.860 and 0.870 at 60°F.

Boiling Point.—Pure turpentine begins to boil at 150°C., and

20

85 per cent. should distill between 155 and 163°C., while the remainder should distill below 183°C.

Polymerization.[1]—Turpentine is converted into a thick brownish-colored liquid by the action of sulphuric acid. The mineral-oil products, since they are practically unaffected by sulfuric acid, will float as a separate layer on top of the polymerized liquid.

Spot Test.—When turpentine is dropped upon a piece of absorbent paper, for example filter paper, it should evaporate completely without leaving a stain and without affecting the water-absorbing property of the paper.

Flash Point.[2]—This should not be under 40°C.

Wood Turpentine.—In contradistinction to wood turpentine, that which is obtained by distillation of the balsam is often called "gum spirits." Wood turpentine is obtained by distillation of the wood, such as stumps, knots, and old trunks of turpentine-producing trees. Its composition differs somewhat from the "gum spirits," but as far as its physical effects in paints and varnishes are concerned it is as suitable as the turpentine obtained from the balsam. However, it contains 2 or 3 per cent. of impurities, such as pyroligneous and formic acids, creosote and pyridine bases, and unless carefully refined to remove these, it has a disagreeable odor that makes it somewhat undesirable for interior use. The wood turpentine now on the market is generally quite free from these impurities.

Turpentine Substitute, "Painter's Spirits."—This is a mineral-oil product having a density of about 62 to 52°Bé. For the preparation and general properties of the light petroleum products, see pages 38 to 40.

There is much difference of opinion concerning the relative merits of turpentine and the petroleum products in paints and varnishes. The objection held against the petroleum products is that they are too light and volatile—that they allow the film to set before it has acquired a smooth surface. This is without doubt true when gasoline or any of the volatile liquids lighter than benzine are used, but the objection does not hold with grades that are sufficiently low in volatility. Toch says[3] that

[1] See footnote, p. 296.
[2] See p. 351.
[3] T“ch: Chemistry and Technology of Mixed Paints, p. 111.

a substitute equal in solvent power, speed of evaporation and viscosity to turpentine can be made by passing certain paraffine oils with wood turpentine over red-hot coke. And as has been said before these substitution products cost much less than pure turpentine.

Ethyl Alcohol (*Grain Alcohol, Spirits of Wine*), C_2H_5OH.— This is produced by the yeast fermentation of the sugars and starches of fruits and grains. Also, it is obtained from various waste products. For example, sawdust is treated with sulfurous acid and water under pressure so that its cellulose is converted into dextrin and glucose, which substances are then fermented in the usual way.

Ethyl alcohol is the alcohol found in intoxicating beverages, and when pure is a colorless, almost odorless, mobile liquid, boiling at 78.4°C. (173.1°F.). It freezes at −130.5°C. It has a great affinity for water, absorbing it when exposed to the air. It mixes with water and ether in all proportions, but not with gasoline. By ordinary distillation all but about 4 per cent. of the water can be removed, the distilled product being known as 96 per cent. alcohol. By reflux distillation over quicklime the water can be removed entirely, the anhydrous distillate being known as *absolute alcohol*, which is a great deal more costly. Besides being stated in per cent. by volume, as above, the strength of alcohol may also be designated in United States proof degrees. One such degree corresponds to 0.5 per cent. by volume. For example 180° proof is equal to 90 per cent. alcohol by volume.

When sufficiently free from water it dissolves gums, resins and similar substances very readily; also, it burns with a very hot, pale blue flame.

The cost of production of the 94 per cent. product is 35 cts. per gallon, which is about half the cost of production of wood alcohol. The United States Government, however, imposes an internal revenue tax on all alcohol which amounts to $2.08 on the 94 per cent. product. This makes it too expensive for a great many uses to which it would otherwise be put. In 1906 Congress removed the tax on denatured alcohol, which can now be obtained for about 70 cts. a gallon.

Denatured alcohol is that which has been rendered unfit for use as a beverage, or as an ingredient of medicines, by mixing with it suitable denaturing materials. According to the regulations issued by the Commissioner of the United States Internal Revenue, the following substances are allowed as denaturants: Wood alcohol, benzine, pyridin bases, camphor, castor oil, caustic soda, nicotin, ether and acetone. The indiscriminate use of all of these materials is not allowed. Indeed, most of them are to be used in manufacturing "specially denatured alcohol," the sale and use of which is permitted only under special restrictions. For the manufacture of "completely denatured alcohol," the variety which may be bought freely by the public, only wood alcohol, benzine and pyridin may be used. However, on account of the relatively high price of the pyridin, the following formula is now altogether employed in making completely denatured alcohol in the United States:

To every 100 *parts of ethyl alcohol of the desired proof (not less than* 180°), *there shall be added* 10 *parts by volume of approved methyl alcohol and one-half of* 1 *part of approved benzine.*

In this manner the alcohol is rendered unfit for liquors but its general usefulness in the industrial arts is unimpaired.

Methyl Alcohol (*Wood Alcohol, Wood Spirit*), CH_3OH.— This is a by-product of the destructive distillation of wood in the manufacture of charcoal, forming about 1 per cent. of the watery distillate obtained. It is a clear, colorless liquid boiling at 66°C. (151°F.). When 99 per cent. pure, its specific gravity at 60°F. is 0.804 (44°Bé.). It mixes with ethyl alcohol, water and ether in all proportions. It is not ordinarily found on the market in the pure state, but contains water, acetone, fusel oil, etc., to which impurities its characteristic odor is due. Pure wood alcohol has scarcely any odor. Administered internally it is poisonous, and this property is greatly increased by its content of fusel oil. Its solvent properties are about equal to those of ethyl alcohol.

Ether, $(C_2H_5)_2O$, is produced by the action of sulphuric acid on ethyl alcohol, and hence is often known as sulfuric ether, although this name is inappropriate. It is a very volatile, colorless liquid. If its vapor is inhaled for some time it produces unconsciousness. The vapors are very heavy, and will come from an unstoppered bottle and spread out over a table or floor to great distances, and since they form a violently explosive mixture with air, ether must never be exposed in the neighborhood of a light or fire. Its boiling point is 34.5°C. (94°F.); its specific gravity at 60°F. is 0.7195 (64°Bc.). It mixes very slightly with water, by far the greater quantity floating as a separate layer· on the surface. With alcohol, chloroform, benzole, petroleum naphtha, petroleum oils, fixed and volatile oils, it mixes in all proportions. The commercial ether contains a considerable quantity of alcohol which carries a little water. It is a very good general solvent for many organic substances, such as oils, resins, fats, pitches, tar and waxes.

Chloroform, $CHCl_3$, is a clear, colorless liquid with a peculiar, sweetish odor. It is made by distilling dilute alcohol with bleaching powder. It is not inflammable, and is scarcely at all soluble in water, but sinks to the bottom, forming a well-defined layer. Its specific gravity at 60°F. is 1.515 (49°Bé.), and its boiling point is 61°C. (142°F.). It is an excellent organic solvent.

Carbon tetrachloride, CCl_4, is a colorless liquid boiling at 77°C. It is made by treating carbon disulfide with chlorine. It is very volatile and its vapors are not inflammable, nor are they able to support combustion. On this account it is much used as a fire extinguisher, and is sold under the trade names "pyrene" and "carbona." It is a very good general solvent.

Amyl Alcohol (*Fusel Oil*), $C_5H_{11}OH$.—This alcohol is otherwise known as pentyl alcohol, being fifth in the series in which each alcohol contains a CH_2 more than the preceding, as methyl, CH_3OH; ethyl, C_2H_5OH; propyl, C_3H_7OH; butyl, C_4H_9OH; and pentyl, $C_5H_{11}OH$.

The word amyl means starch, and this alcohol is obtained together with the ethyl alcohol from the fermentation of grains, and· is separated from the ethyl when the alcohol is rectified. It is a liquid almost insoluble in water, is moderately volatile, has a disagreeable odor, and is quite poisonous. It has an oily appearance and boils at about 130°C. At 60°F. its specific gravity is 0.819 (41°Bé.). *Fusel oil* is almost entirely amyl alcohol. Because it is found in grain alcohol, it is known also as grain oil. The word *fusel* is taken from the German and means "spirits of inferior quality." Fusel oil is also found in wood alcohol in considerable amount, and it is to the fusel oil content that wood alcohol owes, in part, its poisonous property.

Amyl Acetate (*Banana Oil*), $C_5H_{11}(C_2H_3O_2)$.—This is obtained by treating amyl alcohol with glacial acetic acid in the presence of sulfuric acid. It is a thin, water-white, volatile liquid, boiling at 150°C., and has a very pleasant odor resembling that of bananas, hence one of its names. It is much used as a solvent in making collodion lacquers and many varnishes. The thick, viscous, dark-colored liquid which is often referred to as banana oil, but which should be called "bronzing liquid," is in reality a collodion or pyroxyline lacquer. It is made by dissolving nitrated cotton, $C_{12}H_{16}O_6(NO_3)_4$, in the amyl acetate in quantity sufficient to increase the viscosity to such a degree that it will be able to carry the metallic powders, such as aluminum bronze or flake aluminum, thus making the so-called gilts or bronzes so much used in coating radiators and steam pipes.[1]

Acetone, $(CH_3)_2CO$, is obtained from the watery distillate produced in the manufacture of charcoal. It is a clear, colorless, extremely volatile liquid, boiling at 56°C. (133°F.). Its specific gravity at 20°C. is 0.792 (47°Bé.). It mixes in every proportion with water, alcohol, ether and chloroform. The presence of water in it may be determined by shaking a sample with dry crystals of calcium chloride. If the crystals deliquesce, water is present in the acetone.

Benzole (*Benzene*), C_6H_6. This is obtained by distillation from light coal-tar oils. It is the basis for the formation of anilin and coal-tar dyes, by the substitution of certain groups for some or all of its six hydrogen atoms. It is only very slightly soluble in water. It boils at 80.5° and solidifies at 5.4°C. At 15.5°C. its specific gravity is 0.8841 (28.5°Bé.). Because of its penetrative powers it is much used in paint and varnish removers.

PAINTS

As was pointed out in the beginning of this chapter, the characteristics of the pigments and oils determine the properties of the paints made from them. Now, since the properties of the pigments and oils have been to some extent explained, it will be possible to discuss the properties of paints, both with reference to the liquid paint used and the film after it has dried. But,

[1] See p. 329.

before proceeding with the discussion, it may be advisable to review briefly the characteristics that a good paint is required to possess.

General Requirements of a Good Paint.—The paint should possess the proper *spreading quality.* This varies in direct proportion to the fineness of the pigment used. If 1 gal. covers 300 to 400 sq. ft. on a fair surface, it is a good average. The paint should oppose a certain amount of resistance to spreading and not flow excessively so that "slicking" results. However, at the same time it must not be difficult to spread and it must flow sufficiently so that brush marks are not retained. Fine pigments such as zinc oxide and lithopone have a tendency toward excessive flow, while the coarser pigments as white lead, barytes and asbestine, tend to resist spreading. The paint should have sufficient *opacity* so that an excessive amount is not required to hide the underlying surface. Basic carbonate white lead has the greatest hiding power, volume for volume, while zinc oxide excels weight for weight. The paint should have sufficient *penetrative power* that it can enter into the pores of the wood to aid it in sticking, otherwise it may peel. This happens frequently on very hard woods. Penetration may be increased by the use of thinners. When the paint is stirred up it must *not settle* rapidly, so that a surface may receive a uniform application of the pigment in all parts. Also, while standing in the warehouse or on the dealer's shelves the pigment should not settle to a hard compact mass in the bottom of the package. White lead has a pronounced tendency to settle. When exposed in the paint film, the pigments should be as *chemically inert or stable* as possible. They should have no tendency to bring about the destruction of the linoxyn film or to react with the atmospheric gases. There should be the least possible tendency to *check, crack or peel* because of a lack of elasticity of the film.

Exclusion of Voids in a Paint Film.—A coating of pure linoxyn is very soft and does not possess moisture-excluding properties. On this account mineral substances, that is pigments, are used to fill up the pores, harden the surface and strengthen the film. If a pigment consisting of large particles only were used, they would be sufficiently hard and strong in themselves but they would not be well joined together. Also, the interlying surface of linoxyn, known as voids, would still be soft and easily destroyed. Now if the spaces between the large particles are filled with a pigment consisting of extremely fine particles only, the surface of linoxyn exposed would, of course, be reduced, but owing to the numerous small joints the strength would not be great and the moisture could still find its way through the joints. But if there had secondly been used, particles of a medium size, strong in themselves, to fill in between the largest particles, and then thirdly, the very fine particles to fill the remaining

Fig. 58.—Showing the condensing effect of milling a paint pigment with oil. The filled barrel of pigment on the right with oil forms one-third barrel of paste as shown on the left. (Heath.)

spaces, maximum strength with maximum excluding power and minimum surface of exposed linoxyn would be secured.

Then, by the use of at least three pigments: (1) with large particles, like basic carbonate white lead, or barytes; (2) a pigment with medium-sized particles, like calcium carbonate or blanc fixe, including here the paint-coat strengtheners as asbestine and China clay; (3) a pigment with very fine particles like zinc oxide, or sublimed white lead, less linoxyn and more pigment, which is the indestructible part of the paint, will be exposed, and the most compact paint film will be secured, just as by the use of crushed rock, gravel and sand the best concrete is produced.

The fact that two pigments are better than one, and three better than two for the purpose of excluding moisture and gas has been shown by test. The decrease in porosity is in direct proportion to the decrease in voids.

On three test fences, one at Atlantic City, one at Pittsburgh, and one in North Dakota, the above statements were borne out.[1]

Frequently certain percentages of varnishes are added to paints. According to tests it appears that this is not good practice when painting on wood. The coat is thus made non-porous to too great a degree. There is always moisture in wood, even in kiln-dried lumber, and this with certain gases arising from decomposition in the wood must find a way out. If the paint coat is too impervious, blistering results. However, for painting on iron the addition of a good Kauri varnish is beneficial.

Single Pigment and Composite Paints.[2]—Each of the pigments used in paints has, as a rule, certain good properties for which it is noted. But, with practically no exception, each has certain defects as well that must not be overlooked. Single-pigment paints have been much used, especially that containing the basic carbonate white lead, and although this confers upon the paint many qualities that a good paint should have, it gives to it certain serious defects also. In a like manner the other pigments are desirable or not desirable, depending upon the standpoint from which they are considered. But by a proper selection of pigments a paint can be made in which each pigment will tend to offset the defects of the other. This is illustrated in the following discussion in which the basic carbonate white lead and zinc oxide are considered.

If a paint were made up using the basic carbonate white lead alone, it would be very opaque, would be quite compact and free from voids and when it failed, as all paints do in time, it would possess a uniformly, finely roughened surface, characteristic of chalking, that would be easily repainted. But, on the other hand,

---

[1] These tests were conducted by the Scientific Section of the Paint Manufacturers Association of the United States, and from their reports the above discussion has been derived.

[2] From *bulletins* of the Scientific Section of the Paint Manufacturers Association.

this chalking begins at a rather early period and because of the peculiarly roughened surface that results, the paint coating holds very tenaciously any dust that settles on it so that it soon is caused to appear unsightly. Also white lead is blackened in the presence of the rather prevalent atmospheric hydrogen sulfide, because of the formation of the black lead sulfide. Further, a basic carbonate white lead paint lacks spreading power, affects delicate colors with which it may be tinted, and settles badly.

A paint containing zinc oxide only as a pigment would be very white and glossy, would not retain dust particles and darken on this account, nor would it darken because of the action of hydrogen sulfide, and it would have great opacity and spreading qualities. But it dries to a very hard, inelastic film and cracks and peels badly. On this account it would be impossible to produce a uniformly smooth surface in repainting unless the old paint were removed.

Now, in a good composite white paint a certain amount of basic carbonate white lead must be used to contribute opacity, elasticity, and to counterbalance the spreading quality of zinc oxide and to offset its tendency to peel. A considerable amount of zinc oxide is necessary to contribute spreading power which the white lead lacks, to increase the ease of application, to increase the resistance to the darkening influence of hydrogen sulfide, and above all to offset the chalking tendency of the white lead. Also the fine particles of the zinc oxide help to fill up the voids between the larger pigment particles of the white lead and thus decrease the spaces that would be otherwise filled only with oil. On this account, the surface of the paint will have greater wearing qualities. In addition, a small percentage of asbestine or China clay may be advantageously used to prevent settling and strengthen the coat.

The two pigment formulas following gave very good results on a 2-year test in Pittsburgh:

	Per cent.		Per cent.
Basic carbonate white lead....	22	Basic sulfate white lead	60
Zinc oxide.................	50	Zinc oxide..................	27
Whiting....................	2	Asbestine....................	10
China clay and asbestine.......	26	Whiting....................	3

In inspecting these, there was taken into consideration chalking, checking, color, and general condition, partly in respect to repainting.

Paints for Metals.—Concerning paint coatings for metals, practically all that has been said previously holds true. But because no moisture can possibly be in the metal to afterward escape and cause blistering as with wood, the paint is improved by the addition of some fossil resin varnish of good quality to further fill up the pores.

Any metal surface that is to be painted must be *well cleaned*. It should not even be oily, although it is to be painted with a linseed-oil paint. The paint does not mix with the oil layer and so does not adhere well. Galvanized iron is usually oily when new, and in addition is likely to have zinc salts on its surface. Zinc chloride is produced by the action of the ammonium chloride used as a flux in hot galvanizing. And small amounts of zinc salts may be carried by coatings of zinc deposited by the electrolytic method. Zinc chloride especially is very hygroscopic, and by absorbing water underneath the paint it causes peeling. Galvanized iron may be cleaned by washing it well with soapy water, rinsing it thoroughly, and allowing it to dry. Mill scale and rust may be removed from iron by the sand blast, scratch brush, file or by pickling (see page 195).

Corrosive and Anticorrosive Pigments.—As was explained on page 270, the pigments are divisible into groups in accordance with the manner in which they affect the corrosion rate of iron. Those which retard corrosion are called inhibitors; those which hasten corrosion are called stimulators, or accelerators; and those which apparently do neither are called indeterminates. Zinc chromate is a very good inhibitive pigment. No doubt it is active in causing the iron to assume the passive state as explained under "Iron Corrosion." Where indeterminates or other inhibitive pigments are used with it, only a very few per cent. of zinc chromate are required to make a good inhibitive paint. The vehicle itself may also be made inhibitive by introducing into the oil from 5 to 10 per cent. of chromium resinate or linoleate (see metallic resinates under "Driers"). Such a paint would be very good on tinned iron roofs and elsewhere. All of the chrome pigments are inhibitive, as are also the basic carbonate white lead (corroded lead) and the basic sulfate white lead.

The red leads are inhibitive and are much used for first coats on iron and steel where they serve well because they dry quickly to a hard layer that adheres closely and holds well subsequent layers of paint.

Stimulative pigments should never be used in paints for metals. Venetian red because of its calcium sulfate content, which dissociates in water, thus furnishing sulfate ions, is not good for iron. The values of the various pigments for painting on iron are stated in the discussions of the various pigments, which see.

Cement Paints.[1]—Paints prepared for cement and concrete structures usually contain basic carbonate white lead, sublimed white lead, zinc oxide or lithopone as a base pigment combined with a considerable percentage of some inert pigment as barytes, asbestine, silica, China clay, etc., ground in a highly waterproof vehicle prepared from treated Chinese wood oil. In applying paint to fresh cement or concrete surfaces, the alkalinity due to the calcium hydroxide of the cement must be neutralized by washing the surface to be painted with a solution of zinc sulfate.

THE VARNISHES

Varnishes are resinous solutions applied to surfaces to produce hard, lustrous, and generally transparent, protective coatings. The specific solvents or vehicles used, as well as the resins dissolved in them, are quite numerous and varied, but broadly speaking varnishes are classified under two heads according to the general character of the solvent employed, as spirit and oil varnishes. In the spirit varnishes the solvent serves merely to facilitate the spreading, since it practically all evaporates, leaving a thin layer of resin only on the varnished surface. In the oil varnishes, a drying oil is used as the vehicle, and this contributes to the elasticity and durability of the dried varnish film. But in all cases the quality of the varnish depends in a large measure on the properties of the resin employed, hence before proceeding further with the discussion of varnishes, the characteristics of at least the more common resins will be considered.

THE VARNISH RESINS

Source.—The resins are peculiar products of vegetable origin that are secreted in special reservoirs or cells of trees and come to

[1] ROGERS and AUBERT: Industrial Chemistry, p. 342.

the surface either spontaneously or because of incisions. Although their chemical and physical properties vary a great deal, probably all of them are products resulting from the oxidation of the hydrocarbons of the essential, that is, volatile oils.

Gums.—Because of loose usage of the term, the word "gum" has grown to signify in popular language the same as resin. But strictly, a gum is a substance soluble in water but not in alcohol or similar solvents. It belongs to the class of carbohydrates, its typical formula being $(C_6H_{10}O_5)_n$. Examples of true gums are gum arabic and gum tragacanth.

Resins.—True resins are never soluble in water, but are soluble in alcohol or other of the volatile liquids enumerated under "Solvents." However, some of them dissolve only with difficulty in these liquids and some may not be soluble at all unless fused. Resins are not definite chemical compounds with a fixed composition. Different specimens of the same resin may vary considerably in chemical and physical properties. Some resins are neutral substances but many have weak acid characteristics, forming soaps when treated with alkalis or alkaline salts. Some are not susceptible to the action of atmospheric oxygen, while others are considerably affected. They are divisible into two groups: (1) soft or recent resins; and (2) hard or fossil resins.

The *recent resins* are products of trees now living. Examples are dammar, shellac, and rosin, as well as certain of the copals. The *fossil resins* are products of trees growing in other ages and now extinct. The resin being practically non-perishable, is now dug from the earth. In these resins the essential oil has been eliminated and partial oxidation has taken place. They are very hard and many of them are not soluble in the organic solvents unless they are first roasted or fused. During this process, certain molecular changes take place, which in part amount to decomposition with the formation of a by-product of oil. Examples of fossil resins are amber, a great number of the copals and kauri resin. Resins having properties midway between the two preceding groups are known as *semi-fossil resins*.

Balsams.—Belonging to this general class of resinous secretions are a variety of substances known as oleo-resins or balsams. They contain a greater amount of volatile oil than the recent resins, and have usually a syrup-like consistency, although a few are solid or become so on keeping. Examples are Venice turpentine and Canada balsam.

Artificial Resins.—These are synthetic laboratory products, which in some respects resemble the true resins. A very noteworthy example is bakelite.

RECENT RESINS

The recent resins are quite numerous, but the properties of only the most important ones will be discussed here.

Lac.[1]—This resin from which *shellac* is prepared, unlike all the other resins we shall consider, is not a direct secretion produced by a tree or plant, but is a resinous excretion of a parasitic insect that infests various trees, notably certain varieties of fig trees[2] native to northeastern India. Lac is then the digested sap of this tree.

The lac insect is Tachardia Lacca, of the family Coccidæ, and is somewhat similar to the cochineal insect, which belongs to the same family. The word *lac* is the East Indian term for *a hundred thousand* and is without doubt applied to this resin to indicate that such countless numbers of insects are connected with its production.

The young insects appear at two seasons, in June and December, these being known as the swarming periods. When hatched they are about $\frac{1}{40}$ in. long and are of an orange-red color. They migrate from the parent cells in which the eggs are laid, and take up a position on the tender twigs of most recent growth, and in such numbers that the twigs are practically covered. The young insect inserts its proboscis, sucks up the sap and excretes a substance that dries about it in the form of an incrustation which serves as a protective covering. The life of the female insect continues for about 6 months, while that of the male insect is shorter, being completed in about $2\frac{1}{2}$ months. Since the insects are so thickly clustered on the twigs, the incrustations they produce grow together and form a layer that eventually covers the twig entirely.

[1] For a more complete discussion see LIVACHE and MacINTOSH: Manufacture of Varnishes, vol. 3, 285. Also *Wood Craft*, June, 1911.

[2] HEMMING: Moulded Electrical Insulation and Plastics.

Preparation of Shellac.—The twigs containing the resinous incrustation are cut or broken off and these constitute the *stick lac* of commerce. However, the lac is usually not marketed in this condition, but is further refined. Generally the resinous incrustation is separated from the twigs by rolling or crushing in some manner and the resin is then agitated with warm water and the aqueous extract thus obtained is evaporated to dryness. The residue constitutes the *lac dye*. This formerly was of much importance but it is now scarcely more than a commercial curiosity, since it has been replaced by synthetic dyes of coal-tar origin. However, it is necessary to remove the dye as much as possible in order that the resin may be pale. After extraction the resin is taken up from the bottom of the vat and dried by exposure to the atmosphere. In this condition it is known as *seed lac*, being in the form of granules about like grains of wheat.

At this time orpiment and rosin are admixed. Orpiment is arsenious sulfide, As_2S_3, a substance possessing a fine yellow color. The reason given to justify the addition of orpiment is that it confers elasticity, acting in a manner similar to the metallic sulfides in the vulcanization of rubber. Also, it serves to impart to the shellac a fine pale straw color. The original object in the addition of rosin was that the melting point might be lowered, and the lac thus rendered meltable without decomposition. An addition of about 3 per cent. is required for this purpose, and although such an addition is justifiable and probably even necessary, so that little lac is exported that is entirely free from it, advantage has been taken of this fact and shellac is not infrequently found on the market containing as much as 10 to 20 per cent. of rosin or even more. Such quantities, of course, must be classed as adulterations.

When the seed lac is dry, it is melted and strained through fine-mesh cloth. The drippings are caught in warm water or on heated flags. While still warm, the mass is taken up and stretched into thin sheets, which when broken into small pieces furnish the *shellac* in its familiar form. The name applied is without doubt intended to indicate the shell-like or scaly appearance of the lac. It is prepared in this form to increase the surface in proportion to the mass and thus render it more readily soluble. The lac as it is strained may also be poured into a mould, producing that

which is known as *button lac*. Further, the residue left after
straining may be extracted with some solvent, producing that
which is known as *garnet lac* because of its dark red color.

Properties of Shellac.—The melting point of shellac is somewhat indefinite. It begins to soften at about 40°C., emitting a
characteristic odor, but it cannot be melted to a thin fluid without decomposition. It is soluble in both ethyl and methyl alcohol. However, it contains about 6 per cent. of a certain wax
that is not soluble in these solvents, so that the alcoholic solution is rather turbid. On the other hand, this wax is soluble in
petroleum naphtha, and the resin proper is insoluble in this
solvent. The resin is only partly soluble in turpentine, ether
and chloroform.

Also, it is completely soluble in the alkalis and in solutions of
those salts that ionize to produce alkaline solutions, as borax
and sodium and potassium carbonate. Advantage of this fact
may be taken to free the resin from the wax. To do this the
shellac is boiled in about a 2.5 per cent. solution of sodium carbonate. Upon cooling the wax rises to the surface and may be
skimmed off. The solution is then filtered and acidified, whereupon the resin is precipitated. This wax-free resin when washed
and dried may be dissolved in alcohol with the formation of a
clear solution.

Bleached Shellac.—Shellac may be whitened by bleaching.
This is generally accomplished by treating the shellac dissolved
in alkali with a solution of sodium hypochlorite, $NaClO$, from
which chlorine is liberated. The shellac is thrown out of solution by neutralizing the alkali in which it is dissolved, by the use
of dilute sulfuric acid. The precipitated lac is then repeatedly
washed in water. Or it may be melted under water and kneaded
and pulled, under water, to further increase its whiteness.

Bleached shellac has a marked tendency to polymerize or
undergo other molecular transformation on standing so that it becomes less soluble in the ordinary solvents, that is the alcohols.
The longer it is kept, the less soluble it becomes. The change is
hastened by high temperatures but it cannot be prevented even
under the best of conditions. Consequently, bleached shellac
should be promptly used. However, it is possible to dissolve
shellac that has been thus transformed by standing. If it is

allowed to stand about 24 hrs. in ether, it absorbs ether and then is much more readily soluble in alcohol.

Uses.—Shellac is much used in the making of spirit varnishes. It is the best resin for this purpose, since it has a higher degree of combined hardness and elasticity than any other spirit-soluble resin. It is also used as an electrical insulator and as a binder in many forms of composition goods.

Dammar.—The word dammar is the generic Malay term for all resins that exude from the tree and solidify upon exposure to the air. There are many varieties but the Singapore variety is perhaps the most commonly used.

Dammar resin is used for both spirit and oil varnishes, but it is not of exceptionally high quality for varnish making, especially for oil varnishes, since it is too soft and brittle. Nevertheless it is used in these varnishes because of its color, which is almost white. It is somewhat harder than common rosin but it can be scratched with the finger nail and the heat of the hand renders it sticky. It is soluble in benzol, petroleum spirit, turpentine, and in ethyl alcohol with the aid of heat, but it is only partially soluble in cold alcohol. It is used to a considerable extent for making spirit varnishes, but the varnish coat is rather friable so that if it is rubbed with the fingers it becomes powdery.

Rosin is the residue left in the still after the turpentine has been distilled from the balsam. The term rosin must not be confused with the generic term resin. Rosin is only a particular resin. Colophony is another name for it.

After the turpentine removal,[1] the rosin is freed from any moisture left by the steam distillation of the turpentine by a steady application of heat while yet in the still, and then is filtered under pressure through heavy cloth. When filtered, it is run into moulds or barrels and allowed to solidify. Rosin varies in color from extremely pale, through various shades of yellow and brown, to black. Its value increases as its depth of color diminishes. There are various factors that affect its color. If the temperature during its preparation is too high, it will be darkened due to decomposition products. Also, the balsam yielded the first season the tree produces, furnishes the paler grades of rosin, and the product darkens with each succeeding year. The grades of

[1] See p. 304.

21

rosin are indicated by the letters A, B, C, etc. to N. Following N are WG and WW, indicating "window glass" and "water white." Grade A is poorest and blackest, while WW is palest and best. The grade is determined by inspection of the barrels usually as they stand on the docks of the port of shipment, generally Savannah, Ga.

Properties.—Rosin is slightly heavier than water and melts at from 100° to 140°C. It is insoluble in water but is soluble in about eight times its weight of ethyl or methyl alcohol and is much more easily soluble in amyl alcohol, acetone, ether, chloroform, turpentine, benzol, and the fatty oils. It is very soluble in petroleum naphtha. It consists largely of abietic anhydride, $C_{44}H_{62}O_4$, which is readily converted into abietic acid. By boiling with alkalis rosin is changed into rosin soap, this entering into the composition of many laundry soaps. The rosin soap is known also as a resinate. The so-called "hardened rosin" is the product resulting from the treatment of rosin with lime or zinc oxide. Since rosin is soft and brittle, it is not of great value in the varnishes, although it is sometimes used.

Fossil Resins

In the manufacture of the oil varnishes, the fossil resins are used almost without exception. Some of the varieties enumerated in the following pages occur in both the recent and the fossil form, the fossil form being, of course, the more valuable.

Amber.—This is the hardest of all the known resins, and varnishes made from it are the hardest of the varnishes. It is chiefly the chippings and turnings left after the manufacture of amber ornaments that are used for making varnishes, since the resin is too costly to use to any great extent.

The Copals.—The term copal does not indicate a specific resin, but is used as a generic term, covering a wide range. In the varnish-making industry it is used to apply to all those resins that are used for making oil varnishes.

All of the copals are by no means equal or approximately equal in quality. According to their degree of hardness, they are named: (1) hard; (2) medium hard; (3) soft copals. Copals of the first class are the true copals, but those of the second and third classes are known commercially as copals. Other things being equal, the quality of the varnish is in direct proportion to the hardness of the resin used.

The *hard copals* comprise the excellent East African resins, which are known as the *animé resins*. These include the Zanzibar, Madagascar and Mozambique sorts. The most noted and valuable comes from the coast of Zanzibar, where the fossil is dug from the earth, being found at a depth of about 4 ft. The resin occurs in flat tabular pieces, varying from the size of a pea to that of a man's hand. The surface is peculiarly roughened by small papillæ, giving rise to the trade name of "goose skin." Like the other fossil resins the animis are not soluble without fusion. They are so hard that they cannot be scratched with the thumb nail. They are not affected by boiling with caustic soda solutions or mineral acids, which is a somewhat exceptional property for resins. The animis are probably the standard for excellence among resins. Their market value ranges from $1,000 to $1,700 per ton. The *medium-hard copals* comprise the West African sorts of which the Sierra Leone is the best. It is very hard, especially the smaller pieces known as the pebble form, which is almost equal in hardness to the Zanzibar animé. The other copals of the West African group are less known and less regularly used.

Of the *soft copals* the kauri from New Zealand and Australia, and the Manilla and the South American copals are the most important.

Kauri.—This resin is one of the most widely used of all the oil-varnish resins. It occurs in both the recent and fossil form, the fossil being, of course, the more highly prized. This is dug from the earth at depths varying from a few inches to several feet and in pieces ranging in size from that of a small pebble to others weighing 50 lbs. or more. The recent resin obtained from the living trees is known as "young" kauri; it is very pale, almost water white. The fossil varies from amber to dark brown in color, and resembles the true copals in many of its properties.

Kauri has only recently come into use as a varnish resin but it is very much in favor among varnish makers because no other oil-varnish resin is so easy to manipulate. It assimilates the linseed oil at lower temperatures and more quickly than other similar resins. Because lower temperatures are required, carbonization is less likely to occur, and so a paler varnish may be produced. It produces a hard and glossy varnish that is very durable for interior use, but it does not stand exposure to the

Fig. 59.—Digging kauri resin.

weather so well. In general, the varnish is considered not so good as that made from the animé and Sierra Leone copals.

Fig. 60.—Sorting resin on the field.

Manilla Copals.—This name includes all of the known Asiatic resins except dammar. They are not found in the Philippine

Islands but are called "Manilla" because that is the port of final shipment. Generally they are not extremely valuable, although they vary greatly in quality. Some of them, notably those from Borneo, are almost identical with the excellent kauri fossil resin, from which they grade to the soft resin from Singapore, which is almost identical with the dammar from that region. Usually these resins as a class are employed in making only ordinary grades of varnishes.

BALSAMS

The balsams are relatively unimportant in this connection, *Venice turpentine* being the only one that is much used, and the use of this is largely limited to those cases where it is desired to lessen the brittleness of other resins used in conjunction with it. Venice turpentine is obtained from the branches of the European larch. It comes on the market and is used in the form in which it is obtained from the tree. If has a clear yellow color and is thick and viscous at ordinary temperatures, but increases greatly in fluidity on warming. It contains 15 to 20 per cent. of an essential oil.

ARTIFICIAL RESINS

Bakelite is an artificial resin-like substance discovered by Dr. L. H. Bakeland, who presented the first general account of his work on this substance in 1909. Bakelite is prepared by boiling about molecular weight proportions of phenol (carbolic acid), C_6H_5OH, and formaldehyde, $CHOH$, in the presence of small amounts of ammonia or sodium hydroxide. A condensation reaction occurs and water is eliminated. The product exists in three stages, known as A, B, and C. A is the initial product and may exist in either the liquid, pasty or solid form. The solid form melts and sets again on cooling. It is soluble in caustic soda and in general resembles a true resin. Also it is soluble in alcohol, acetone and similar solvents, and its solutions may be used as a varnish or lacquer. By the proper application of heat, A may be converted into either B or C.

Bakelite B is somewhat similar to solid A, except that it is insoluble. It is softened by heat so that it may be moulded or welded.

Bakelite C is the final form and is made by the application of heat to either A or B. It is infusible and insoluble in the ordinary solvents. It withstands boiling water, oil, strong acids, and other chemicals, excepting caustic soda which most varieties of C do not withstand. Its color is usually transparent or yellow but it may be prepared in other colors.

A great point in favor of bakelite is that it may be used to impregnate or coat various substances in the form of A or may be moulded in the form of B and then be converted afterward into the hard, infusible, insoluble substance C by the application of heat.[1]

SPIRIT VARNISHES

Composition and Properties.—Spirit varnishes are made of the recent resins—shellac, dammar, sandarac and mastic—but shellac is most important. The varnishes are merely solutions of these resins in alcohol or some other volatile solvent.

The solvent only facilitates the spreading, none of it being left in the dried varnish layer. Hence, the dried varnish consists merely of a thin coating of the original resin, and is, therefore, very brittle and liable to crack and flake off. The coating is hard and very brilliant usually, but it must not be applied to any article that will receive rough handling. Also, since the resins used are dissolved and converted into soaps by the action of the alkaline substances such as washing soda, borax, etc., articles coated with spirit varnishes must not be washed with soaps or other cleansing preparations of similar nature.

Because the drying is rapid, spirit varnishes are somewhat difficult to apply, but because they dry rapidly, they are suitable for application to wood patterns. In addition they are good excluders, preventing alteration in the size of the pattern due to moisture absorption, and they furnish a hard smooth coat. Shellac varnish is especially suitable for this purpose.

The brittleness of the spirit-varnish coat may be reduced by mixing with the varnish 1 to 1½ per cent. of castor oil, or some balsam, as Venice turpentine, or better still a small quantity of

---

[1] For additional information on bakelite, see *Jour. Ind. Eng. Chem.*, March, 1909, August, 1909; and December, 1911. Also various pamphlets issued by the General Bakelite Co.

an oil varnish. This, however, lengthens its period of drying. Resins by themselves are very unyielding substances, so that any expansion or contraction of the article to which they are applied causes cracks to appear in a spirit-varnish coat.

A very clear, almost water-white varnish may be prepared by dissolving dammar in alcohol. *Colored varnishes* may be prepared by dissolving with the resin an aniline dye soluble in alcohol, or by the use of dragon's blood, gamboge or cochineal.

Spirit varnishes may be prepared on a small scale by putting the materials into a well-stoppered receptacle and agitating. This agitation may be secured by attaching the receptacle to a slowly revolving shaft. Free flame for heat must not be used on account of the great inflammability of the materials. The proportions are 5 lbs. of resin to 1 gal. of 95 per cent. alcohol.

Lacquers.—It is very difficult to draw any line of distinction between varnishes and lacquers. Lacquers are spirit varnishes generally, and so prepared that they may be applied to metals. They are usually pale, but may also be colored to heighten the natural color of the lacquered article. "Gold" lacquers contain saffron, gamboge, etc., and all the various colors may be prepared by the use of aniline dyes. The article to be lacquered is usually warmed before the application of the lacquer, in order that it may spread out to a thin, even coat.

Varieties of Lacquers.—Lacquers are divisible into three distinct varieties: (1) resin lacquers, (2) collodion, cotton, or pyroxyline lacquers, and (3) combination lacquers.

Resin lacquers are made of the recent resins, including recent kauri, dissolved in one of the solvents, If a spirit lacquer must be thinned, it should be done only with absolute alcohol, since any water present prevents transparency.

Pyroxyline lacquers are the most frequently used. They consist of nitrated cotton, having a lesser degree of nitration than the usual gun cotton, $C_{12}H_{14}O_4(NO_3)_6$, which is not soluble in a mixture of ether and alcohol, amyl acetate, amyl alcohol, acetone and wood alcohol, in which solvents the pyroxyline is soluble. However, its solutions must be filtered to secure transparency, and the pyroxyline lacquers are made in this way. The lacquer is applied generally by spraying or dipping the article, since brushing causes streaking. The formula for pyroxyline is considered to be approximately $C_{12}H_{16}O_6(NO_3)_4$. It is also called *collodion*.

Aside from its use as a lacquer, pyroxyline finds its way into the composition of waterproofing materials, celluloid, photographic films, and it is used in making artificial silk. A solution of it is sold as "liquid court plaster," but this may at times contain other materials as Canada balsam, etc.

Combination lacquers may be made by mixing the first two mentioned. Such a mixture produces a very serviceable lacquer. Lacquers may also contain chromium resinate or chromic acid properly introduced, which act as rust inhibitors.[1]

Japanning.—When lacquers are applied to metals, or sometimes even to wood and other materials, and the article baked in an oven to render the coating harder and more durable, the process is known as japanning. The name is derived from the fact that the process originated long ago in Japan where the natives employ a certain rather thinly fluid balsam that possesses naturally the requisite properties for a most excellent varnish or lacquer. As it exudes from the tree, it has a creamy tint, but it blackens upon exposure to the air.

The baking of a lacquer, that is, japanning, causes the surface to be more lustrous and far more adhesive than when allowed to dry in the ordinary way. The baking is carried out on metals at a temperature of about 130°C. During the baking a certain amount of decomposition takes place and also some oxygen is absorbed.

Enamels.—Sometimes there may be incorporated in the lacquer certain pigments to furnish color to the japanned surface, as in the case of bicycle frames and similar articles. After the colored, so-called ground had been applied and baked, a coating of clear lacquer is applied and then it is again baked. Shellac lacquers are most suitable for the usual enamelling operations.

The black enamels applied to hardware, conduits for electric wiring, and other metallic articles are bituminous coatings, and may or may not be baked. Sometimes the coating consists of asphalt dissolved in benzine, benzole or other solvent. But coatings that consist of this alone after a time crack very much.

---

[1] Directions for making up lacquers may be obtained from HURST and HEATON's Painter's Colors, Oils and Varnishes, Griffin and Co., London.

Better coatings are those in which the brittleness has been overcome by the admixture of drying oils or oil varnishes.

Although the term enamelling is applied to those japanning operations in which the colored grounds are used, the coatings produced in this way must not be confused with the *true enamels*. These latter are porcelain glazes produced on pottery, ironware, etc., by heating thereon silicate substances until fusion or vitrification occurs. Also the baked lacquer or japan enamel must be distinguished from another so-called enamel—the *varnish enamel*. This is similar to a paint, except that the vehicle or medium for the application of the pigment is either a spirit or an oil varnish instead of an oil merely. The resin introduced in this manner confers upon the so-called enamel such properties that it is suitable for use in such locations where the coating is desired to have a high luster, to be waterproof and hence be capable of being cleaned by washing, as for example in kitchens and bathrooms.

Gilts and bronzes may in a sense be considered paints in which the vehicle is about a 3 per cent. solution of pyroxyline in amyl acetate and the pigment a finely divided metal, as brass, bronze, aluminum or aluminum bronze. Also the so-called mosaic gold, stannic sulfide, SnS_2, is used for this purpose. The metals mentioned are beaten into leaf by steam hammers, and are then converted into powder by forcing through a fine-mesh wire sieve, by means of a scratch brush, aided by the admission of oil. The metal is afterward ground in oil, and later separated therefrom by water, and finally is pressed to remove the residual oil. The mosaic gold is prepared by heating a mixture of tin filings, sulfur and ammonium chloride.

OIL VARNISHES

Manufacture.—The process of manufacturing oil varnishes consists of six stages. (1) The resin is melted. Care is necessary, since overheating darkens the varnish and underheating lessens its durability. (2) The oil is boiled. Lead or manganese compounds or both are added to the raw linseed or other oil being used, and the whole is kept at about 260°C. for 1 to 2 hrs. The metallic compounds dissolve in the oil. (3) The materials are mixed. The necessary quantity of boiling oil is poured into

the melted resin and stirred well. (4) The mixture is boiled.
When the mixture is first made, it is cloudy, so that it must be
boiled again to make it transparent and to increase its tenacity
or make it "string." (5) The mixture is thinned. It is first
allowed to cool to about 120°C. and then the necessary amount of
turpentine or other thinner is introduced so that the viscosity is
lessened to the desired degree. (6) The varnish is now allowed

Fig. 61.—Varnish chimney. Showing method of heating resin with oil.

to clear and age. It is allowed to stand until the more insoluble
matter settles out and the top layer becomes clear.

Properties.—Oil varnishes are used wherever durability is required. The durability of a varnish is dependent upon the proportion of oil used in it and the elasticity of its film when dried.
Also it depends upon the hardness of the resin used. Oil varnishes should combine the hardness and brilliance of the resins
with the elasticity of the dried oil films.

The brilliancy and luster also depend upon the proportions of
oil and resin; the greater the amount of resin, the greater the
brilliancy and luster. But beyond a certain point, the greater
amount of resin the less the durability also, consequently the proportions must be determined with both these properties in view.
Where great brilliance is desired, the resin must be no less than

Fig. 62.—Thinning varnish by the addition of turpentine.

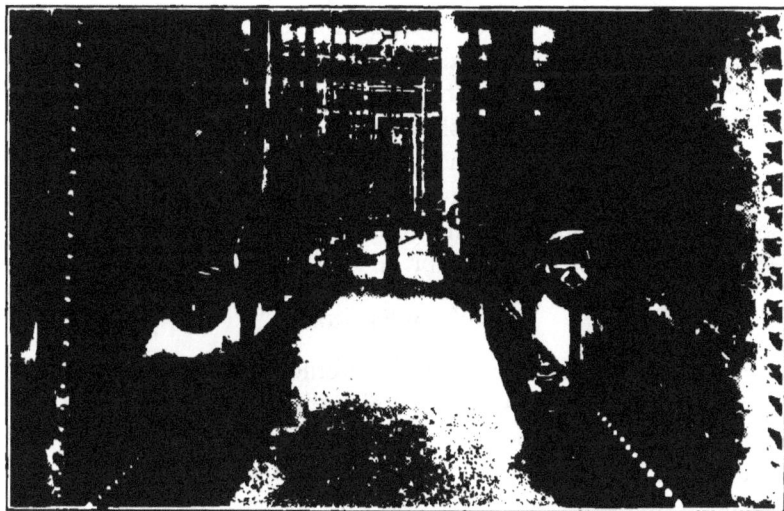

Fig. 63.—Storage tanks.

one-third or one-fourth of the dried coat.[1] If great elasticity is required, the resin must be no more than one-sixth of the dried coat. Also the harder the resin, the greater the brilliancy. Other things being equal, a varnish made from Manilla copal will be less lustrous than one made from Zanzibar animé.[1]

Varnishes should be used only for the purpose for which they are made, since different properties reside in varnishes designed for different uses, as exterior, interior, coach, cabinet, floor, polishing, rubbing varnishes, etc. An outside varnish contains a large percentage of oil to give it toughness and elasticity, hence it dries slowly. Interior varnishes contain less oil in order that they may dry quickly, and be hard and suitable for rubbing and polishing. Floor varnishes must be hard and durable as well as elastic, requiring the use of the hardest of resins in a carefully determined proportion of oil.

Other things being equal, pale varnishes are more expensive than dark ones, since they require the use of special, hand-selected, pale resins. Also, the use of cold-pressed and bleached oil lightens the product, or even the use of oil that has been tanked or aged for a long time is of value in this respect. Overheating during any stage of making, darkens the varnish. On the other hand, varnishes may be rendered lighter in color by underboiling, but this produces an inferior product. Dammar resin is of great aid in lightening the color, but the use of this resin detracts from the hardness and toughness of the varnish.

The drying qualities of the varnish depend upon the resin used as well as upon the kind and amount of drier, and the degree of boiling. Some inferior varnishes contain balsamic or pitchy, resinous materials that prohibit drying.

WOOD STAINS

Practically all the wood used in cabinet making or in finishing the interior of buildings, etc., is stained in some manner, either to augment its natural beauty or to cause a less valuable wood to assume the appearance of a more valuable one.

The staining materials include a great variety of substances. Organic dyes, both of vegetable and coal-tar origin, acids, alkalis,

[1] LIVACHE and MCINTOSH: Varnish Materials and Oil Varnish Making, vol. II, p. 120.

metallic salts and mineral pigments are used. The stains are usually classified according to the solvent or vehicle with which they are applied, as water, spirit, acid and oil stains.

Water stains are the most used. The advantages of water as a solvent are its low cost and the ease with which it may be applied, but water causes the fibers of the wood to rise, producing a roughened surface and this necessitates refinishing. It is said that the raising of the fiber may be minimized by the addition of a small quantity of glycerine to the water. The water stains may consist of solutions of the aniline dyes or of water extractions of logwood, Brazil wood, madder, gall nuts, etc. The colors of these may be modified or changed entirely by the action of alkalis or acids. The aniline dyes are quite susceptible to the action of the substances contained in the wood, especially the tannic acid. These substances may exert a noteworthy reducing action, absorbing oxygen from the molecule of the dyestuff and so modifying its color. Some dyes are more susceptible to this action than others. The action will be most noticeable, of course, on those woods rich in tannic acid, as oaks.

Spirit stains are preparations made by dissolving the coloring materials such as the aniline dyes and other materials in alcohol or in a good spirit varnish. Spirit stains dry very rapidly—so rapidly that it is difficult to apply them properly. Also the spirit stain may "bleed," that is, come through the varnish coat because of the dissolving action of the oil or solvent contained in the varnish. This difficulty may be overcome by selecting a dye that is soluble only in some solvent as nitrobenzol, petroleum spirit, amyl acetate, acetone or other liquid not contained in the varnish that is being applied. Many dyes of the various colors are obtainable, whose solubilities are entirely different, so that they may be chosen with this point in view. All solutions of spirit stains must be clear without sediment or excess crystals. This is important because colors are frequently produced by the dye manufacturer, by mixing two or more distinct dyestuffs. It may be that one of these dyestuffs will be soluble in the solvent chosen and the other not, so that if it be not all dissolved the true value of the dye will not be secured. To fix stains after the manner of a mordant in dyeing cloth, there may be applied after the stain has dried, about a 5 per cent. solution of alum or ferrous sulfate.

Acid stains comprise a great variety of substances. In this connection, the term "acid" is often employed to designate many materials that are not acids in any sense. It includes the true acids such as sulfuric, nitric and picric; salts such as alum, ferrous sulfate or copperas, potassium chromate and dichromate, potassium permanganate, sodium and potassium carbonate; and even the alkalis, sodium and potassium hydroxide themselves.

Wood may be "aged" by the direct application of ammonium hydroxide or by allowing the wood to stand in chambers, containing vessels of ammonium hydroxide from which the ammonia gas may arise and act upon the wood. Oak is especially susceptible to this action, the product being known as *fumed oak*. This treatment with ammonia is useful only with those woods containing considerable gallic or tannic acids, since the color is dependent upon the formation of the ammonium compound of these substances.

Oil stains do not penetrate deeply and are most suitable for woods that are not to be filled. The oil used in the stain, when dried, partly fills the pores of the wood so that proper filling with a suitable filler cannot be accomplished. The oil stains are prepared by adding mineral pigments, such as burnt sienna, umber, chrome yellow, etc., to the oils, together with turpentine and some japan drier. Colors produced by oil stains are less clear and transparent then those produced by acid stains.[1]

Bleaching of Woods.—Dark woods may be bleached by the action of chlorine. A thin paste is made of bleaching powder, $Ca(OCl)_2$, applied and allowed to dry. Then a solution of some acid, as a 5 per cent. solution of hydrochloric or acetic, must be applied. The acid reacts with the hypochlorite and liberates nascent chlorine which causes the organic coloring matter of the wood to be destroyed. If the action is insufficient, the operation may be repeated.

WOOD FILLERS

Fillers are used to close the pores in open-grained woods. If not filled, capillary action would occur in the pores and produce

[1] For further information on wood stains and methods of use, see *Wood Craft*, September, 1906; July and August, 1907; March, 1908; August, 1910; February and August, 1911; and March and April, 1912. Also SCHMIDT: Problems of the Finishing Room.

"sinking of the varnish." There are two kinds; paste and liquid fillers.

Paste Fillers.—The best grade of these consist of silica or barytes, boiled oil, and japan drier. Both silica and barytes become practically transparent in oil and so produce a transparent filling. Fillers of poorer quality consist of corn starch, whiting, boiled oil, turpentine and drier. These have a tendency to shrink after drying, especially if made of the starch, thereby defeating the object for which the filler was applied. If intended to produce some color effect, the filler may also contain a coloring matter. The paste fillers are used in conjunction with the liquid fillers which are applied subsequently. The oaks, the walnuts, mahogany, ash, chestnut and red gum are open-grained woods, and are filled, with both paste and liquid fillers.

Liquid fillers consist of practically the same materials as the paste fillers, but in these the liquid constituents preponderate. Liquid fillers are used only on close-grained woods, as cherry and birch. When using fillers in conjunction with stains, the filler must be applied after the stain. If the filler contains whiting or marble dust, and the stain is an acid stain, then a thin coating of shellac should be applied over the stain before filling, to prevent the acid from acting on the calcium carbonate of which these substances consist.

Putty usually consists of whiting and raw linseed oil kneaded together. It may also consist occasionally of whiting, lead carbonate, white clay or other similar materials with linseed oil.[1]

Paint and Varnish Removers.—There are a number of liquids that exert a solvent action on dried paint and varnish films, but being highly volatile also, they do not remain in place long enough to become effective. Consequently, if the liquids are entangled in some medium that will serve to prevent their evaporation and the whole is applied in a paste form they may be advantageously used.

A solvent paste of this sort may be prepared as follows: Dissolve with the aid of heat 8 parts of finely divided paraffine wax in 8 parts of benzole. (Care! Benzole is inflammable.) To this solution while still warm, add 7 parts of methyl alcohol. The

---

[1] For a description of various putties see *Furniture Manufacturer and Artisan*, February, 1914, p. 93.

alcohol is miscible with the benzole, but is not itself a solvent for the wax, consequently the wax is precipitated out in the form of a gelatinous mass that holds the liquids entangled. The solvent power of the paste may be increased by stirring into it about 1 part of acetone or about one-half of 1 part of amyl acetate.

A quite active preparation of an entirely different sort is the following, used also in paste form: caustic soda, 1 part, fresh powdered quicklime, 3 parts; and whiting 4 parts, with sufficient water to form a paste. This is applied and allowed to remain for a short time. This preparation is active because it converts the linoxyn into soap. Or a 10 to 12 per cent. solution of sodium carbonate may be used, as may also a moderately strong solution of ammonium hydroxide. These liquids are used with an abrasive as steel wool, or a stiff-bristled brush. After the removal of paint or varnish by any one of these alkaline substances, the residual alkali on the surface of the wood must be corrected with some weak acid, as for example, about a 5 per cent. solution of oxalic acid, which both neutralizes the alkali and acts as a bleaching agent.

References

A

SABIN: Technology of Paint and Varnish, New York, 1906.

HALL: The Chemistry of Paint and Paint Vehicles, New York, 1906.

TOCH: Technology of Mixed Paints, New York, 1907.

MAIRE: Modern Pigments and Their Vehicles, New York, 1908.

ENNIS: Linseed Oil, and Other Seed Oils, New York, 1909.

HOLLEY: Lead and Zinc Pigments, New York, 1909.

CHURCH: The Chemistry of Paints and Painting, London, 1910.

FRIEND: An Introduction to the Chemistry of Paints, London, 1910.

LIVACHE and McINTOSH: Manufacture of Varnishes, London, 1904–1911.

BOTTLER and SABIN: German and American Varnish Making, New York, 1912.

ALLEN: Commercial Organic Analysis, Philadelphia, 1909–1913.

HURST and HEATON: Painters' Colors, Oils and Varnishes, London, 1913.

LEWKOWITSCH: Chemical Technology and Analysis of Oils, Fats and Waxes, London, 1913–1915.

SCHMIDT: Problems of the Finishing Room, Grand Rapids, 1916.

B

DUNLAP and SCHENK: Oxidation of Linseed Oil, *Jour. Am. Chem. Soc.*, **25**, 826 (1903).

BAKELAND: The Synthesis, Constitution, and Uses of Bakelite, *Jour. Ind. Eng. Chem.*, March, 1909; August, 1909, and December, 1911.

——*Bulletins* of the Scientific Section of the Paint Manufacturers' Association of the United States, 1909 and 1910.

BAKELAND: Bakelite and Its Uses, *Jour. Frnkl. Inst.* (1910).

THOMPSON: Scientific Preparation and Application of Paints, *Jour. Ind. Eng. Chem.*, **2**, 87 (1910).

BOTTLER: Hardened Resins, *Chem. Abst.*, **5**, 1198 (1911).

TOCH: Fish Oil as a Paint Vehicle, *Jour. Ind. Eng. Chem.*, **3**, 267 (1911).

——Removal of Old Varnish, *Wood Craft*, August, 1911.

COSTE and NASH: Turpentine Substitutes, *Analyst*, **36**, 207 (1911).

——The Kauri Gum Industry of New Zealand, *Chem. Trade Jour.*, **49**, 487 (1911).

——Driers. *Jour. Frnkl. Inst.*, January, 1911.

VEITCH and DONK: Wood Turpentine; Its Production, Refining, Properties and Uses, U. S. Department of Agriculture, Bureau of Chemistry, *Bull.* 144 (1911).

CABOT: Carbon Black and Lamp Black, *Proc.* 8th Inter. Cong. App. Chem., **20**, 59 (1912).

——Chalking of Paints, *Jour. Frankl. Inst.*, January, 1912.

TOCH: Soya Bean Oil as a Substitute for Linseed Oil in Paints, *Jour. Soc. Chem. Ind.*, **31**, 572 (1912).

TOCH: Paint as an Engineering Material, *Jour. Ind. Eng. Chem.*, **5**, 366 (1913).

WALKER and VOORHEES: Some Tests of Paints for Steel, *Jour. Ind. Eng. Chem.*, **5**, 899 (1913).

GARDNER: The Toxic and Antiseptic Properties of Paints, *Drugs, Oils and Paints*, **30**, 10 (1914).

THOMPSON: Painting Defects, Their Causes and Prevention, *Drugs, Oils and Paints*, **30**, 253 (1914).

WALKER ET AL.: Report of Committee D-1 on Preservative Coatings for Structural Materials, *Proc. Am. Soc. Test Matls*, **15**, Pt. 1, 186 (1915).

BOUGHTON: The Effect of Certain Pigments on Linseed Oil, U. S. Bureau of Standards, *Tech. Paper*, 71 (1916).

WARE and CHRISTMAN: A Study of the Effect of Storage on Mixed Paints, *Jour. Ind. Eng. Chem.*, **8**, 879 (1916).

CHAPTER XIV

LUBRICANTS

Object.—When two solid bodies are caused to slide in contact with one another there is more or less resistance to motion, which results in an absorption of energy that is manifested as heat. The resistance to motion is due somewhat to molecular cohesion, but chiefly it is due to interlocking of projecting portions on the uneven surfaces of the bodies. It is impossible to prepare a perfectly smooth surface. Consequently, when it is desired to lessen friction, as in machinery bearings, a lubricant, which is generally an oil, is used to keep the metallic surfaces apart. In perfect lubrication the metallic surfaces do not contact at all, but this degree is usually not attained in practice.

A Fundamental Requirement.—Gill[1] says: "the cardinal principle underlying all lubrication is to *use the thinnest (or least viscous) oil that will stay in place and do the work.*" In all lubrication the lubricant should adhere to the moving solid parts with a consequent shearing of the lubricating layer. A good lubricant is, then, one that offers but little resistance to shearing, or in other words, that has but little friction between its own moving particles. The term viscosity is used in designating the amount of this internal friction. An oil that flows sluggishly is said to be high in viscosity. A lubricating oil should have sufficient adhesion and viscosity to enable it to stay in place, but if it possesses greater viscosity than this, power is wasted in shearing the oil. In brief, as long as it can resist the force that tends to squeeze it out, the value of a lubricant increases with its mobility.[2]

[1] A. H. GILL: ROGERS and AUBERT's Industrial Chemistry, p. 602.

[2] In experimental work, air has been used as a lubricant, furnishing almost perfect lubrication. A steel cylinder 6 in. in diameter and $6\frac{1}{4}$ in. long, weighing $50\frac{1}{2}$ lbs. was fitted into a cast-iron journal with a fit of $\frac{1}{2,000}$ in., and supplied with a detachable crank-rotating mechanism. After a speed of 500 revolutions per minute had been obtained, the harsh grating noise ceased, and the cylinder ran smoothly of its own accord for 4 or 5 min.

338

Chemical Stability Essential.—It is also of fundamental importance that the oils used for lubrication be as chemically inert as possible. The lubricating oils belong to two major classes: (1) the fatty oils, which are of both animal and vegetable origin; and (2) mineral oils. The former are relatively active, and the latter inert.

Chemical Activity of Fatty Oils.—Fatty oils in general are more or less affected by exposure to the atmosphere. Some of them are quite susceptible to *oxidation*, and are converted into a dry hardened mass. These are the paint oils, and this property which makes them suitable for paints manifestly renders them useless for lubrication. Others of these oils that do not oxidize sufficiently to form a hardened mass, become thickened to such an extent that friction would be much increased by their use. To some extent the fatty oils react with moisture in the air, the reaction which they undergo being known as *hydrolysis*. The structure of the fatty oils[1] bears a close analogy to the structure of metallic salts. They consist of mixtures of compounds made up of a trivalent basic radicle, C_3H_5, called glyceryl, united with various fatty radicles, which may be typified by the radicle of oleic acid, $H(O_2C_{18}H_{33})$. Thus glyceryl oleate, called olein, is $C_3H_5(O_2C_{18}H_{33})_3$. The hydrolysis that occurs upon exposure to the air is small in amount compared to that which occurs in steam cylinders where fatty oils are often used. Hydrolysis is represented by the following equation which shows the reaction between water and stearin, the products being glycerine and stearic acid.

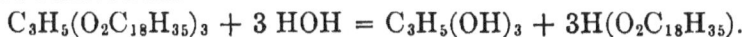

$$C_3H_5(O_2C_{18}H_{35})_3 + 3\ HOH = C_3H_5(OH)_3 + 3H(O_2C_{18}H_{35}).$$

The fatty acids thus developed attack the metallic bearings, particularly when the metal is alternately exposed to the acidified oil and then to the air.

Mineral Oils Inert.—The mineral oils, on the other hand, are relatively quite indifferent to the action of the atmosphere. As

---

Attempts to establish electrical connection between the journal and shaft failed until after the cylinder had slowed down, thereby showing insulation due to the supporting film of air. No grease or oil was used in this experiment (Lubrication and Lubricants, by ARCHBUTT and DEELEY, 1912 ed., p. 90).

[1] For a more complete discussion see p. 288.

was stated under the subject of "Fuels," it is chiefly the paraffine-base oils that furnish the best of the refined petroleum products, as petroleum spirit, lamp oil, lubricating oils, etc.   This *paraffine series* of hydrocarbons is especially inert.   It is called a saturated series; that is, the molecules composing it have no "double bondings" between the carbon atoms, and so are not susceptible to the formation of addition products.   It is because of its inertness that it is called the paraffine series—from the Latin *parum*, little, and *affinis*, affinity—indicating that these compounds have but little chemical affinity for other substances.   These oils are, therefore, as far as this point is concerned, eminently suitable for lubricants.

Classification of Lubricants.—As has been shown, lubricants may consist of either fatty or mineral oils.   Mineral oils are most used, for aside from their chemical stability they are the cheapest of all the oils.   Still for certain uses, they are improved by an admixture of certain animal and vegetable oils.   For example, in steam-cylinder lubrication an admixture of a fatty oil helps to cause the oil to adhere to the metal.   These mixtures are called compounded oils.   Fatty oils are only to a comparatively small extent used alone.   For especially heavy burdens, the oils are thickened to form a grease by means of soaps or lime and these may be further reinforced with mica, graphite, or the talcose variety of soapstone.   For convenience in discussion, then, the lubricants may be separated into the following classes:

(I) Mineral oils.

(II) Fatty oils.

(III) Greases.

I. The Mineral Lubricating Oils.—Preparation.—After the naphthas and lamp oils have been removed from the paraffine-base crude oil as described under "Fuels" on pages 37 to 43, the lubricating oils are derived from the residue remaining in the still. If these lighter portions were removed by the *destructive distillation* process, only about 10 to 12 per cent. of residue or "tar" will be left, and from it the paraffine lubricating oils are obtained.   If the lighter portions were removed by distillation with the aid of *steam* about 15 to 18 per cent. of residue remains in the still. This is known as "cylinder stock" and from it are obtained the better grades of lubricating oils—spindle oils, steam and gas-engine cylinder oils, etc.

Destructive Distillation (continued from page 43).—The 10 to 12 per cent. of "tar" that remains in the "crude still" is first subjected to straight (not fractional) distillation without steam at a relatively high heat either in the crude still itself or in a smaller "tar still." At·the temperature used, some "cracking" of the heavier molecules result. The process is continued until nothing remains in the still but coke, which is formed by the cracking. The distillate is known as the tar distillate. It contains the wax and the paraffine lubricating oils. It is then treated with sulfuric acid to remove traces of asphaltic or tarry substance, after which it is washed with water and then it is treated with a sodium hydroxide solution to neutralize any unseparated acid. After this it is washed again and then chilled to about 20 to 24°F. by means of cold brine. The chilled mass is now filtered through canvas in filter presses. The wax, which remains in the press, is called "*slack wax*" and still contains about 30 to 40 per cent. of oil.

The filtrate from the wax presses is first "reduced" with steam; *i.e.*, it is heated in steam stills until the more readily voltatile products are driven off and the fire test of that which remains in the still is thereby raised to the desired point. These more or less volatile products, which are removed by the steam, resulted from cracking in distilling the "tar." In this "steam reduction" the first fraction that distils off belongs with the low fire-test lamp oils. The second fraction having a specific gravity too great for a lamp oil but having practically no lubricating value is used for a fuel oil. The third fraction constitutes a low viscosity lubricating oil, *i.e.*, a spindle oil. Sometimes this fraction is filtered through bone-black or fuller's earth and then exposed in shallow vats to the action of sunlight and air. In this way the greenish fluorescence or "bloom" that is characteristic of mineral oils is removed, and it is then known as a "*de-bloomed*" oil. The "bloom" may also be removed by treating the oil with about 1 per cent. of nitric acid, nitronaphthalene or dinitrobenzol. It is said that the bloom returns after a time when removed by these reagents. This de-blooming does not in any way improve the lubricating properties of the oil. It is de-bloomed in order that it may be used with lessened probability of detection to adulterate the fatty oils, such as linseed, cottonseed, neatsfoot, lard oil, etc. An admixture of as much as 40 to 50 per cent. of de-bloomed oil does, however, lessen the risk of spontaneous ignition of these fatty oils when they are soaked up in cotton waste. The de-bloomed oils are also sometimes known as "*neutral oils*," but all so-called neutral oils are not de-bloomed. The term neutral is used in general to indicate an oil having about the composition of this third fraction, in that it is free from the volatile products on the one hand and from the heavy or wax-like products on the other. The fourth fraction constitutes the medium viscosity lubricating oil, and the fifth, which is the residue remaining in the retort, is the heavy lubricating oil. This is purified by the acid and alkali process, producing engine or machinery oil.

Steam Distillation (continued from page 43).—The 15 to 18 per cent. of cylinder stock remaining in still after the removal of the naphthas and lamp oils by steam distillation is now further fractionated by the use of steam into spindle oils, steam-cylinder oils and gas-engine oils. An effort is made to completely avoid cracking, since cracking increases the amount of asphaltic or tarry substances, which are very undesirable, especially in the oils designed for use in the internal-combustion engines, where the high heat causes a deposition of carbon from such compounds. After being fractionated the oils are brought to the fire test desired by heating them in steam stills in the same manner as the tar distillate after the wax has been separated from it.

The spindle oils and gas-engine oils produced in this way are clarified by filtering through bone-black or fuller's earth to remove the asphaltic matter which is always present in small quantity. The first oil coming through the filter is practically colorless, but the filtering material loses its efficiency gradually and the oil that comes through later is not so clear. The cylinder oils may be sold as *unfiltered cylinder oils* or they also may be filtered. The unfiltered oils are a dark green, while the *filtered cylinder oils* are likely to be a dark red. The filtering process is very slow and much increases the cost.

Reduced Natural Lubricating Oils.—A limited quantity of crude petroleum found in the United States in the vicinity of such places as Erie, Greensburg and Franklin, Pa.; Charlestown, W. Va.; Mecca, O.; and Monticello, Ky., are so rich in lubricating hydrocarbons that all that is necessary to prepare them is to place them in open vats partly filled with water heated to about 110°C. with steam coils. (Sometimes heat is furnished by the sun only.) The naphthas are removed by evaporation and the suspended mineral matter is removed by settling.

If the naphtha content in these natural oils is sufficiently great to make its recovery profitable, it may be removed by steam or vacuum distillation with a careful avoidance of cracking.

Oils obtained by these methods are free from decomposition products due to heating, and from traces of the chemicals such as are used in treating the paraffine lubricating oils. Reduced natural oils are considered very fine lubricants but the quantity on the market is only small.

Paraffine Wax.—The "slack wax" expressed from the tar distillate may be freed from the residual oil it contains by washing it with naphtha, after which it is again chilled and pressed. Or, the oil may be removed by warming the caked wax to a definite temperature at which the oil sweats out and drains off. The wax is then melted and filtered through heated bone-black or fuller's earth to whiten it. Paraffine wax is practically inert under the influence of all chemical reagents. It burns when heated to its kindling point, but aside from this is very stable chemically. It consists of that portion of the hydrocarbon series extending from about $C_{24}H_{50}$ to $C_{35}H_{72}$.

Vaseline or petrolatum is made by the acid and alkali treatment of the residue left by the careful vacuum distillation of crude oil. The treated residue is washed and then filtered through bone-black to clarify it. After filtering it is sometimes redistilled *in vacuo.*

II. The Fatty Oils.—The fatty oils used for lubrication are selected from those of this class that are the most stable, although all fatty oils oxidize somewhat when exposed to the air, and are therefore to some degree susceptible to "gumming" when in use. However, the fatty oils have an advantage over the mineral oils in that their viscosities are lowered to a less degree by warming. Also, as has been stated before, their adhesion to the moist metallic surfaces in steam-cylinder lubrication is greater than that of the mineral oils. The fatty oils that are most commonly used for lubricants are castor, lard oil, tallow oil, rape seed, neatsfoot and sperm oil. Sperm oil is the least viscous of these, and castor oil the most.

Castor oil is obtained from the seeds of *Ricinus Communis*, a plant native to India, but grown quite commonly elsewhere. The oil that is obtained by cold pressure from the ground seeds is practically colorless, and is used medicinally. A second quality of oil, darker in color, is obtained by heating the once-pressed mass and subjecting it to further pressure. This lower grade of oil is used as a lubricant as well as for other purposes, such as for belt dressings and soap making.

Castor oil consists chiefly of the glyceride of ricinoleic acid, but it also contains a little palmatin which separates out in cold weather, although the oil as a whole does not solidify until about $-18°C.$ is reached. It is not entirely free from gumming tendencies, but it is much less susceptible to this action than most of the fatty oils, and it does not turn rancid as readily as the majority of them do.

Unlike other fatty oils, castor oil is easily soluble in absolute alcohol. Because of this, its adulterants may be detected by means of absolute alcohol if they are present to as much as 10 per cent., since a turbidity will then show itself in the alcoholic solution. Also, unlike the others, it cannot be mixed with mineral oils without separating out again on standing, although it will make a permanent homogeneous mixture with other fatty oils. And when mixed with, or adulterated with a small quantity of some other fatty oil, it will then remain dissolved in mineral oil. It does not dissolve very well in the light petroleum products in which the other fatty oils are readily soluble.

At ordinary temperatures castor oil is the most viscous of all the fatty oils. It is suitable for the lubrication of heavy machinery running at relatively high speeds.

The common adulterants of castor oil are linseed, rape seed, cottonseed or rosin oil that have been thickened by blowing air through them while warm.

Rape seed or colza oil is obtained from the seeds of several varieties of *Brassica Campestris*, a plant belonging to the mustard family. It is cultivated extensively in Europe but not in America. The expressed oil is refined by the sulfuric acid and alkali treatment. The refined oil has a pale yellow color, characteristic odor, and rather unpleasant taste. It thickens noticeably when exposed to the air.

It has a moderately high viscosity and is often used as a standard for comparison in testing the viscosity of other oils. It is much used as a lubricant in European countries, but is not much so used in America, because of the high price of the imported oil. The more common adulterants are cottonseed, hemp and refined fish oils.

Lard oil consists chiefly of olein, which is separated from the solid fats of lard by pressure. Melted lard is allowed to cool slowly to about 30°C. and then is kept at this temperature for about 24 hrs. Under these conditions the stearin and palmatin largely solidify in granular form. The mass is then subjected to a pressure of about 4 tons per square inch in a filter press, and the more fluid olein is filtered out. About 40 to 60 per cent. of oil is obtained from the lard. It is chiefly olein but contains some other glycerides, especially palmatin. The solidifying point or cold test can to a degree be regulated by the temperature at which the mass is filtered. The oil used as a lubricant will usually bear a temperature of 0°C. without solidifying.

The market price of lard oil varies largely according to color, which ranges from a pale or greenish yellow to a reddish brown. The odor varies from practically none to a very pronounced odor in the brown or poorer grades. The best grades contain only a little free fatty acid, usually less than 1 per cent., but the poorer grades contain a considerable quantity.

Lard oil is used as a lubricant when mixed with mineral oils, and also as a cooling agent for cutting tools. Its more common adulterants are corn, cottonseed and neutral petroleum oils.

Tallow oil is obtained from tallow in a manner similar to that by which lard oil is obtained from lard. Its odor is characteristic of tallow. As a lubricant it is used chiefly with mineral oils in steam-cylinder lubrication.

Neatsfoot Oil.—As the name indicates this oil is obtained from the feet of cattle. After the hide and hoofs have been removed the feet are boiled in water with the result that an emulsion forms from which the oil ultimately separates. The oil is pale yellow, and consists chiefly of olein, but contains also a small quantity of stearin which separates out on standing.

It is a very good lubricant having but little tendency to "gum," and may be used alone or with mineral oils. Also, it is much used as a leather dressing. It is subject to adulteration because it is relatively high in price. Cottonseed, corn, rape seed and refined fish or mineral oils are used for this purpose.

Sperm Oil.—This is not a fatty oil but a liquid wax. Its composition has not been definitely ascertained, although it is known to contain a considerable amount of cetyl palmitate, $C_{16}H_{33}(O \cdot OC_{16}H_{31})$. It contains practically no glycerides.

The true sperm oil is obtained from the head cavity or "case" of the sperm whale, *Physeter macrocephalus*. A great many barrels of free oil[1] are not infrequently obtained from the head of a single specimen. The name, sperm oil, is also applied to the body or blubber oil, and very often the head and body oils are mixed prior to refining. Aside from these oils, the oil obtained from the blubber or body fat of the Arctic sperm or bottle-nose whale is also called sperm oil.

When first obtained from the animal the true sperm oils contain much solid wax. This is separated by chilling and filtering with pressure. Different grades of oil are thus obtained depending upon the temperature at which the oil is pressed. The solid wax which is retained by the filter, yields when refined that which is commonly known as *spermaceti wax*. It is white, lustrous and very brittle, and is used chiefly for the making of candles.

Sperm oil has a lower viscosity than any fatty oil, and its viscosity is much less decreased by increase of temperature than is that of the true fatty oils. Also, it does not gum nor become rancid. It is therefore an excellent lubricant especially for light machinery running at high speeds.

Sperm oil is much subject to adulteration; it may contain mineral oils, other whale oils and fish oils.

"Blown" or Thickened Oils.—These oils are thick and viscous, having been made so by blowing a current of air through the oil while heated. It is generally cottonseed, rape seed, and lard oils that are treated in this way. Mineral oils are mixed with about 20 to 30 per cent. of the blown oils and the mixture is used as a substitute for castor oil.

III. Greases.—For very heavy pressures and slow speeds,

---

[1] HURST, in his Lubricating Oils, Fats and Greases, says the head cavity will hold as much as 200 bbl. of oil.

greases are often used as lubricants. According to Gillett,[1] the commercial greases may be classified as follows:

(*A*) **The Tallow Type.**—These greases are made up of tallow and more or less of an alkali soap, commonly the sodium and potassium soaps of palm oil, mixed with a smaller amount of mineral oil. These were the common type many years ago, but are now less used than greases of type (*B*).

(*B*) **The Soap-thickened Mineral-oil Type.**—These are the most common journal greases today. They are composed of mineral oils of various grades, made solid by the addition of calcium or sodium soaps. Calcium soap is more used than sodium soap. The former is cheaper, producing a stiffer grease with a higher melting point, but the latter is the better lubricant, probably because it is more soluble in the oil and produces a smoother grease.

(*C*) Types (*A*) or (*B*) with the addition of a *mineral lubricant*, usually graphite, mica or talc.

(*D*) **The Rosin-oil Type.**—These consist of rosin oil thickened by lime or less commonly litharge, to which is added a certain amount of mineral oil, either paraffine or asphalt oils being used. These greases are sticky, usually contain 20 to 30 per cent. of water, and find their chief application as gear greases where true lubrication is not so essential as the prevention of the rattling of the gears. Some very heavy bearings are occasionally lubricated with this type of grease. Tar, pitch, asphaltum, graphite and such fillers as wood pulp and ground cork are often put into these greases. The high content of water is not objectionable since it serves to lessen the stickiness.

(*E*) **Non-fluid Oils.**—These are thin greases stiffened to some extent with aluminum oleate, or a mixture of soaps, as of sodium and calcium. They are sometimes known as soap-thickened oils.

(*F*) **Special greases,** such as mixtures of wood pulp and graphite; thin greases of any of the above types containing wool or cotton fibers, etc.

Under test the rosin greases showed a high friction at first, but after the bearing had warmed up compared well with the more expensive greases.

[1] Analysis and Friction Tests of Greases, *Jour. Ind. Eng. Chem.*, June, 1909.

Gillett further says that since friction cannot be reduced until the temperature of the bearing is high enough to melt the grease, or at least to render it sufficiently fluid that it can flow over the surface, it follows that other things being equal, the highest melting point greases will produce the highest coefficient of friction. The lowest melting point greases are, therefore, the best.　In other words, an oil instead of a grease should be used whenever possible.

Graphite.—Benson[1] states that graphite serves as a lubricant for cast-iron bearings where it acts as a surface evener of the porous metal, but that on finer surfaces it does not work so well, having a tendency to collect in such quantities as to seriously score or abrade the journal and bearing.　To serve as a lubricant under these conditions Acheson has prepared graphite in an extremely fine condition, which is known as deflocculated graphite.

Deflocculated graphite is so extremely fine that it will remain suspended indefinitely in water, and will pass readily through the finest filter paper.　It is no doubt in colloidal solution.　It is prepared[2] by first heating carbon in the electric furnace to 7,500°F.　All material other than the carbon is volatilized, the carbon thus attaining a purity of 99.9 per cent.　This purified carbon is then masticated with a solution of tannin, whereby it is deflocculated.　After allowing this to stand until any suspended matter settles out, it is filtered under pressure through canvas impregnated with thin films of rubber, by which process a large part of the water is separated and a carbon paste is formed.　To this has been given the commercial name "aquadag" (aqueous deflocculated Acheson graphite).　By working the "aquadag" with oil in a mixing or pugging machine, the water is separated and a paste with oil is left, which bears the commercial name "oildag."

Deflocculated graphite seems to have a very remarkable value as a lubricant.　It is said to surpass ordinary graphite in unctuous quality, and the ease with which it remains suspended in water and purified oils renders it very convenient to use.　It is very important that any medium into which it is introduced be

[1] Industrial Chemistry, p. 185.
[2] *The Chemical Engineer*, March, 1915, p. 127.

free from any substance capable of ionizing, or in other words, that will act as an electrolyte, since carbon in colloidal solution is precipitated by such substances. Hence, it cannot be used with an oil that contains any acid either mineral or fatty.

Oildag has been found specially suitable for automobile lubrication.[1] Dr. Acheson himself found that with a four-cylinder car, 0.35 per cent. of deflocculated graphite, which is the amount usually specified for use in oils, added to the cylinder oil reduced the consumption of oil from 1 gal. per 200 miles to 1 gal. per 750 miles. The graphite film which is formed on the piston rings and inside surfaces of the cylinder is said to produce such a perfect fit that the compression is increased.

Aquadag is said to be useful in place of the oil and soap mixture for lathe and die threading, for boring, reaming, planing and general machine shop work. It is also said to be useful for air-compressor lubrication, since it eliminates the risk of explosion from the inflammable vapors generated from lubricating oil as the cylinder becomes heated.

EXAMINATION OF LUBRICANTS

Mineral oils may be distinguished, as a rule, from fatty oils by means of the greenish fluorescence which mineral oils show. Distinction on this basis cannot always be relied upon, since mineral oils may be de-bloomed, and because in rare cases fatty oils also show a certain sort of fluorescence. The saponification test is a far more certain test for the detection of such admixtures, but it must be remembered that rosin oil also is largely unsaponifiable.

The color of an oil is not a reliable aid in determining the kind of oil since oils may be colored as desired by means of the oleates and butyrates of iron and copper. An oil may be identified by a careful determination of its iodine number, saponification value, specific gravity and other properties, but the determination of the nature of mixtures of fatty oils is complex and difficult. The odor and taste of oils, particularly when warm, yield much information to one having had some experience with them, since many have a decidedly characteristic odor and taste. It is not

[1] ARCHBUTT and DEELEY, pp. 150–152.

the oil proper that possesses this odor and taste but certain characteristic impurities that are always present in the oil.

Aside from tests for identification, lubricating oils are generally examined for specific gravity, viscosity, flash point and fire test, tendency to "gum," solidifying point, volatility, acidity, the presence of soap, and if it be a solid lubricant or grease, its melting point.

Specific Gravity.—This is sometimes determined by means of the hydrostatic balance, but more commonly by means of the hydrometer. Hydrometers may be divided into two classes, those from which the specific gravity may be read directly and those graduated with arbitrary scales, of which the Baumé is the most common example in this country. For a description of hydrometers and their methods of use, see Chapter XX.

Viscosity.—A thick or gummy liquid is said to be viscous while at the same time it may be either high or low in specific gravity. Viscosity is then the measure or degree of fluidity of an oil or other liquid.

A very rough idea of the relative viscosities of two oils may be gotten by shaking them in bottles and noting the time required for bubbles of equal size to rise. The more sluggish the movement of the bubbles, the greater the viscosity. It is necessary, of course, to allow the oils to stand side by side for an hour or more in order that they may be at the same temperature.

Viscosity is usually represented by the number of seconds required for a stated amount of oil at a definite temperature to pass through a small orifice. This is in reality also a relative test since the time must be compared to the time required for the flow of an equal quantity of water under the same conditions, although sometimes it is compared to the rate of flow of standard oils, as sperm or rape. There are a great many varieties of viscosimeters, differing essentially in the method employed for stirring the sample and keeping it at a uniform temperature. Temperature is a very important factor in the determination of viscosity, even the tenth of a degree producing a noticeable variation. Unfortunately the various viscosimeters do not conform to any fixed standard and the values obtained on each cannot be compared directly to those of another instrument, although tables have been prepared by the Bureau of Standards by which comparisons can be made. The Engler[1] viscosimeter has been adopted as the standard instrument for accurate scientific work. The Saybolt and Tagliabue instruments are commonly used in the oil trade, the Saybolt Universal being recommended for American practice by the American Society for Testing Materials.

A rough quantitative method that may be of service for commercial use consists of filling a 100-c.c. pipette with the oil and then counting the number of seconds required for the oil to flow from a mark on the upper stem to another on the lower stem. This must be repeated with some standard oil and, of course, to secure good results the oils must be at the same temperature. Oils may be quickly graded by this method.

Viscosity is an important factor in choosing an oil for any special use; also, it is of great value in comparing an oil with another that has been found to give good results in practice. Since viscosities of different oils change at different rates with changes of temperature, Gill[2] says that the viscosity should not be taken at any arbitrary temperature, as 70°, 100°, 212°F., but at the temperature or between the temperatures *at which the oil is used.* It often happens that one oil is less viscous than another at one of these temperatures and at the temperature of use the reverse is the case.

Flash Point and Burning Point or "Fire Test."—These points are not determined in order to furnish any information concerning the lubricating value, but in order that safety may be ensured while the oil is in use. The *flash point* is the lowest temperature at which the oil will give off sufficient vapor to support a momentary flash when a small flame is brought to within about ¼ in. of the surface of the oil. The *burning or ignition point* is the lowest temperature at which an oil will give off vapor at a rate sufficiently rapid that it will burn continuously when ignited. The flash point is always lower than the burning point or fire test and both are below the boiling point.

These points are determined[3] commonly by either the "open"

[1] For method of using the Engler viscosimeter see Lewkowitsch, vol. 1, 1913 ed., p. 354 or Sherman: Organic Analysis, 1912 ed., p. 226.

[2] Gill: Oil Analysis, p. 26.

[3] For methods see Lewkowitsch, vol. 3, 1909 ed., p. 57; or Thorp: Outlines of Industrial Chemistry, p. 311; or Hurst: Lubricating Oils, Fats and Greases, p. 245.

or "closed tester." In the closed tester the vapors do not escape so readily by dissemination into the air, and consequently the results obtained on an instrument of this type are always lower, perhaps from 12° to 20°. Also, its results are likely to be more uniform and accurate. Then, in stating the flash point or fire test, the type of instrument used should be specified.

The Gumming Test.—In order to determine the amount of change a mineral oil will undergo while in use Gill[1] treats the sample with nitrosulfuric acid, which is prepared by saturating sulfuric acid of 1.47 specific gravity with nitric oxide, NO, at 0°C. A certain amount of tarry matter is produced and this is said to be a measure of the amount of resinification that will occur on the bearing and thus increase friction when the oil is in use. Also, the test is used to determine the amount of oil that will carbonize in a gas-engine cylinder.

Cloud Test[2] (*"Cold Test"*).—The cloud test indicates the temperature at which waxes, such as paraffine, or solid fats, such as stearin or palmatin, separate from solution in the oil. The point at which the sample becomes opaque under specified conditions is known as the cloud test.

Pour Test[2] (*"Cold Test"*).—The pour test indicates the temperature at which a sample of oil in a cylindrical container of specified diameter and length will just flow under given conditions. This temperature is determined by placing the sample in a refrigerating mixture and noting the temperature at which the oil ceases to flow.[3]

Acidity.—Occasionally lubricants contain small amounts of free acid, and this corrodes the metallic parts with which the oil may come into contact. If any acid exists in mineral oil it is more than likely sulfuric which was not removed because of lack of care in refining, but properly refined mineral oils should contain no acid. In fatty oils very often fatty acids are present because of hydrolysis and oxidation, this being especially true of old oils.

The presence of free acid may be determined by shaking about 5 c.c. of the sample in a test-tube together with about an equal

[1] Oil Analysis, 1909 ed., p. 45.

[2] *Proc.* Am. Soc. Test. Matls., **15**, 281.

[3] See STILLMAN: Examination of Lubricating Oils, p. 10.

quantity of water to which a drop of phenolphthalein solution has been added and only *just* enough sodium hydroxide to color it faintly pink. If too much of the alkali is used, then even though acid be present, it may not exist in sufficient quantity to neutralize all of the alkali. If the pink tint is discharged, acid is indicated. But this test does not distinguish between mineral and fatty acids. If it is desired to make this distinction, the following method will suffice. Shake about 5 c.c. of the oil with an equal quantity of hot water in a test-tube for about 2 min. Allow the resultant emulsion to cool and separate, and then draw off the water by means of a pipette and transfer it to an evaporating dish. Add to this a few drops of methyl orange solution, which is unaffected in the presence of fatty acids but in the presence of mineral acids turns red.[1]

The Soap Test.—When an oil has an unnatural viscosity, or tends to form threads upon dropping or when the stopper is removed from its container, it should be examined for soap. The test given by Gill[2] depends upon the fact that the metaphosphates of the metals used as soap bases are insoluble in absolute alcohol. Also, the oil may be ignited and its ashes examined.

The thickening of oil with soap is generally considered as very bad practice. The soap or "oil pulp" greatly increases the viscosity of the oil, but the disadvantages[2] are that it causes the oil to chill more readily and to emulsify, thus increasing the friction. Furthermore, it is precipitated by water and steam thus causing clogging of the machinery. Also, it separates from the oil and clogs the oil inlets, thus causing uneven lubrication. This results in a heated bearing, and when a soap-thickened oil is heated its acquired viscosity vanishes. The oil will not then stay in place and the final result is disastrous.

THE SELECTION OF LUBRICANTS

General Requirements.—An oil must be especially adapted to the purpose to which it is put. The amount of pressure is a very important determining factor, the heavier the pressure the more viscous the oil and *vice versa*. At the temperature at which

[1] For methods of determining the amount of acid see LEWKOWITSCH, 1913 ed., vol. 2, p. 437, and STILLMAN: *Loc. cit.*, p. 46.

[2] Oil Analysis, p. 42.

23

the oil is used it must have sufficient viscosity to resist being forced out by the maximum load that is placed upon the bearing, but any viscosity above this means only a waste of power. On general principles the most fluid oil that will stay in place must be used.

Pressure and speed are closely associated. High speeds are usually accompanied by light pressures and heavy pressures by slow speeds. The higher the speed the less is the density and the viscosity of the oil required. If a viscous oil were used for high speeds, as for spindles rotating at 6,000 to 8,000 revolutions per minute, the cohesion of the particles would be too great to be overcome with sufficient rapidity. Thus there would be a tendency to retard the motion. On the other hand, an oil that would be suitable for a rapidly rotating spindle would be squeezed out at a slow speed.

Also, as has been indicated, it is very important to consider the temperature at which the oil is used. If the oil solidifies at a relatively high temperature, the friction would be unduly increased if used in exposed situations in cold weather. And an oil that would work satisfactorily in a cold location would not be equally satisfactory for the same speed and pressure in a warm room, for under this condition its viscosity would likely be too low.

The oil should not thicken when exposed to the air or while in use, either because of chemical alteration, or because of the evaporation of a volatile portion that would leave a more viscous residue. The oil should not lose more than 4 per cent. by weight when exposed for a working day (8 to 10 hrs.) at the temperature of the bearing. The flash point should be above 300°F. The fire test is usually from 30 to 50°F. above the flash point. Viscosity usually increases with increase of flash point and fire test.

The oil should contain no free acid, for this would corrode and roughen the bearings. Fatty acids form soaps with the metal thus causing the oil to "gum."

Special Requirements.—The characteristics of the oils suitable for the various purposes indicated in the following paragraphs are largely as stated by Gill.[1]   In some cases extracts from specifications are given to serve as examples of the requirements demanded for these special uses.

[1] ROGERS and AUBERT: Industrial Chemistry, p. 606.

Spindle Oils.—The oils known by this name have usually a low specific gravity and may be either thinly fluid or quite viscous. They are used for a variety of light machinery operating both at low and high speeds. The spindles in textile mills, sewing machines and typewriters may be considered examples.

The density varies from 35 to 27°Bé.; flash point from 320 to 430°F.; viscosity 30 to 400 sec. Saybolt at 70°F.; and the evaporation test should not be over 4 per cent.

Engine Oils.—These are classed as light and heavy oils. Aside from being used on engines they are applicable to shafting and machinery in general.

The density varies from 32 to 23°Bé., flash point from 300 to 430°F., and the viscosity from 50 to 400 sec. Saybolt at 70°F. For heavy pressures or rough bearings they may contain about 20 per cent. of animal oil, as lard or whale oil.

Electric Motor and Dynamo Oils.—These oils are subject to very high speed, medium pressure, and low temperature. The following specifications are issued by the S. T. Chase Machine Co.[1]

The specific gravity should be between 32 and 24°Bé., the specific viscosity[2] taken at 70°F. should be between 3.50 and 7.50 with a preferred viscosity of about 5.25; flashing point (closed tester) must not be less than 330°F., and the burning point (open tester) should not be less than 420°F.; acidity must not exceed 0.10 per cent. of free mineral acid or more than 1.50 per cent. of fatty acid calculated as stearic. The congealing point must be below 28°F.

Crank-case Oils.—An oil having the following characteristics has proved to be very satisfactory: density 27 to 26°Bé., flash point 455°F., viscosity Saybolt 100 sec. at 212°F.

Steam-cylinder oils are usually compounded oils, unless the cylinder is operated with superheated steam in which case it is advisable to use a pure mineral oil. The compounded oils are prepared by mixing with the mineral-oil cylinder stock, from 5 to

---

[1] Quoted by STILLMAN: Examination of Lubricating Oils, p. 82.

[2] Specific viscosity is obtained by dividing the time of efflux of the sample by the time of efflux of pure water. But specific viscosity obtained in this way is not a constant factor. The type of viscosimeter used affects the value.

25 per cent. of fatty oil, preferably tallow oil that is as acid-free as possible. The advantage in using the mixture depends upon the fact that fatty oils adhere to the moist metal to a greater degree than mineral oils, and also that they enable the lubricant to retain a greater viscosity at the higher temperatures. However, in the high-pressure cylinders the fatty oil is not necessary since the metal in this case is absolutely dry. Besides, with the superheated steam the fatty oil would be more completely hydrolyzed with the production of fatty acids that would corrode the bearings. Also in either high- or low-pressure operation, if it is desired to use the condensed steam for feed water, or for ice making or similar purposes, it is advisable to use only pure mineral oil since the fatty oils produce emulsions with water that do not very readily separate. Since the working temperature in steam cylinders is not sufficiently high to decompose the asphaltic or other similar hydrocarbons in a mineral oil, the use of filtered oils is not so essential as it is for gas-engine lubrication.

The mineral oil stocks which may be used for compounding have the following characteristics: density 28 to 23°Bé., flash point 500 to 630°F.; viscosity 100 to 230 sec. Saybolt at 212°F.

Gas-engine Cylinder Oils.—It is obvious that an oil that is suitable for the lubrication of a steam cylinder would not be equally suitable for a gas-engine cylinder. The temperature within the steam cylinder ranges usually between 300 and 400°F., while the temperature of a gas-engine cylinder when in operation lies in the neighborhood of 1,200°F. Aside from having a relatively high viscosity at a high temperature these oils must be free from any material that will leave a carbonaceous deposit when burned. Practically all of the oil that enters the cylinder is destroyed by combustion.

In order to remove the asphaltic and similar matter which seems to have a pronounced tendency to yield a carbonaceous deposit, these oils are usually carefully filtered through bone-black or China clay in the refining process. The removal of this asphaltic matter lessens the viscosity.

It is said that other unsaturated hydrocarbons such as the olefines, which are produced in a certain amount by the process of destructive distillation, as shown on page 36, also tend to yield a carbonaceous deposit. Assuming this, it follows that oils whether of paraffine or asphaltic base, if refined by the "cracking" process are not so suitable for gas-engine cylinders as are the paraffine-base oils refined by steam distillation. These olefine and similar hydrocarbons are not removable by filtering.

It has been found that the carbonizing tendency of the mineral oils is indicated by the amount of tarry matter yielded in the gumming test.

For automobile engines, oils having the following characteristics are indicated: density 33 to 30°Bé., flash point 380 to 450°F., and viscosity 180 to 185 sec. Saybolt at 100°F. For larger-sized gas engines a heavier oil is used: density 28 to 26°Bé., flash point 400 to 475°F.; viscosity 250 sec. at 70°F.

Steam-turbine Oils.—Turbine bearings are lubricated by a forced or circulating pump system, hence the oil used should be comparatively low in viscosity. Also, because the working temperatures are frequently rather high, a thin oil is more desirable since such an oil loses its viscosity to a less degree than a more viscous one. A thin oil is also demanded by the high speed of the turbine. Another important requirement is that it shall separate readily from water, consequently the oil should be purely mineral. Gill says an oil of 30°Bé., 150 sec. Saybolt at 70°F. and 420°F. flash point has given good results.

Cooling Liquids and Lubricants for Cutting Tools.—To prevent an injury to the tool by the heat developed during use, a stream of liquid is directed against it. Because of its high specific heat, water serves well for this purpose, but because plain water readily rusts clean iron surfaces soaps are dissolved in it, either sodium (hard) or potassium (soft) soap or salts such as borax or washing soda, all of which ionize to produce an alkaline reaction, and thus offset the corroding tendency.

Also it is advantageous and frequently necessary that a lubricant be introduced between the tool and the work. Then for this double purpose the liquid used commonly consists of mixtures of watery solutions with oils. A fatty oil, often lard oil, is more commonly used, at least in part, because it emulsifies readily with the watery solution. Aside from their corrosion-inhibitive properties the soaps aid in the lubrication.

That the lubricant penetrates to the immediate neighborhood

of the cutting edge of the tool is shown[1] by the increased smoothness of the finished surface and the greater force required to keep the tool against the work. The lubricant finds its way in through the gap between the work and the beveled side of the tool, probably by the aid of capillarity. The edge of the tool being lubricated best on the beveled side nearest the work, is unable to dig in and tear the metal, which is, therefore, finished smooth and partially burnished.

References

A

Davis: Friction and Lubrication, Pittsburgh, 1904.
Gill: Oil Analysis, Philadelphia, 1909.
Hurst: Lubricating Oils, Fats, and Greases, London, 1911.
Alford: Bearings and Their Lubrication, New York, 1911.
Archbutt and Deeley: Lubrication and Lubricants, London, 1912.
Thorp: Outlines of Industrial Chemistry, New York, 1912.
Southcombe: Chemistry of the Oil Industries, London, 1913.
Stillman: Examination of Lubricating Oils, Easton, 1914.
Rogers and Aubert: Industrial Chemistry, New York, 1915.

B

Acheson: A New Lubricant, Deflocculated Graphite, *Eng. News*, **58**, 127 (1907).
Mabery and Mathews: Viscosity and Lubrication, *Jour. Am. Chem. Soc.*, **30**, 992 (1908).
Gillett: Analysis and Friction Tests of Lubricating Greases, *Jour. Ind. Eng. Chem.*, **1**, 351 (1909).
Conradson: Laboratory Tests of Lubricants, *Jour. Ind. Eng. Chem.*, **2**, 171 (1910).
Mabery: Lubrication and Lubricants, *Jour. Ind. Eng. Chem.*, **2**, 115 (1910).
Fils: Manufacture of Lubricating Oils, *Petroleum*, **5**, 1165 (1910).
Waters: Action of Sunlight and Air Upon Some Lubricating Oils, *Jour. Ind. Eng. Chem.*, **2**, 451 (1910).
Waters: Effect of Added Fatty and Other Oils upon the Carbonization of Mineral Lubricating Oils, *Jour. Ind. Eng. Chem.*, **3**, 812 (1911).
Nastyukov: Chemical Composition of Lubricating Oils, *Jour. Soc. Chem. Ind.*, **30**, 201 (1911).
Southcombe: Deposits Resulting From the Lubrication of Engines, *Jour. Soc. Chem. Ind.*, **30**, 261 (1911).
Ellis: Relation Between the Temperature and Viscosity of Lubricants, *Met. Chem. Eng.*, **10**, 546 (1912).
Michael: The Critical Examination of Lubricating Oils, *Petroleum*, **6**, 2238 (1912).

[1] Archbutt and Deeley: Lubrication and Lubricants, p. 546.

JONES: Automobile Greases, *Nat. Petroleum News*, **5**, No. 12, 31 (1913).
FAWCETT: Lubricating Oils, *Nat. Petroleum News*, **6**, No. 2, 37 (1914).
ACHESON: Colloidal Graphite Lubrication, *Chem. Eng.*, March, 1915.
PHILIP: Demulsification Values of Mineral Lubricating Oil for Steam Turbines, *Jour. Soc. Chem. Ind.*, **34**, 697 (1915).
MOORE: Lubricating Oil for Diesel Engines and Air Compressors, *Engineering*, **120**, 176 (1915).

CHAPTER XV

GLUE

Glue is a product obtained by boiling with water suitably prepared animal matter, such as hides, bones, sinews, etc., and then drying the resultant solution. It is generally believed that the boiling brings about a process of hydrolysis; that is, a definite chemical combination between water and the glue-yielding substance. Certainly the process accomplishes more than the mere extraction of a substance or substances that exist naturally in the tissue, for the material used contains no glue as such.

Composition of Glue.—Glue is a very complex material. It consists of several more definite substances, such as gelatine, chondrin, keratin, etc. The gelatine makes up the greater bulk of the glue and is itself a substance from which a great number of compounds have been isolated.[1] But even then, the chemical constitution of gelatine is very imperfectly known. Still less is known of the more complex glue.

Distinction between Glue and Gelatine.—It is rather difficult to make an exact distinction between glue and gelatine. Commercially they are distinguished by their appearance; gelatines being usually transparent, producing very clear solutions and jellies, while glues are darker. But gelatine is obtained like glue by boiling hides, bones, etc., although the stock is much more carefully selected and washed, and the process is carried out under more sanitary conditions. In a broad sense, glue may be considered as an impure gelatine, or gelatine as a refined glue. It should be noted that the boiling process in the production of glue is carried out at higher temperatures, and glue is generally considered to be more hydrolyzed than the gelatine. The higher degree of heat employed in the production of glue makes it more liquid and lessens its gelatinizing power, probably because of the greater hydrolysis. Although glue has a lower gelatinizing power than gelatine, its adhesive power is greater.

[1] ROGERS and AUBERT: Industrial Chemistry, p. 961.

Colloidal State of Glue and Gelatine.—Glue and gelatine are excellent examples of substances that exist naturally in the colloidal state. In fact, from the Greek κόλλα (glue), the term colloid has been derived. Substances that exist in the colloidal state are distinguished primarily from those that exist in the crystalloidal state, but most substances can be caused to assume either state by suitably adjusting conditions.

Colloidal Sols and Gels.—When colloids are added to water they may appear to dissolve and the resulting fluid state is commonly called a "solution," as for example, a "glue solution." In reality, the degree of dispersion or physical subdivision is not so fine as in the case of a true solution such as is produced when crystalloids dissolve, and the so-called solution is more properly considered as a state of extremely fine suspension. In colloidal chemistry this state of fine suspension is known as the *disperse phase* or *sol* of the colloid. Thus the fresh white of egg, just as it comes from the shell, is a colloidal sol of albumin. Although the suspended particles in the sol are considerably larger than the dissolved particles in a true solution, they are nevertheless sufficiently fine that they do not readily settle out as do the particles of coarser suspensions, such as muddy water, for example. Neither can they be filtered out by ordinary filters. However, they are too coarse to pass through parchment paper or other dialyzing membranes that are permeable to particles in true solution.

When water is extracted from a sol its viscosity increases and eventually a solid, jelly-like mass is produced. To this form the name *gel* has been applied. In the cases of glue and gelatine, this swelled, jelly-like state is also assumed when the dry substance is allowed to stand in cold water. The swelling is due to the water imbibed and the amount taken up may amount to several times the weight of the dry substance, but glue and gelatine do not form sols in cold water, although they very readily assume this state when the softened gel is gently warmed. When the sol is allowed to cool the gel forms again and the process may be repeated. Of course, it is possible to prepare sols of these substances so weak that they will not form gels on cooling, but Rideal[1] says that if the strength of the gelatine sol is not under

[1] Glue and Glue Testing, p. 6.

1 per cent., a gel will be formed when the sol is allowed to cool.

It must be understood that gels are not always soft substances like the water-saturated glue jelly. The degree of softness or hardness depends upon the degree of swelling. Gels may become very hard when they are allowed to dry out, but they are still gels. Thus the familiar, hard-flake or sheet glue is a hard gel produced by drying a voluminous, swelled jelly.

Reversibility.—All dried gels are not able to be transformed into sols again in the manner just described for glue. Upon the basis of whether or not they can be thus transformed, colloidal substances are divided into two classes known as *reversible* and *irreversible colloids*. Glue and gelatine are excellent examples of the reversible class, as has been shown. But those colloids that become impermeable to water as they dry cannot be caused to swell or to be directly transformed into the sol state and are known as irreversible colloids. The colloidal carbon of drawing inks and silicic acid are examples of this class.

Coagulation is the term applied to the formation of a precipitate in sols by the application of heat or by the addition of precipitating agents. Thus, the albumin sol of natural egg-white is coagulated by heating to the temperature of boiling water for a short time, the product being known as "hard-boiled white of egg." The same effect may be produced by the addition of nitric acid to the albumin sol. Just what constitutes the change during coagulation is not exactly understood, but there is evidence to show that it partakes of the nature of a chemical reaction.

Glue and gelatin sols are coagulated by formaldehyde, tannin and alum, the coagulated product being "irreversible" in that it is immune to the action of water. The same effect is produced by chromium salts upon exposure to light. This effect is of importance in using glue where it is required to resist being softened by water. Methods of preparing solutions of glue for this purpose are shown at the end of the chapter.

Materials Used for the Manufacture of Glue.—The stock used in the manufacture of glue consists chiefly of the hides, bones, and sinews of cattle, the hides of horses, the skins of the sheep, goat, deer, and pig, together with the bones of some of these. Also, the air bladders and other portions of fish are used. Aside from these, glue is made also from the rabbit skins left after the hair has been removed for the manufacture of hats, from clippings of parchment and gloves and from any other form of hide waste. The calf, cattle, goat, deer, pig, and horse skins are valued in about the order given. Of these hides and skins, it is only the portions that are unsuitable for leather that are used for glue. Tanned hides are not available for glue stock because the tannin renders the glue-yielding substance insoluble.

Preparation of Stock.—If the stock to be used consists of skins, it is first necessary to treat it with lime water, except in those cases where it has been delivered from the tannery and is already sufficiently limed. The lime treatment is necessary to remove the hair and fleshy portions, and it also causes the hide to swell and makes the extraction more easy. The length of time required for the liming process depends upon the stock, varying generally from 1 to 3 days, but for very heavy pieces several weeks are necessary. Care must be taken to prevent both over- and underliming, since either will produce defects in the stock that cannot be prevented from producing defects in the glue. After liming is finished, the hides may be merely washed, or the remaining lime in the skins may be neutralized with sulfuric acid. Calcium sulfate is formed in this way, and this being insoluble is carried into the finished glue, producing a certain cloudiness. For very clear glues hydrochloric acid is used, which produces the very soluble calcium chloride, and which may be more readily washed out. Depending upon the treatment at this stage, that is, whether the lime is removed as much as possible by washing or more completely by treatment with acid, the finished product may be acid or alkaline in reaction. Fernbach says[1] that this is important since the bacteria which bring about the decomposition of a glue multiply rapidly when it is alkaline, but have a much harder struggle for existence in a glue having an acid reaction. Hence the latter keep longer than the former in the form of a solution or jelly without the use of preservatives.

If the stock consists of bones, these are generally ground and the grease removed by steaming or extraction with benzine. After degreasing, the mineral matter, that is, the calcium phosphate, is removed by leaching with dilute hydrochloric or sul-

[1] Glues and Gelatine, p. 7.

furic acid and washing with water. A soft cartilaginous mass, equal to about one-third the weight of the original bone, is left. This, when dried, is commercially known as *ossein*. From ossein, glue may be made by boiling. The term "acid-treated bone glue" is used to indicate glues made in the manner just described, in order to differentiate them from glues made from untreated bones. Mixtures are often made of bone and hide stock or of sheep, goat or ox stock, thereby producing better glues than the straight glue would be.

Boiling.—After the stock has been prepared as desired, it is boiled with water to convert the glue-yielding substance into glue. The change produced by the boiling is generally believed to consist essentially of a process of hydrolysis, but it is very probable that other changes occur also; the transition process, as a whole, from glue-yielding substance to glue being not very well understood.

According to the nature of the stock and the preference of the manufacturer, the glue may be obtained from the stock in one operation or in several "runs." If the latter method is used, the extraction is begun using only a small quantity of water and the lowest temperature by which glue may be produced. After this extraction has been drawn off, more water is added and the boiling is continued for a longer time and at a higher temperature. Four or five extractions may be made in this way, the temperature employed for the last extraction being about 100°C. The higher the temperature and the longer it is applied, the lower will be the quality of the glue. Thus, the glue obtained by the first extraction has the greatest gelatinizing power and the greatest strength, while these qualities in the later "runs" grow less. Also, in producing glue from untreated bones, pressure tanks are used and the glue produced in this way has usually less strength than that produced by the open-tank method because of the higher temperature employed.

Preservatives.—During the extraction, if it is thought necessary, preservatives are added. There are at least three of these that may be used: formaldehyde, phenol, and boric acid. Rideal says[1] that the minimum strengths that will prevent putrefaction are as follows:

[1] Glue and Glue Testing, p. 40.

Formaldehyde...................... 1 part in 20,000
Phenol............................ 1 part in 1,000
Boric acid........................ 1 part in   200

Formaldehyde is a good preservative but has a tendency to cause
the glue sol to coagulate when the glue is prepared for use.
Phenol, or carbolic acid, is barred to some extent on account of
its odor. Boric acid, although low in antiseptic power, necessi-
tating the use of greater amounts than the other preservatives,
is found to be very satisfactory.

Clear Glues.—Calf and acid-treated bone stock generally yield
very clear glues. Clear glues may be produced also by bleaching
and clarification, while the glue is still in the liquid form. Bleaching is generally accomplished by the use of sulfur dioxide, after
which the glue solution is treated with milk of lime to neutralize the
sulfurous acid formed. Depending upon whether an insufficient
amount or an excess of the lime is used, the final reaction may
be either acid or alkaline, and this should be remembered in
testing. Or the coloring matter may be swept from the solution
by producing therein flocculent precipitates which upon settling
carry down the coloring matter. Alum, albumin, and other substances are used for this purpose. Fernbach says that alum has
a tendency to make the glue solution very stringy and low in
penetrative power.

Opaque or Moulding Glues.—The appearance of the glue may
also be modified by the addition of zinc white, ZnO; chalk, $CaCO_3$;
barytes, $BaSO_4$; or talc, a hydrated magnesium silicate. A limited amount of such additions produces that which is known as
a "colored" glue, while greater amounts produce a white or
"opaque" glue. The latter is also known as *moulding glue.*
Its price is always somewhat higher than natural glues of the
same strength.

The Glue Jelly.—After the glue solution has been treated as
desired, it is run into pans and allowed to cool, or is chilled artificially, in order that it may gelatinize. When the jelly has
formed it is removed from the pans, placed in frames and sliced
by wires into sections averaging ½ in. in thickness. These slices
are then placed on nets of twine or galvanized wire and dried,
either artificially or by natural air currents. The glue pieces
often retain the impression of these nets, the marks being readily
seen in the sheet and flake glues of commerce.

Commercial Forms of Glue.—Since the glue jelly when placed on the drying nets consists largely of water, the drying process greatly diminishes the thickness, so that the slices take on the form of more or less thin sheets. The amount of the decrease is, as a rule, greater in the glues of better quality, since in the formation of the jellies of these, less glue by weight was required to produce a jelly firm enough to handle. The dried glue may be marketed in the form of sheets, but usually the sheets are broken into small pieces, which constitute the familiar flake glues. The flakes may also be ground to varying degrees of fineness. There are other forms, as strip, ribbon glue, etc.

Grades of Glue.—When the glue is marketed it must be classified or graded in order that its price may be fixed. The grades are determined by comparison with accepted standards. The comparison is based on the determination of certain physical factors, of which the viscosity of the solution and the strength of the jelly are considered most important. Other factors such as odor, keeping qualities, acidity or alkalinity, presence of grease, tendency to foam, etc., may also be ascertained. Only by determining the grade of the glue can the manufacturer properly assign a price. Cost of stock and cost of production cannot govern this entirely, since variations in the conditions of production, perhaps accidental or uncontrollable, may markedly affect the quality. In the same manner the consumer should determine the grade in buying. A more or less fixed price pertains to a definite strength of glue, and by determining the strength it can be determined whether or not its quality is commensurate with the price paid.

There is no absolutely fixed system of standards in use, but the eleven grades originally employed by Peter Cooper, both because they were among the very first manufactured in this country, and because they have maintained a certain reliability, have been more or less commonly chosen as standards by both manufacturer and consumer. Ranging from strongest to weakest they are: A Extra; 1 Extra; No 1; IX (one cross); $1\frac{1}{4}$; $1\frac{3}{8}$; $1\frac{1}{2}$; $1\frac{5}{8}$; $1\frac{3}{4}$; $1\frac{7}{8}$; and No. 2. Although these standards are chosen for discussion here, it is not meant to indicate that they only are reliable. Grades exceeding the strength of the best of the series are regularly manufactured, as well as grades that may lie between any two.

Glue Testing and Information Derived from Tests.[1]—As has been indicated, the tests applied to glue are concerned chiefly with the determination of certain physical characteristics of the glue solution and the jelly formed from it. In addition, a preliminary examination of the dry glue is of value. Although the tests applied are largely only comparisons, they are of sufficient accuracy to furnish quite a definite idea of the strength of the glue and the price that should be paid for it.

Preliminary Examination of the Dried Glue. Cut and Flexibility.—The cut of a glue may give some indication of its quality. Thin-cut glues are obtained by making a jelly that consists largely of water and cutting in slices varying from $\frac{1}{4}$ to $\frac{1}{2}$ in. thick. These reduce to about $\frac{1}{16}$ in. upon drying. If an attempt is made to cut the glue into such thin sections directly from a solid jelly, it curls badly. It is only a very good glue that can stand the large excess of water and still form a jelly solid enough to cut. A thinly cut glue that is flexible enough to be bent into the shape of a horseshoe is a good glue. Moisture tends to increase the flexibility of glues and this should be remembered in testing.

Fracture.—When a piece of glue is broken, if the fracture is splintery, a strong glue is indicated. A glassy fracture indicates a brittle glue of less strength. Glues made from bone stock generally break with a clean fracture, as do also those that have been overheated in manufacturing.

Bubbles.—The surface of the piece of glue, examined in a good light, should be free from bubbles. Bubbles are generally considered to be evidence of putrefaction, which began in the glue stock and continued in the jelly while on the drying nets. Further information may be obtained by moistening the glue and noting the odor. Sometimes bubbles may occur in the glue beneath the surface due to the incorporation of air while the glue solution was being poured into the mould to gelatinize. Such bubbles are apt to be found in good glues whose solutions have higher viscosity and hence allow bubbles of air to separate less readily. Air bubbles are generally small and more or less regular

---

[1] The discussion given here on methods of testing and information derived from tests is largely as given in FERNBACH'S Glues and Gelatines. For detailed directions for carrying out tests, see this volume, p. 50.

in outline, while bubbles resulting from decomposition are larger and more irregular in shape. The marks of the drying net must not be mistaken for surface bubbles.

Mixtures.—Flake, and particularly ground glues, may consist of mixtures. This may be determined by visual examination in a good light. If the glue is ground, it is necessary to compare pieces of the same size, since smaller pieces possess apparently a lighter color. Glues may be mixed either to cause the jelly strength of the mixture to accord with the specifications or perhaps to lower the price to suit the requirement by the introduction of a cheaper glue.

Examination of the Glue Solution. Odor.—The odor of the solution will vary according to the stock from which the glue was made, but there should be no trace of putrefaction.

Acidity and Alkalinity.—Strictly neutral glues are very seldom found. Enough acid or alkali will likely be present to show a reaction with litmus paper. Correction of overliming, bleaching, and acid treatment to remove mineral matter in the making of bone glue, will cause the finished product to be acid. Since "colored" and "opaque" glues are generally made from bleached stock, these glues may be acid unless the whitening agent used was some "anti-acid" as chalk or zinc white. Overlimed stock that has been washed only will be alkaline. If a glue has an earthy smell, it quite likely has been overlimed.

Viscosity.—This term refers to the rate of flow of a glue solution through a specified opening, as compared with a standard, usually water. A solution that flows sluggishly is said to be high in viscosity and *vice versa*. The viscosity determination is of great value but is not alone sufficient for grading glues. Acid-treated bone glues are much lower in viscosity than would be expected according to their true strength. Opaque and colored glues are slightly higher in viscosity than natural glues of the same strength. The majority of foreign glues, having been clarified, are low in viscosity, unless clarified by alum, when the reverse is the case. Very clear glues that are exceedingly high in viscosity—so that they are "stringy"—are likely to be alum-clarified.

Jelly Strength.—This is the most important and determinative of all the tests. The strength of a glue jelly may be determined by especially designed apparatus, as plungers weighted with shot and by other devices, but for all practical purposes it can be well enough measured in comparison with standards by its resistance to the pressure of the finger.[1] The real strength of a glue bears a fixed relation to the strength of its jelly.

Foreign Glues.—Fernbach says that the foreign glues are generally made from more carefully selected stock than are the domestic glues, the stock being frequently hand-picked, a process that is out of the question for the American glue maker because of the cost of labor. And further, that the glue liquor is generally clarified, which causes the glue to be lower in viscosity than is usual for glues of their jelly strength, just as is the case with domestic acid-treated bone glues. But on the other hand, the attempt to remove the viscous, non-adhesive elements by clarification is apt to result in the loss of adhesive material as well. Foreign glues appear usually in the form of more or less thick cakes.

Fish glues are prepared from the heads, bones, swimming bladders, and other portions of fish. With the exception of isinglass, they are not prepared in the dry state, but in the form of heavy solutions or pastes, which have the characteristic odor of fish. An effort is generally made to mask this odor by the use of essential oils, such as sassafras and wintergreen. Fish glues in general possess high adhesive properties. Unlike the other fish glues, isinglass occurs in various solid forms, as sheet, lump, etc. It is very hard, transparent and practically colorless. It is prepared from the swimming bladder of the Russian sturgeon and possesses the highest adhesive properties of all the glues.

Selection of Glues for Wood Joints.—Other factors being satisfactory, glues ranging in strength from 1⅜ to A Extra are suitable for wood joints, the exact grade being determined by the character of the work. The viscosity of the solution should be as low as possible, since the adhesive power of the glue is dependent upon the degree to which it enters the pores of the wood. On this account, thin solutions produce stronger joints than thick ones. A well-made, acid-treated bone glue will make a strong joint because its solution will be thin. Hence it is necessary to consider both viscosity and jelly strength. Clarified glues, unless

[1] For method of conducting test, see FERNBACH: *Loc. cit.*, p. 50.

24

clarified by alum, have lower viscosity and, therefore, greater penetrating power. Foreign glues are generally clarified and, therefore, of a foreign and domestic glue of the same jelly strength, the former will likely produce the stronger joint because it will penetrate the wood more readily than the latter. But, if a foreign and a domestic glue of equal price are chosen, the domestic will be by far the stronger. Therefore, more glue solution of a specified quality can be made from a pound of domestic glue than from a pound of foreign glue costing the same amount.[1]

Preparation of Glue Solutions for Use.—To prepare the glue solution for use it is very important to first soak the dry glue in cold water. Then, after it has been entirely softened by soaking, it should be warmed in a water-jacketed container at a temperature not exceeding 80°C., using a thermometer to determine the temperature. Hot or even warm water must never be poured upon dry glue. If this is done, a layer of partially dissolved glue is formed on the surface that is practically impermeable to water and then the glue cannot be dissolved without the application of heat to either a high degree or for a long period. Such excess heating greatly injures the glue. The damage done is generally considered to be due to an increased hydrolysis brought about by the overheating, the change being described as comparable to the softening sought after in the cooking of meats to prepare them for food. Glue solutions should never be boiled.

For the preliminary soaking it is advisable to have the glue as finely divided as possible. A heavy mortar with a cardboard cover, having a hole through which the pestle may pass, can be used very advantageously to crush the glue if it is not as fine as desired. The soaking process will require from 4 to 12 hrs., or even over night, depending upon the degree of fineness. Good glues will generally absorb from one and a half to two and a half times their weight of cold water. It should be seen to that all pieces of glue are covered during the soaking process. If portions are not sufficiently softened, prolonged heating is necessary to effect the solution of these and then the whole of the contents of the vessel is injured. Thin, voluminous, flake glues may be softened without increasing the amount of water, by adding only portions at a time. Being thin, they absorb water rapidly and

[1] FERNBACH: *Loc. cit.*, p. 108.

recede quickly. During soaking it is advantageous to stir the mass, at least at first, to prevent sticking, since if this occurs a prolonged period of soaking is necessary, and if sufficient time is not allowed protracted heating is required. If ground glue is used, particles will float and stick to the sides of the container, with the result that they will be only partially soaked and hence insoluble without undue heating. To prevent this, only about half the proposed amount of water is added at first, and when partial soaking has occurred so that the particles will no longer float, the remainder of the water is added. In this way, the original surface line where the sticking occurred will be submerged.[1]

When the glue has been softened by proper soaking it may be readily dissolved by warming as has been described. Since glue solutions cannot be used cold, but must be continuously heated to keep them in a condition suitable for use, the strength of the joints will be far greater if successive small portions of glue are dissolved and used, rather than to prepare large amounts that must be kept heated for long periods.

The glue container should be very frequently cleaned. Glue solutions and jellies form excellent breeding media for liquefying bacteria and other microörganisms.

Flexible glues consist of glue, glycerine and water. The glue is dissolved in the ordinary way and an amount of glycerine about equal to the weight of the glue is added. Flexible glue is used for book binding, for preparing writing pads, etc.

Liquid glues may be fish glues as was stated under that head, or may be animal glues that have been kept from gelatinizing by the addition of certain salts or acids. For this purpose weak animal glues, that possess the gelatinizing power in low degree, are generally chosen. Consequently, their strength is far less than the fish glues they may be designed to imitate. Acetic and nitric are the acids most used.

Some so-called "liquid glues" contain no glue whatever, being merely solutions of corn dextrine, preserved generally with formaldehyde. This adhesive is used in manufacturing paper boxes, postage stamps, etc.

Insoluble or Waterproof Glues.—Glue may be rendered insoluble by the action of potassium dichromate in the presence of

[1] FERNBACH: p. 137.

light. The glue solution is prepared and then the dichromate, previously dissolved, is stirred in. The amount of the salt required is about equal to 2 per cent. of the glue solution. When applied and allowed to dry the glue becomes practically insoluble if exposed to the light, even for a brief period. About 5 per cent. of alum may be used in place of the chromium salt if the glue is to be applied where the light cannot penetrate, as in wood joints. Also formaldehyde and tannin are capable of rendering the glue insoluble under certain conditions.

Marine "Glue."—This rather well-known substance has a varying composition, but it never contains glue. It generally consists of rubber and some other substance as shellac, asphaltum, or some form of pitch dissolved in benzine.

References

RIDEAL: Glue and Glue Testing, London, 1900.
LAMBERT: Glue, Gelatin, and Their Allied Products, London, 1905.
KAHRS: Glue Handling, New York, 1906.
FERNBACH: Glues and Gelatine, New York, 1906.
TAGGART: The Glue Book, Toledo, 1913.
ROGERS and AUBERT: Industrial Chemistry, New York, 1915.

CHAPTER XVI

RUBBER

Rubber, or caoutchouc, seems to have first become known to Europeans about the year 1525, at which time it was mentioned in the writings of d'Anghiera, a Spanish traveller in Mexico. The name "rubber," as applied to this substance, was not thought of until about two and a half centuries later. In 1770, Priestley, the discoverer of oxygen, recommended it as a substance suitable for the erasing of pencil marks—hence a rubber. Thus from a specific use, its general name was derived.

Source.—Rubber is obtained from a considerable variety of vines, shrubs and trees that grow in moist climates of tropical and subtropical countries. The best grade has been obtained from a tree of the botanical name, *Hevea Braziliensis*, which is native to the Amazon district of South America. And because the shipments were made from the port of Para, Brazil, this rubber has become known as Para rubber. Comparatively recently, plantations stocked principally with Hevea have been established in various parts of the world, notably in Ceylon and Malaya, and a considerable portion of the world's rubber output has been produced by these, but for some rather obscure reason the plantation Hevea does not possess all the excellence of the wild Hevea of the Amazon district. Rubber is also produced by other botanical species in various countries, as in portions of Africa, India, Mexico, Central America, etc.

The Latex.—The juice of the plant from which the rubber is obtained is called latex. It is not the sap. It is white, resembling an oil emulsion, being in fact, a great deal like the juice of the common milkweed. The latex is obtained by the means of incisions made in the bark, the flowing juice being caught in small cups attached to the tree by the means of moist clay or in some similar manner. The amount of fluid collected from a single tapping varies greatly, but it is on the average about 1 oz.

And this is by no means pure rubber. According to Perkin[1] a typical latex shows the following composition:

	Per cent.
Pure rubber	32
Albuminoid and mineral matter	12
Water	56

Coagulation of the Latex.—In order that it may be exported, the latex must be treated so that the rubber globules are caused to coalesce and the albuminoid matter to be rendered immune to fermentation. Also the water must be evaporated to a large extent. To accomplish these ends the latex is subjected to a process known as coagulation, and there are many methods by which this is brought about. The latex may be smoked, treated with chemical reagents, boiled or evaporated, either by spreading on the ground or as is done in the Congo, by spreading on the human body. In the Para district the latex is coagulated largely by smoking. A paddle is dipped into the latex and then is rotated in the smoke of a smouldering fire, the process being repeated until a large ball or "biscuit" is formed. The methods employing chemical reagents are used largely on plantations where facilities are at hand for carrying them out.

Cleaning.—After coagulation, the rubber is ready for the market. But before it can be used it must be cleaned to remove earthy material, bark, twigs, etc. This is done by slicing and macerating it between rolls and washing it with water. Much loss occurs in the washing process, the amount varying from 10 to 50 per cent. After washing, the rubber is passed through rolls and formed into thin sheets, which because of their appearance are known as *crepe*. These sheets are then carefully dried, frequently in a vacuum drying apparatus.

Vulcanizing.—Pure rubber does not possess the properties that cause the familiar manufactured form to be considered so valuable. At the temperature of very hot summer weather, pure rubber becomes soft and sticky; during very cold weather, it becomes hard and brittle. In either of these conditions it is useless for the purposes to which it is commonly put. It is upon the discovery of the effect of heating the rubber with sulfur that the evolution of the present-time product rests. The sulfur effect

---

[1] F. Mollwo Perkin: *The Chemical Engineer*, August, 1913.

was first discovered by Goodyear at New Haven, Conn., in 1839, and 4 years later it was noted independently by Hancock of London. In 1846, Parkes of Birmingham, Eng., observed that the same effect is produced when thin sheets of rubber are immersed in a solution of sulfur monochloride, S_2Cl_2. The process of treating rubber with sulfur is known as vulcanization. It is not necessary to enumerate here all the effects produced by the treatment, but some of the more important are these: the rubber is rendered much less sensitive to changes of temperature; it acquires increased elasticity and tensile strength, it is more durable when exposed to the weather, and more resistant to the action of chemical reagents.

In vulcanizing by the Goodyear process, the crude rubber is very finely divided and intimately and uniformly mixed with the sulfur by continuously passing the mixture through rolls. For ordinary vulcanization, from 5 to 7 lbs. of sulfur are used for every 100 lbs. of rubber. However, because of other materials that are also used in the mixture as fillers, the amount of sulfur in the finished product does not usually exceed 2 or 3 per cent. After mixing, the mass is heated to about 120°C., usually with steam, this being known as the "steam cure."

The manner in which the sulfur reacts with the rubber is not definitely understood, but the rubber molecule is not broken down in any way, the sulfur uniting with it where certain "double bonds" exist to form addition products. All of the sulfur does not combine, a rather small percentage of it remaining as a mere mixture.

In the Parkes process, which is known as the "cold cure," thin sheets of rubber are immersed in a solution consisting of 2 or 3 per cent. of sulfur monochloride in some one of the more or less active rubber solvents, such as carbon disulfid, benzole, etc. Also, vulcanization may be brought about by exposing the rubber to the vapors of sulfur monochloride.

The properly vulcanized product containing 40 per cent. of high-grade rubber will stretch to seven times its length and return to its original condition. If when drawn out it remains partly stretched, or if a strong indentation leaves a permanent hollow in the solid piece, the rubber has been poorly or incompletely vulcanized.

Other Materials Used in Rubber.—Even though rubber is greatly improved by vulcanization it does not even then possess the properties required for most of its uses. On this account, at the same time that the sulfur is incorporated there is usually mixed with the rubber a certain amount of other materials. These added materials are employed even in high-grade rubber; in fact, pure vulcanized rubber has a very limited use. Of course, it not infrequently happens that many foreign substances are introduced into rubber merely to increase the weight and bulk and thus to cheapen, but if the materials are properly chosen and used, they increase the durability of the rubber by lessening its oxidation, mechanical abrasion, etc. All materials used in this way are admixed in the form of a very fine powder.

According to Schidrowitz[1] the materials employed in the manufacture of rubber articles, aside from rubber and sulfur, may be classified as follows:

"(a) 'Fillers' or 'cheapeners' pure and simple.

"(b) Materials employed for imparting a certain quality or qualities, as hardness, toughness, resistance to steam, etc.

"(c) Pigments.

"(a) **Fillers.**—Powdered chalk, barytes and ground rubber waste may be regarded as typical 'cheapeners,' possessing little if any valuable specific quality. In many cases such substances as zinc oxide, lithopone, and oil substitutes are also used solely as cheapeners, but each of these materials may, if rationally employed, serve some useful specific purpose aside from the lowering of cost. Thus zinc oxide hardens and toughens a mixing, and increases its resistance to cutting and abrading forces, if used within certain limits and together with other suitable materials. The same remarks apply to lithopone. The oil 'substitute' serves to produce rubber goods in which cheapness combined with a low specific gravity and an absence of hardness are required. Good 'reclaimed' rubber is a valuable ingredient in mixings that have to be made at a moderate price and in which a large quantity of 'minerals' cannot be employed.

"(b) **Materials Used for Specific Purposes.**—1. *For increasing mechanical strength,* or, in other words, for hardening and toughening the material, minerals, such as zinc oxide, lithopone,

---

[1] Rubber; Its Production and Industrial Uses.

magnesium oxide, lime, litharge and for special purposes (for instance, for 'eraser' rubber) ground glass and talc are used. Certain organic substances, for instance balata—which contains a hydrocarbon similar to that of rubber—also possesses a toughening effect, without, however, speaking broadly, decreasing the elasticity or rather distensibility of the material to the same extent as 'minerals.'

"2. *For making dense, i.e., non-porous, mixings for increasing resistance to water and improving the dielectric properties,* organic materials such as asphalt, bitumen and coal-tar pitch are exceedingly useful.

"3. *For softening* harsh mixtures, *i.e.,* for cheap goods containing a large quantity of minerals, such as chalk or barytes, certain oils and more particularly vaseline are employed.

"4. *For hastening vulcanization or improving vulcanizing conditions,* such materials as magnesia, quicklime, and antimony sulfide (golden sulfide) are in many cases essential.

"(c) **Pigments.**—The most important white pigments are zinc oxide, lithopone, barytes, and white lead. Antimony pentasulfide and mercuric sulfide (vermilion, cinnabar) are very largely employed as red pigments, the latter chiefly for hard rubber, mainly in dental work. Pigments and dyes other than white and red, are employed only for special purposes, as for toys, flooring tiles, imitation mozaics, fancy waterproof cloth, etc."

Manufacture.—In the manufacture of rubber articles of the so-called "mechanical" type—pneumatic tires, for example— the washed and dried rubber is mixed with the desired amount of sulfur and filling materials and the resultant "dough" is rolled out into sheets of the required thickness by means of smooth rolls, and attached to cloth to prevent sticking. The sheets are then cut and made up according to the method best adapted to the article being produced. Sometimes strips made from the sheets are wound in a manner suitable to form the article, or they may be superimposed in layers after having been cut to the desired form. After being formed into shape they are vulcanized, or "cured" as it is called, by applying heat and pressure, or heat only. The heat may be furnished by the means of steam or in some other manner, but the best results seem to be produced by

steam generally.   In either case the vulcanization may be carried
out in a mould or in the open.   Washers, heels and valves are
stamped from sheets and then vulcanized in moulds.

Properties of Soft Vulcanized Rubber.—In contact with the
air, rubber is slowly oxidized, with the formation first of a plastic,
sticky substance, which by further oxidation is gradually changed
into a dry and readily pulverent form.   This oxidizing action is
rapidly hastened by the ultraviolet or "chemical" rays of the
sun.   Hence the better the rubber is protected from light, the
longer its life will be.   Rubber articles when not in use, should
be stored in a cool, dark place.   Articles exposed in shop windows
may deteriorate more than those in use.   Vulcanized rubber
that is undergoing oxidation has quite a distinct odor from that
which is sound.   Rubber that has become hardened by oxidation may be softened again by soaking for several hours in about
a 3 per cent. solution of carbolic acid (phenol) in water.

All rubber resists in a high degree the passage of heat and the
electric current.   However, the changes brought about in it
upon being exposed to the atmosphere, cause it after a time to
become a relatively good conductor of both.

Dilute acids and caustic alkalis affect rubber but little.   Concentrated hydrochloric acid, both liquid and gaseous, attacks it.
Nitric acid acts feebly when cold but energetically when hot.
Sulfuric acid acts in about the same way as it does upon cork,
seeming to char it.

Rubber is soluble in benzole, petroleum ether, carbon disulfide, carbon tetrachloride, chloroform and turpentine, but only
the first three mentioned are much used as commercial solvents,
as in making solutions for rubber cements.   Rubber, being a
true colloid, does not dissolve sharply as do salts in water.   The
solvent is first absorbed by the rubber, whereby it is caused to
swell and form a jelly-like mass.   In a sense the solvent may be
said to have dissolved in the rubber.   Finally, when enough of
the solvent has been absorbed, the mass assumes the liquid state.

Grease or oil of any kind is decidedly injurious to rubber,
softening and partially dissolving it to form a sticky mass.

It has been observed that copper and its alloys have a destructive action on rubber.   Also its oxides and salts affect it in a
similar manner.   The reason does not seem to be very clear.

Of course, copper reacts with the sulfur contained in the rubber, but also a rapid hardening takes place. This has been attributed to oxidation induced by the oxygen occluded or combined with the copper in the form of cuprous oxide, or in the case of the compounds, by their catalytic action. The action is greatly hastened by grease or any substance capable of exerting a solvent action on the rubber.

Ebonite or Hard Rubber.—In producing this substance, approximately the same materials are used in the mixture as were used for the soft vulcanized rubber, but more sulfur is employed, there being about 30 lbs. of sulfur for every 100 lbs. of rubber. Also the temperature of vulcanization is higher, being from 150 to 160°C. Like the soft rubber, it is a non-conductor of electricity, but it is much more resistant to atmospheric action and is acidproof. It lacks extensibility, being similar to gutta percha in this respect.

Reclaimed Rubber.—It is possible to recover or reclaim rubber in a usable condition from worn-out articles and rubber waste from the factories. Although usable, it must not be understood that it has been devulcanized and returned to the condition of crude rubber, for to the present time, no one has been able to accomplish this very desirable end.

There are several processes by which rubber may be reclaimed, but that which is most widely used is known as the *alkali process*. In this process, the rubber after having been separated from the metal and fiber as much as possible, is heated in a closed iron vessel with an aqueous solution of an alkali, generally sodium hydroxide. This treatment removes the remainder of the fabric, and all the free sulfur as an alkaline sulfide. Any rubber substitutes present are saponified and dissolved out. The rubber thus treated is then carefully washed and dried and may be incorporated with a lot of fresh rubber.

Reclaimed rubber is sometimes known as "rubber shoddy," and in general possesses much lower tensile strength, elasticity, and wearing power than new rubber. Nevertheless, some of it is of extremely good quality, being, in fact, superior to some of the poorer grades of crude or virgin rubber. Its value seems to be to a certain extent dependent upon the degree to which the reclaiming process has regenerated its plasticity, but in any event its use with the fresh rubber is said to make a product that is less susceptible to oxidation than all new rubber would be, because the old rubber has already been largely oxidized. On the other hand, the elasticity of such mixtures must be in the same measure sacrificed.

Synthetic or Artificial Rubber.—Rubber is a hydrocarbon of complex structure, the formula usually ascribed to it being $(C_{10}H_{16})_x$. When rubber is subjected to destructive distillation, its complex structure is broken down and certain liquids result. Among these is isoprene, C_5H_8, a liquid boiling at 32°C. It was early noted that isoprene can be caused to condense or polymerize[1] with the information of true rubber. Also erythrene, C_4H_6, a gas, when cooled with a freezing mixture, can under certain conditions be caused to polymerize to form rubber. Therefore, it is necessary only to produce these substances cheaply and in sufficient amounts to produce rubber in commercial quantities.

Isoprene has been made from turpentine, from isoamyl alcohol, which is a constituent of fusel alcohol, and from starch, sugar and sawdust. Erythrene has been made from acetone, butyl alcohol and other substances. The polymerization has been brought about by temperature modifications, or by contact, under suitable conditions, with such substances as acids, alkalis, starch, glycerine, blood serum, albumen, etc. Even the polymerized substance itself can bring about the polymerization of isoprene and erythrene.[2]

However, either because of the smallness of the yield, or because of the cost of the raw material, as in the case of turpentine, or because of the expense of the processes necessary for conversion, or because of all of these, synthetic rubber has not become a commercial success. Dannerth says[3] that the synthetic rubber research received its first great stimulus in 1910 when fine quality Para rubber was selling for $2.50 per pound, but that now when high-grade plantation rubber is selling in the

[1] See footnote, p. 296.
[2] For literature on the synthesis of rubber, see:
  Jour. Am. Chem. Soc., 1914, 165.
  Jour. Ind. Eng. Chem., **3**, 279, May, 1911.
  Chem. Eng., **18**, No. 2, 61, August, 1913.
[3] ROGERS and AUBERT: Industrial Chemistry, p. 720.

New York market for 65 cts. per pound, there is little hope for any laboratory product, unless, of course, it can be made at a price considerably lower than that asked for the natural rubber.

Rubber Substitutes.—A great many, more or less rubber-like substitutes for rubber have been prepared, but these must not be confused with the synthetic rubber previously discussed. The most important of these substitutes is known as *factis* and is made by "sulfurizing." unsaturated fatty oils. There are two kinds of this factis: white, and brown or black.

The white modification is made by stirring from 20 to 25 per cent. of sulfur monochloride, S_2Cl_2, into linseed, rapeseed, corn, cottonseed or like oil, that has previously been mixed with a little petroleum naphtha. The product is more or less elastic. The brown or black modification is made of "blown" oil. Linseed or rapeseed oil is heated and a current of air is forced through it until it is quite thick and viscous. This blown oil is then vulcanized by heating with sulfur in a manner similar to rubber. The brown factis is quite elastic and is more stable than the white.

These substitutes are largely used to mix with rubber of a more or less inferior quality, or with some resin, heavy hydrocarbon, asphalt, etc., for cheap compositions. They are assimilated very thoroughly by heated rubber, but can be readily detected when in composition with rubber by treating the sample with alcoholic caustic potash. The substitute is saponified while the true rubber is not.

References

A

TERRY: India Rubber and Its Manufacture, London, 1907.
SCHIDROWITZ: Rubber; Its Production, and Industrial Uses, London, 1911.
POTTS: The Rubber Industry, London, 1912.
CLOUTH: Rubber, Gutta Percha, and Balata.
HEIL and ESCH: The Manufacture of Rubber Goods.

B

SHARPLESS: Utilization of Waste India Rubber, *Trans.* Am. Inst. Chem. Eng., **2**, 229 (1909).
FRANK: Preparation of Raw Caoutchouc, *Chem. Ind.*, **35**, 339 (1912).
FRANKENBERG: History of the Commercial Development of India Rubber, *Jour. Soc. Chem. Ind.*, **31**, 416 (1912).
PERKIN: Natural and Synthetic Rubber, *Chem. Eng.*, August, 1913.
Testing of Mechanical Rubber Goods, U. S. Bureau of Standards, *Cir.* 38 (1915).

CHAPTER XVII

ELECTRICAL INSULATING MATERIALS

The range of materials used for electrical insulation is very wide; for, aside from the primary requisite of possessing an insulating value, the material used must also have the necessary physical and chemical properties to enable it to maintain this value under the conditions of use. Obviously, the same material cannot be equally suitable in all locations. In some cases a high flexibility is necessary, while in others, rigidity and a high mechanical strength are desired. There are, however, some properties that all insulators should possess in common; for example, all must be able to exclude moisture, must not be easily affected by temperature changes, and should be as chemically inert as possible.

In this treatment it will be possible to mention only the more important of the insulating materials. Some of them have been described under the head of other uses and will be but briefly referred to here.

Rubber, in the soft vulcanized state, is used largely as a wire covering, although it is used also for impregnating tapes, and for preparing various insulating compounds.

In the manufacture of rubber-covered wires, a mixture consisting of rubber and sulfur, together with the desired amount of filling materials, is prepared and the whole is kneaded into the form of a "dough." The filling materials include a great variety of substances, and are added for various reasons. To harden and toughen the mass, inorganic materials, such as zinc oxide, lithopone, lime, litharge, magnesium oxide, etc., are employed; while other substances of an organic nature, such as ozokerite, asphalt, and other bitumens, are added to make it more waterproof and to increase its dielectric properties. This mixture may be applied to the wire by either the "seam" or "seamless" method. Since copper reacts with the sulfur contained in the covering, and this eventually affects its insulating properties, the wire is first tinned. In the "seam" method, strips of rubber are pressed about the wire as it passes through specially grooved rolls. In the "seamless" method, the insulating material, while still plastic, is forced through a die by means of a hollow screw, through the center of which the wire runs. After application, the rubber is vulcanized in a closed chamber by the means of steam.

In the form of hard rubber or ebonite, rubber is used for bushings, parts of switch gear, etc. Curtis says[1] that when hard rubber is exposed to light in the presence of air, complicated chemical reactions are set up, which are equivalent to oxidizing the sulfur of the rubber to sulfuric acid. The acid may then take up ammonia from the air to form ammonium sulfate. Or it may form sulfates by reacting with the mineral fillers of the rubber. Many of the compounds that the acid produces in this way are hygroscopic and form a surface layer of dissolved salts that noticeably lowers the insulating value of the rubber. He says, also, that such rubber can be renovated by suspending it in distilled water for about 2 days to remove the soluble salts, then wiping dry, and coating it with a thin film of light oil. Hard rubber cannot withstand any considerable degree of heat.

Other properties of both the soft vulcanized and hard rubber have been described in the preceding chapter, which see.

Gutta percha has a high dielectric strength and a better insulating value even than rubber. It is very extensively used for making submarine and subway cables. In many respects, it is similar to rubber, being obtained like it from the latex of certain tropical trees. Unlike rubber, however, it finds its chief use in the natural or unvulcanized state. Still, it can be vulcanized, this treatment rendering it harder and less plastic when heated.

Although gutta percha is plastic, it is not elastic, as is rubber. When moulded by an application of force, gutta percha retains the shape given it, while rubber returns to its original form when the moulding force is removed. At temperatures between 90 and 100°C. gutta percha is sufficiently plastic that articles may be formed of it by pressing it in moulds. It melts at about 120°C. In appearance gutta percha very closely resembles ebonite,

[1] Insulating Properties of Solid Dielectrics, U. S. Bureau of Standards, *Scientific Paper* 234, 1915.

but it can be readily distinguished from it by the fact that it is readily softened in boiling water, which property ebonite does not have.

Gutta percha is soluble in carbon disulfide, carbon tetrachloride, chloroform and warm benzene. It is not affected by alkalis and dilute acids, nor even by strong hydrochloric, but is attacked by strong sulfuric and nitric acids, as well as by other strong oxidizing agents. It is not especially resistant to atmospheric action, being rather easily oxidized and in this condition it is brittle.

A material that is often used in place of gutta percha for some purposes is *ozokerite*, called also *ceresin* or *mineral wax*. It is a hydrocarbon found upon the earth in the vicinity of oil wells. Upon distillation, it yields about 15 per cent. of paraffine oil. It is a good dielectric, and atmospheric and chemical agents have no action upon it. *Nigrite* is another substitute made by fusing together at the very lowest temperature, india rubber and the residuum obtained by the distillation of ozokerite.

Bakelite has a high dielectric value and is used either alone to impregnate windings, or as a moulded composition with wood fiber, or as a filling material for blotting paper and similar substances. For a discussion of its preparation and general properties see page 325.

Varnishes.—As a rule, the oil varnishes are used to saturate paper and fabrics, while the spirit varnishes are used as a cementing material. Both oil and spirit varnishes are used to a large extent for the insulation of windings. The spirit varnishes dry more quickly, but are less flexible than oil varnishes. Shellac varnish is the most widely used of the spirit-varnish class, and is particularly suitable for low-voltage windings. The composition and general properties of both classes of varnishes have been discussed in Chapter XIII.

Oils.—The use of insulating oils is confined chiefly to transformers, rheostats and switches. For transformers and rheostats, the main function is that of cooling, although the insulating action is very important. Mineral oil is generally employed, and it should have a high dielectric strength, high flash point, low viscosity, and be free from moisture, acid, alkali, sediment and impurities.

The insulation value of the oil depends upon the absence of water. A very delicate test for its presence consists of shaking a warmed sample of the oil in a test-tube with anhydrous copper sulfate. If water is present, the white, anhydrous salt will take it up and become blue. For this test the sample should be taken from the bottom of the barrel or other container, since if water is present, it will likely be found there.

Fibers are used largely for the wedges of armature windings, the most widely used being known as vulcanized fiber. Vulcanized fiber is prepared by treating wood pulp with zinc chloride, under the action of which the pulp becomes gelatinous. The product is then kneaded and rolled to wash out the zinc salt, but the removal is difficult and is generally not complete. Zinc chloride is a very hygroscopic salt, so the finished material has a tendency to absorb moisture. It swells when it becomes wet, but returns to its original size upon drying. Vulcanized fiber is insoluble in the mineral oils, turpentine, alcohol and other organic solvents. Preparations known by the trade names of leatheroid, fish paper, horn fiber, etc., are very similar to vulcanized fiber, but are more flexible.

Mica is one of the most widely used of the solid insulating materials. The variety generally employed is muscovite. It is obtained from the mines in the form of rough cakes, which are split and trimmed into sheets. The harder varieties can be split economically as thin as 0.0005 in.[1]

Mica is unaffected by heat until a temperature of several hundred degrees Centigrade is reached. Then the laminæ of the harder varieties separate, imparting a very soft and flexible character to the mica, but which also gives it a tendency to disintegrate into scales and flakes. Mica is capable of absorbing considerable moisture between its laminæ, and this increases its flexibility. Also, it will take up oil and other fluids in the same way. The presence of oil seems to lower its insulating value somewhat.

To lessen cost by using waste flakes, and sometimes for the sake of the flexibility that may be secured, that which is known as *reconstructed mica* is prepared. Layers of mica are cemented together by the use of an insulating varnish, as shellac, and are

[1] FLEMING and JOHNSON: Insulation and Design of Electrical Windings, p. 81.

25

then formed into the desired shape by hot pressure. Stiff, flat plates prepared in this way and milled to uniform thickness are known to the trade as *micanite*. Flexible sheets are prepared in a similar way by using a non-drying varnish, but in this case the hot pressing is omitted.

Asbestos is a hydrated magnesium silicate, $3MgO \cdot 2SiO_2 \cdot 2H_2O$. It does not naturally possess very high insulating properties, but is valuable in electrical work because of its non-combustible nature and its resistance to heat. Sheets are prepared by boiling asbestos in a solution of sodium silicate and then subjecting the mass to heavy pressure. Sheets prepared in this way are hard and mechanically durable, but are hygroscopic and have a low insulating value. Also, they have a tendency to disintegrate when wet. Attempts have been made to render the silicate binder less soluble and hygroscopic by treating it with calcium and aluminum compounds, but reaction products are always left, that are hygroscopic in much the same way as the sodium silicate.

Talc, like asbestos, is a hydrated magnesium silicate, but is more acid than asbestos, having the formula, $3MgO \cdot 4SiO_2 \cdot 2H_2O$. It is soft and is easily drilled and machined. After it has been fashioned as desired, it may be rendered exceedingly hard by heating to about 1,100°C. Stone prepared in this way is marketed under the name of *"lava."* Another product known as *"lava composition"* is also made from talc by preparing a mixture of the finely powdered stone and a suitable binding material, which is pressed into the forms desired by hydraulic pressure. These forms are dried in air and then heated in kilns until all moisture is removed. The semi-hardened material is now machined to suit the requirements and is then finally hardened by heating to 1,100°C., as in the case of the natural stone.[1] Both the natural and the artificial material are heat-resistant and possess good insulating properties.

Marble is used chiefly for switchboard work, being desirable for this purpose because it is non-combustible, easily worked, and possesses a pleasing appearance. Its insulating properties may be faulty because of the presence of metallic veins, which are difficult to locate, except by high-voltage tests. Fleming and

[1] BENSON: Industrial Chemistry, p. 369.

Johnson say[1] that marble is very hygroscopic and normally contains so much water that it is necessary to bush the holes through which conductors pass. And further, that this is all the more necessary, because the water usually employed for drilling is absorbed by the stone and cannot be effectually removed, except by baking for a considerable time. If the water is allowed to remain, and if it is not suitably bushed, local heating will occur that may be sufficient to fracture the stone. Because marble absorbs dissolved salts produced by the corrosion of metals with which it may be in contact, it becomes stained and discolored. Such stains are practically impossible to remove, because of the distance to which they have penetrated. For further information concerning marble, see page 220.

Slate, like marble, is used for switchboard work, and in other locations, where a non-combustible material, having fair insulating properties is necessary. It is harder and less hygroscopic than marble, but is much more liable to contain metallic veins, so that its insulating value is uncertain.[1] For switchboards, slate is very suitable in those locations where marble would be very rapidly discolored. For a further discussion of the properties of slate, see page 215.

Porcelain possesses a high insulating value and is used largely for high-tension work. It is hard and chemically durable, but is very brittle. The methods of porcelain manufacture have been described on page 265, which see. The hard or single-fired porcelain is most suitable for electrical work, because it has practically no tendency to absorb moisture. Good-quality porcelains of this sort have a clean, glassy appearance. The less desirable varieties are porous and readily absorb moisture. They are, therefore, quite unsuitable for high-tension work.

References

FLEMING and JOHNSON: Insulation and Design of Electrical Windings, London, 1913.

HEMMING: Moulded Electrical Insulation and Plastics, New York, 1914.

WATTS: Porcelain for Electrical Purposes, *Trans.* Am. Ceramic Soc., **4,** 86 (1902) and **6, 600** (1907).

BOGGS: Chemical Specifications for Rubber-covered Wires and Cables, *Elect. World,* **56,** 1234 (1910).

[1] FLEMING and JOHNSON: *Loc. cit.,* p. 101.

KORBER and DILTSCH: Ceramic Materials of High Electrical Resistance, *Chem. Abst.*, **6**, 2679 (1912).

CURTIS: Insulating Properties of Solid Dielectrics, U. S. Bureau of Standards, *Scientific Paper* 234 (1915).

CREIGHTON and HOSEGOOD: Testing Electrical Porcelain, *Gen. Elec. Rev.*, **19**, 479 (1916).

CHAPTER XVIII

PRIMARY ELECTRIC CELLS

Definition.—A primary electric cell is a system of elements and chemical compounds so arranged that the energy of the chemical reactions that occur in the system, is directly transformed into electrical energy.

Theory of Cell Action.—It has been observed that chemical compounds of only one class react in such a manner that they are suitable for use in electric cells. Such compounds are those that when dissolved, dissociate into ions, these being considered to be subdivisions of the molecule carrying electrical charges. It is by means of ions that the current is assumed to be carried through the cell. In fact, the whole of the present conception of primary-cell action rests upon the ionic theory; therefore, in order to logically develop this subject, it will be necessary to consider the experimental work by which the ionic theory has been confirmed. This confirmation has been arrived at by a study of various phenomena manifested by solutions, important among which are osmotic pressure and the effect of dissolved substances upon the freezing point and boiling point of water. Because of their significance in this connection, a discussion of these topics will be taken up at this point.

Osmotic Pressure.—If a concentrated aqueous solution of some salt, copper sulfate, for example, is placed in the bottom of a tall cylinder, and then pure water is poured on in such a manner that it floats above the solution, producing a sharply defined layer, it will be found, after a time, that a part of the dissolved salt has passed upward against gravity and entered that which was formerly the pure water. It is apparent that there is some force similar to a pressure driving the dissolved salt from one layer into the other. This pressure is called osmotic pressure. By the use of semi-permeable membranes, the osmotic pressure of dissolved substances may be rendered more evident, and may even be measured. Semi-permeable membranes are such as

389

are permeable to water, but are practically impermeable to the molecule of the dissolved substance, and hence, in a sense, act as a filter for these, retaining them as the water passes through. At first, before better methods were devised, osmotic pressure was demonstrated by filling an animal bladder with an aqueous solution of alcohol, and immersing the bladder in pure water. The water could pass into the bladder, but the alcohol could pass out to only a slight extent. In this way, sufficient pressure was finally developed within the bladder to cause it to burst.

A partial measurement of osmotic pressure may be made in the following manner. A piece of animal parchment is carefully tied over the mouth of a small glass funnel or thistle tube, and the interior of the funnel and stem are filled with a rather strong solution of cane sugar. A long piece of capillary tubing is now joined to the stem of the funnel by a sound, well-fitting rubber connection, and the funnel is placed mouth downward in a vessel of pure water, the capillary tube being supported in a vertical position. After standing quietly for several hours, it will be noted that the solution rises in the capillary tube, and it may rise 2 or 3 ft. By the use of stronger and generally more satisfactory membranes, prepared artificially, it is not difficult to show a pressure of 30 ft. or more of sugar solution. Pressures as high as 31.4 atmospheres have been measured.[1]

Various hypotheses have been advanced to show what is the cause of osmotic pressure, but it has not yet been satisfactorily explained.

It is worthy of note that osmotic pressure obeys the laws for gases. For example, it has been found that the osmotic pressure of different solutions at the same temperature and of equimolecular concentration, made so by dissolving a number of grams equal to their molecular weights in equal amounts of water, exert equal osmotic pressures; thus showing that the pressure depends upon the number of molecules in solution. Avogadro's hypothesis for gases states that equal volumes of gases measured under the same conditions of temperature and pressure contain the same number of molecules.

In a similar manner the law of Boyle applies to osmotic pressure. Boyle's law states that if the temperature remains con-

[1] Jones: Principles of Inorganic Chemistry, p. 103.

stant, the volume of a gas varies inversely as the pressure upon it; which is equivalent to saying that the pressure of a gas varies directly with its concentration. Observation has shown this to be true also of osmotic pressure.

The law of Gay-Lussac pertains in that the osmotic pressure of the solution increases in proportion to its absolute temperature, and the rate of increase is approximately $\frac{1}{273}$ for each degree rise in temperature.

Osmotic Pressure of Electrolytes.—It was early observed, that for equimolecular concentrations, certain substances when dissolved, exerted an osmotic pressure far greater than that produced by other substances. In fact, chemical compounds were found to be divisible on this basis into two classes: first, a class in which the pressure was called normal, and second, a class in which it was either *two or three times as great*. In the first class are substances like cane sugar, glycerine, and the alcohols, in general, substances that are not very active chemically. In the second class are the acids, bases, and salts, substances that are chemically the most active. Moreover, it is only the latter class of substances that conduct the electric current, and are decomposed by it, these being, therefore, called *electrolytes*. The former do not conduct the current and are not decomposed by it, and hence are called *non-electrolytes*.

The Lowering of the Freezing Point.—When any substance is dissolved in water the freezing point of the solution is lower than that of the pure solvent. It has been shown that the amount of lowering is dependent upon the ratio between the number of the particles of the dissolved substance and the number of the particles of the solvent, in other words, upon the concentration of the solution. In the case of the electrolytes in water the lowering of the freezing point is either two or three times as great as for a water solution of a non-electrolyte of equivalent concentration; for example, a gram molecule per liter in each case.

Boiling Points Affected.—In a similar manner, the vapor tension of the solution is lowered, or in other words, the boiling point is raised[1] either two or three times as much in a solution of

---

[1] Certain exceptions must be noted. With some dissolved substances, as for example, alcohol, a low-boiling-point liquid, the boiling point of the solution is not raised, but lowered below the boiling point of pure water.

an electrolyte as in a solution of a non-electrolyte of equivalent concentration.

.**Summary.**—All three of these properties of solutions—osmotic pressure, lowering of the freezing point, and raising of the boiling point—have been shown in the case of the non-electrolytes to be dependent upon the number of particles of dissolved substance, that is, upon the concentration of the solution. Electrolytes behave as though they contain two or three times as many particles as their concentration warrants. This observed difference between electrolytes and non-electrolytes is accounted for by the theory of electrolytic dissociation advanced by the Swedish chemist, Arrhenius.

Theory of Electrolytic Dissociation.—This theory states that when acids, bases, and salts are dissolved in water, a certain number of their molecules break down or dissociate into two or more parts. The exact extent to which dissociation proceeds, depends upon the particular substance considered. For example, some acids and some bases dissociate almost completely, such being called strong acids and bases. Examples are hydrochloric and nitric among the acids, and sodium and potassium hydroxide among the bases. Others that dissociate to only a slight extent are called weak, as carbonic acid and ferrous hydroxide. Also, the degree of dilution has much to do with the degree of dissociation, the percentage of molecules dissociated being comparatively low in concentrated solutions and increasing as the solution grows more dilute.

As the molecules dissociate, the separated parts are charged with an electric charge, one atom or group of atoms being charged positively and the other negatively, the total amount of the positive being equal to the total amount of the negative. These charged parts are called *ions*, and the process by which they are formed is known as *ionization*. The positively charged ions are usually simple atoms, as atoms of the metals or hydrogen, although sometimes they are radicals, as in the case of ammonium. The negatively charged ions are generally radicals, for example, SO_4, OH, and NO_3, or they may also be charged atoms, as atoms of chlorine and bromine. All acids have the common positive ion hydrogen, and all bases have the common negative ion OH, the other ion in each case depending upon the particular acid or base considered. The plus sign, $+$, is used to show a positive charge, while the minus sign, $-$, is used to show that the ion carries a negative charge. The number of the charges carried by each ion is equivalent to the valence of the atom or radical, the ions being called monovalent and divalent, as $\overset{+}{N}a$ and $\overset{--}{SO_4}$.

Ionization may be represented in the following manner, the examples shown being respectively, an acid, a base, and a salt:

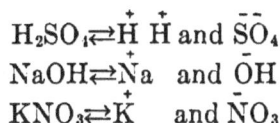

$$H_2SO_4 \rightleftharpoons \overset{+}{H}\ \overset{+}{H} \text{ and } \overset{--}{SO_4}$$
$$NaOH \rightleftharpoons \overset{+}{N}a \text{ and } \overset{-}{OH}$$
$$KNO_3 \rightleftharpoons \overset{+}{K} \text{ and } \overset{-}{N}O_3$$

It is very important to keep in mind that dissociation is a reversible process as the equations show. The dissociated compound readily reverts to the undissociated state, as for example, when water is evaporated and the solution made more concentrated.

The reason why electrolytes produce two and three times as much effect as non-electrolytes upon the freezing point, boiling point and osmotic pressure of solutions is now apparent. With the former, there may be two and three times as many particles as with the latter in solutions of equal concentrations. The size of the particles is not determinative.

Solution Tension.—Whenever a metal is immersed in water, in a dilute acid, or in a solution of one of its salts, it exerts a certain tension or pressure that tends to throw metallic ions into solution. This is known as the solution tension of the metal. Nernst, who proposed the theory, suggested that it is analogous to the vapor tension of liquids that causes them to evaporate. Now it is well known that if the vapor produced from a liquid is confined above it, there will be developed ultimately for a given temperature, a definite vapor pressure that will prevent the amount of vapor from increasing. The vapor tension, of course, continues to manifest itself, and to drive molecules from the liquid into the vapor above; but under the influence of the vapor pressure that has developed, molecules pass back into the liquid at the same rate at which molecules from the liquid enter the vapor. That is, there is a constant and equal interchange.

In this manner a condition known as equilibrium is established.

In a similar manner, a state of equilibrium may be reached between the solution tension of a metal and the osmotic pressure of its dissolved ions. When enough ions have passed into solution so that their osmotic pressure is equal to the solution tension of the metal, as many ions will separate or precipitate upon the metal in a given period as pass into solution.

However, the analogy between the vapor tension of liquids and the solution tension of metals is not entirely exact. In the one case molecules are dealt with, and in the other, ions. When solution tension forces an ion from the surface of the metal into solution, it carries positive charges with it, and leaves behind an equal number of negative charges on the metal. Thus, the solution becomes positively charged, and the metal negatively charged. On the other hand, when osmotic pressure drives ions out of solution onto the metal, they can no longer remain ions but become atoms; and their charges are given up to the metal, thus charging it positively.

Then, whether or not a metal immersed in a solution of one of its salts will become positively or negatively charged, depends upon which is the greater, the osmotic pressure of the dissolved ions or the solution tension of the metal. If the solution tension is the greater, the metal will become negatively charged; if the osmotic pressure is the greater, it will become positively charged; if the two forces are equal, nothing will happen.

In selecting metals for primary cells, two metals are chosen, one having a high, and the other a low solution pressure. When put into suitable electrolytes, the former will become negatively, and the latter positively charged. In this condition, a current will flow between them when they are suitably connected.

The table[1] on the opposite page shows the comparative solution tensions of some of the more common metals.

Jones says also[2] that the solution tensions of magnesium, zinc, aluminum, cadmium, iron, cobalt, and nickel, are always greater than the osmotic pressures of the metallic ions in any solutions of their salts that can be prepared. Hence, they are always negative when immersed in their own salt solutions. On the

---

[1] JONES: Principles of Inorganic Chemistry, p. 393.
[2] Outlines of Electrochemistry, p. 82.

Metal	Solution tension in atmospheres
Magnesium...............................	10^{44}
Zinc......................................	10^{18}
Aluminum................................	10^{13}
Cadmium................................	3×10^6
Iron.....................................	10^4
Cobalt...................................	2×10^0
Nickel...................................	1×10^0
Lead....................................	10^{-3}
Mercury.................................	10^{-16}
Silver...................................	10^{-17}
Copper..................................	10^{-20}

other hand, mercury, silver, and copper have such low solution tensions, that they are less than the osmotic pressure of their dissolved ions; and will be positive when immersed in solutions of their salts, unless the solutions are extremely dilute.

Because zinc has a very high solution tension, and is more common and cheaper than magnesium, it is very generally chosen for the negative element of primary cells. In one of the very common

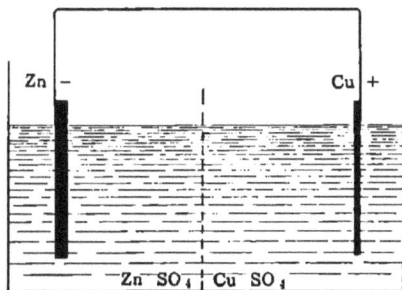

FIG. 64.—A two-metal, two-fluid, primary, electric cell.

forms, copper, because of its low solution tension, is used as the positive element.

A Two-metal, Two-fluid Cell.—The action that occurs in a cell of this type can be very readily shown by the use of copper and zinc in solutions of their salts as shown in Fig. 64. Because of the high solution tension of zinc, it becomes negatively charged by projecting zinc ions into the zinc sulfate solution. Because of the low solution tension of copper, copper ions from the solution precipitate upon the plate, and charge the plate positively. Because of the oppositively charged states of the two metals, a current will flow through the connecting wire when the two plates are joined. Since the opposing charges are thus neutralized, more zinc ions can enter solution and more copper ions can precipitate upon the copper, the whole process being continuously repeated. Thus, the latent energy of the dissolving zinc is manifested as a current of electricity passing through the system.

Local Action.—It frequently happens in primary cells that a considerable part of the energy theoretically rendered available by the metal dissolving in the cell, is not manifested as a current of electricity in the external circuit. Sometimes more than half of it is used up in maintaining numerous, small currents that flow from part to part of the dissolving metal itself. This is due to lack of uniformity in the composition of the metal, or in other words, to the presence of impurities. In zinc, which is used in all the common cells, the impurities are iron, lead and cadmium. These metals have lower solution pressures than zinc, and so can fill the same function as copper in the cell just discussed. Zinc ions enter the solution at the points where the metal is relatively pure; and other ions, generally hydrogen, precipitate at the impure points, where the solution pressure is low. Thus, local currents are maintained between portions of the metal and electrolyte, zinc being consumed to no purpose. Such action is termed local action.

Aside from the presence of impurities in the zinc, local action may be engendered by the presence in the solution of ions of metals having low solution pressures. For example, copper ions in solution, when they come into contact with zinc, give their charges to the zinc and become atoms of copper, while zinc ions enter solution; for as has been explained, zinc has a high, and copper a low solution pressure. This deposited copper acts as a discharge point for hydrogen ions, thus causing local action.

Amalgamation.—Local action may be prevented by amalgamating the zinc; that is, by producing a mercurial alloy on its surface. Amalgamation may be accomplished by rubbing the well-cleaned metal with a tuft of cotton moistened with dilute acid and containing a few entangled globules of mercury.

It is thought that the surface amalgam is effective in preventing local action, because it acts, in a sense, as a filter, forming an alloy with the zinc, but less readily with the impurities. As zinc ions escape from the outer layer, the mercury takes up more zinc from the metal on the inner side; but since the impurities do not alloy so well with the mercury, they cannot come into contact with the electrolyte to the same extent. Or it has been suggested that the prevention of local action is brought about by the mercurial layer merely because it makes the surface uniform, and the solution pressure equal at all points. That is, it may dissolve even the impurities, both those in the zinc and those that precipitate from the electrolyte, thus preventing them from becoming discharge points. Also, the amalgam presents a very smooth surface, and on this account it is more difficult for hydrogen ions to discharge to it.

Polarization.—When hydrogen ions precipitate upon the metal or element that conducts the current from the cell, they become atoms and adhere as a layer of gas. When this has happened, the production of current is very largely cut down, and the cell is said to be *polarized.* The decreased activity of the cell in this state is partly due to the great non-conductivity of the gaseous layer, but it is also partly due to the fact that the hydrogen plate (a plate covered with hydrogen) has a higher solution pressure than the underlying metal that was intended to serve as this electrode. The result is that there is a tendency to set up a counter-current, opposing the main current.

Polarization is overcome by using suitable oxidizing agents, as oxides, peroxides, etc., which are able to oxidize the hydrogen to water.

Electrochemical Terms.—Before proceeding further with the discussion of primary cells, it will be necessary to define certain terms commonly used in reference to them.

As was indicated earlier in the chapter, the chemically active solution used in the cell is called the *electrolyte.* This may be a single solution, or at times, two solutions kept separate in some convenient manner. The metals employed in the cell are generically known as *plates*, although they may not always be in plate form. The *surfaces of contact* between these plates and the electrolyte are known as *electrodes*. The electrode on which ions precipitate and to which positive charges are given up, in the regular action of the cell, is known as the *positive electrode*, or *kathode*. The electrode where metallic ions enter solution, and which is, therefore, negatively charged, is the *negative electrode* or *anode*. The ions that move toward the kathode as the current passes through the cell are known as *kations*, and those that move toward the anode, the *anions*. Kations carry positive, and anions negative charges. The path by which the current flows from kathode plate to the anode plate, outside the cell, is known as the *external circuit*. The junctions of the external circuit with these plates are known as the *poles* of the cell or battery, the junction with the kathode being the *positive pole*, and the junction with the anode, the *negative pole*. The cells themselves are frequently called *voltaic* or *galvanic cells*, in honor of Volta and Galvani, who were pioneer investigators in this field.

The unit of resistance offered to the flow of the electric current is called the *ohm*. It is defined as the resistance at 0°C. of a column of mercury 1 sq. mm. in sectional area, and 106.300 cm. long. The unit of current strength is the *ampere*. It is the current that, when passed through silver nitrate under specified conditions, will deposit silver (see "Electrolysis," page 408) at the rate of 0.0011180 gram per second. The unit of potential difference or electromotive force is called the *volt*. It is the driving force or electrical pressure that causes the current to flow. It may be defined relatively as the amount of this force that will cause a current of 1 amp. to flow through a resistance of 1 ohm.

Kinds of Cells.—There are many kinds of primary cells, and generally several modifications of each kind, but only those that are the most widely used will be discussed here. These are the Daniell, the Lalande, and the Leclanché, of which the dry cell is a modification.

THE DANIELL CELL

The Daniell cell is the oldest of the three types mentioned, having been brought out in 1836. In respect to mechanical construction, it may be of either the porous-cup, or gravity type.

The porous-cup type consists of a jar containing a saturated solution of copper sulfate, in which is immersed a partially closed cylinder of copper sheet. Within the cylinder of copper sheet stands a porous clay cup, containing either a dilute solution of sulfuric acid or zinc sulfate. In this solution stands a rather heavy zinc grid. The cup allows ions to travel through its pores, but still prevents a rapid mixing of the two solutions.

The gravity type differs from the porous-cup form in that

the two solutions are kept separated only by the difference of the specific gravities of the two solutions. Also, the shapes of the electrodes are modified.

The *electrochemical action* in the Daniell cell may be sketched graphically. Before the circuit has been closed, the constituents may be considered to bear the following relation to each other:

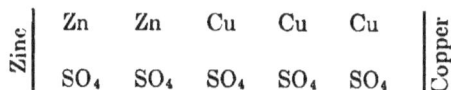

$$\text{Zinc} \left| \begin{array}{ccccc} Zn & Zn & Cu & Cu & Cu \\ SO_4 & SO_4 & SO_4 & SO_4 & SO_4 \end{array} \right| \text{Copper}$$

But immediately after the circuit is closed the following ionic changes occur:

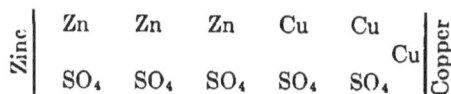

$$\text{Zinc} \left| \begin{array}{ccccc} Zn & Zn & Zn & Cu & Cu \\ SO_4 & SO_4 & SO_4 & SO_4 & SO_4 \end{array} \right. Cu \left| \text{Copper} \right.$$

Zinc ions enter solution and copper ions leave it, being converted into metallic copper. In this way electricity is carried through the cell at a slow but definite rate.

As a result of this migration of ions a change in concentration is brought about in the cell, as sketched in Fig. 65. Since the anode is dissolving, the solution there becomes continually more concentrated in zinc ions, and the osmotic pressure proportionally increased. Consequently, the solution pressure of the zinc meets with more resistance in sending ions into solution and the speed of the

FIG. 65.—Concentration changes in a Daniell cell.

cell is lessened. In a similar manner the osmotic pressure of the copper ions in the copper sulfate solution becomes less as ions precipitate upon the metal, and the solution becomes less concentrated. In caring for a Daniell cell, attention must be paid to these concentration changes. To keep up the activity of the cell, copper ions must be constantly supplied to the solution by adding more copper sulfate, and zinc ions must be removed from the solution by removing the concentrated zinc sulfate solution.

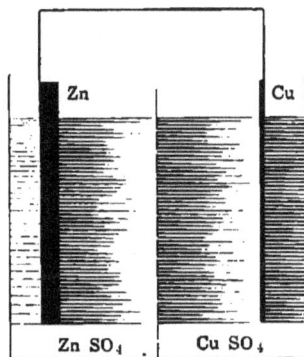

The E.m.f. of the Cell.—In the Daniell cell, the zinc is about 0.8 volt negative to the zinc sulfate solution, and the copper is about 0.3 volt positive to the solution of copper sulfate, consequently, the electromotive force of the cell is about 1.1 volts. In this cell, as in all cells, the electromotive force is not dependent upon the size of plates but on the amount of solution tension and osmotic pressure.   A small battery will produce the same electromotive force as a large one of the same kind, if the solutions used are the same, and of the same concentration.   But, the current strength depends upon the size of the plates.   In the ordinary-sized Daniell cell, the current strength does not usually exceed 0.5 amp.

Depolarization.—In this cell, polarization is prevented by substituting another ion, that is, the copper ion for hydrogen, so that when the cell is in good condition, the hydrogen ion does not reach the kathode plate.

Defects.—When the cell is allowed to stand idle, the copper sulfate diffuses through the porous cup wall to the zinc.   Because of the high solution tension of the zinc and the low solution tension of copper, zinc ions enter solution and copper ions precipitate on the metal in exchange.   In extreme cases the zinc may be entirely covered with a layer of copper.   This replacement of zinc by copper, of course, uses up zinc, but the interchange is especially objectionable because the deposited copper serves as kathodes for local currents.   In all Daniell cells, local action is likely to be very great.   According to Haber, quoted by Allmand,[1] even under favorable conditions, only 30 per cent. of the zinc is electrochemically utilized.   When the cell is working, the migration of the kations toward the kathode overcomes the diffusion toward the zinc.

Another objectionable feature is that the porous cup increases the internal resistance of the cell.   To aid in keeping this resistance as low as possible, the copper kathode is generally shaped to fit very closely about the cup, so that the distance between the anode and kathode may be as little as possible.   But this arrangement causes very little copper sulfate solution to be available, so that its concentration soon becomes low, and the resistance increases on this account.   To keep up the concentration,

[1] *Principles of Applied Electrochemistry*, p. 201.

crystals of copper sulfate are sometimes packed between the copper plate and the cup, but this is objectionable because the crystals also have a high resistance. Since all of these objections are either directly or indirectly due to the porous cup, a type of cell was devised in which the cup could be omitted. This is known as the gravity type, and is now the most common form of Daniell cell.

The Gravity Type.—In this type the two solutions are kept from mixing merely by the differences in their densities. Such an arrangement decidedly lessens the internal resistance, but also introduces a disadvantage in that the cell cannot be readily transported, because any shaking tends to mix the solutions. Further, when standing idle, the solutions are free to diffuse into each other, so that copper ions may reach the zinc. To prevent this diffusion, the gravity cell should always be kept working, even though current from it is not desired for use. Of course, in such case, the current taken from the cell should be but a small quantity—enough only that the kationic migration will overcome the diffusion. The current may be regulated by placing a few ohms resistance in the external circuit.

To Set Up a Daniell Cell.—For the kathode electrolyte, a saturated solution of copper sulfate is prepared. At 20°C., which is about room temperature, the crystals are soluble to the extent of 42.3 in 100 parts of water. For the anode electrolyte, either dilute sulfuric acid or a solution of zinc sulfate may be used, and there are advantages pertaining to each. With the salt solution, the amalgamation of the zinc is not so imperative, since local action is less pronounced. On the other hand, the internal resistance is less when the acid is used, and the output of the cell is greater. However, the current strength is less constant with the acid, especially during the first 2 or 3 hrs. of discharge, or until considerable zinc sulfate is formed.

If the acid is decided upon, a convenient strength is about 20 per cent. The strength of the zinc sulfate may vary considerably—from 10 to 30 per cent. For short periods of service it is advantageous to use the stronger solution, since its conductivity is greater; and up to a certain point, this seems to somewhat more than counterbalance the detrimental effect of the greater osmotic pressure. But on the other hand, the greater the initial concentration, the shorter will be the time during which the cell can work without showing lessened activity due to zinc ion concentration. In general, for longer periods of service a weaker solution is desired.

THE EDISON-LALANDE CELL

This cell was devised in. 1883 by De Lalande, but has been improved by Edison and others. It consists of a copper oxide plate between two zinc plates, suspended from a lid, in a 20 per cent. solution of sodium hydroxide contained in a porcelain jar. The zinc plates are heavily amalgamated. The copper oxide plate is made by mixing finely divided copper oxide with sodium hydroxide solution, and then forming the mass into cakes under heavy pressure. These are dried and baked at a bright red temperature. After cooling, the plate is reduced to metallic copper on the surface, to give it better conductance at the start.

The electrolyte in this cell is kept covered with a layer of heavy mineral oil, since if it were exposed to the air, it would absorb carbon dioxide and be converted into sodium carbonate, which salt "creeps" badly. There is a disadvantage in having this layer of oil on the electrolyte, since the plates cannot be withdrawn without becoming coated with a film of it, and this film prevents them from being chemically active. In setting up the cell, the elements should be immersed before the oil is poured on.

Electrochemical Action.—The electrochemical reactions that occur in the Lalande cell may be represented graphically as follows:

Before closing the circuit.

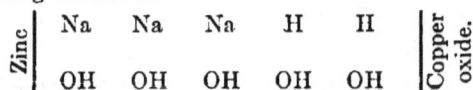

Zinc	Na	Na	Na	H	II	Copper oxide.
	OH	OH	OH	OH	OH	

After closing the circuit.

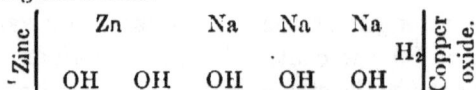

Zinc	Zn		Na	Na	Na		Copper oxide.
	OH	OH	OH	OH	OH	H_2	

Zinc ions enter solution and sodium ions pass over to the cathode, but the hydrogen ions, present as the result of dissociated water, give up their charges more readily than sodium ions, therefore, hydrogen ions separate out. The zinc hydroxide that is formed in this way is immediately dissolved by the excess of sodium hydroxide present, with the formation of sodium zincate as follows:

$$Zn(OH)_2 + 2NaOH \rightarrow Na_2ZnO_2 + 2H_2O$$

Depolarization.—The hydrogen separated at the kathode would soon produce polarization were it not that it is oxidized to water by the copper oxide, which is thus reduced to metallic copper, as:

$$CuO + 2H \rightarrow Cu + H_2O$$

The copper oxide is quite active in this way. When it has been reduced to such an extent that its effectiveness is lost, it may be regenerated by heating the plate to incipient redness in the air, over an iron plate to protect it from the direct contact with the flame. Before heating, the sodium hydroxide must be removed from the plate by prolonged soaking in water. In using a regenerated plate, the cell should be short-circuited for a time to reduce the oxide on the surface and increase the conductance. A regenerated plate is never quite as good as a new one.

E.m.f. of the Lalande Cell.—The internal resistance of the Lalande cell is very low, ranging from 0.03 to 0.05 ohm. The electromotive force of the 500-amp.-hr. type is 0.62 volt per cell, when discharging at the rate of 1 amp. When discharged at a more rapid rate the voltage drops, being 0.58 volt per cell at 2 amp., 0.55 volt per cell at 3 amp., etc.[1]

Efficiency.—During the action of the cell, practically all of the zinc is utilized electrochemically; that is, there is very little local action, and while standing the zinc consumption is practically negligible. In this respect, it probably excels any other type of cell. Also, it requires very little attention.

THE LECLANCHÉ CELL

This cell in its original form was introduced in 1868, but a great many modifications have been brought out since.

---

[1] Information and Directions for Use, furnished by the Edison Co.

The cell consists of a zinc anode and a carbon cathode, suspended in about a 20 per cent. solution of ammonium chloride. To the electrolyte may be added sometimes, certain hygroscopic substances, as zinc chloride, calcium chloride, glycerine, etc., to lessen the loss of water by evaporation. The depolarizer consists of manganese dioxide, this being kept in position about the kathode by various devices. In one form, blocks are made of the manganese dioxide, which has been previously mixed with carbon to increase the conductance, and the blocks are then clamped to the sides of the cathode. In another form there are two concentric carbon cylinders, with manganese and carbon packed between them and sealed in, the zinc being in the center.

The electrochemical action is substantially as follows: The zinc goes into solution as zinc chloride, and ammonium ions move over to the kathode; but the hydrogen ions of the dissociated water retain their charges less powerfully than the ammonium ions, hence the former separate out, as the following sketches show:

Before closing the circuit.

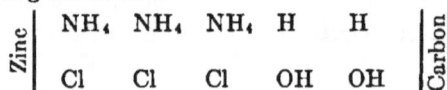

Zinc	NH$_4$	NH$_4$	NH$_4$	H	H	Carbon
	Cl	Cl	Cl	OH	OH	

After closing the circuit.

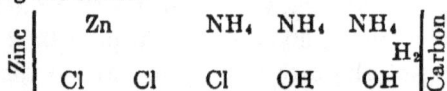

Zinc	Zn		NH$_4$	NH$_4$	NH$_4$	Carbon
					H$_2$	
	Cl	Cl	Cl	OH	OH	

The result of this action is that undissociated ammonium hydroxide[1] is formed, which in turn furnishes ammonia gas. Some of this gas escapes into the air, but the greater part of it remains dissolved and diffuses into the zinc chloride solution, with which it reacts, producing the double salt, $Zn(NH_3)Cl_2$.[2] This compound is but slightly soluble and gradually separates out. Also, by the action of the ammonium hydroxide, insoluble zinc hydroxide is produced. Some of this precipitates within the pores of the depolarizer and greatly increases its resistance.

---

[1] Ammonium hydroxide dissociates very little, only about one-fortieth as much as potassium hydroxide, for example. See JONES: Outlines of Electrochemistry, p. 27.

[2] ALLMAND: Principles of Applied Electrochemistry, p. 207.

The conductivity of the depolarizer of an exhausted cell is very low.

Concentration of Effect.—Because of concentration changes that take place in the electrolyte, there is a certain amount of local action in this cell. The zinc chloride formed at the anode is heavier than the other constituents, and sinks to the bottom, making the concentration of the zinc ions greater here than at the top. It was pointed out in the discussion of osmotic pressure that the concentration of dissolved ions lessened the rate at which more ions could enter. In this case, then, the zinc anode can dissolve more readily at the top than at the bottom. In fact, the action is that of a concentration cell. Zinc ions enter solution at the top of the anode and hydrogen ions separate from solution at the bottom. Sometimes the effect is so pronounced that the rod becomes noticeably "necked" at the top and may even be entirely separated into two parts.

Depolarization.—The depolarizing reaction of the Leclanché cell may be represented by the following equation:

$$2MnO_2 + 2H \rightarrow Mn_2O_3 + H_2O$$

The manganese dioxide does not react nearly so rapidly as the copper oxide of the Lalande cell; in fact, depolarization is very slow. Consequently, but little current can be taken from the cell at a time without the voltage dropping greatly, but depolarization takes place eventually and the cell recovers. Because of this slow action of the depolarizer, Leclanché cells are most suitable where only an intermittent current is desired.

E.m.f. of the Cell.—Depending upon the form of the cell, the initial voltage may vary from 1 to 1.8. The internal resistance is about 3 ohms. When exhausted, the cell may be renewed with fresh electrolyte, but the e.m.f. is not so high as that of a new cell.

"Creeping" of Salt.—Leclanché cells are sometimes troublesome because of the "creeping" of the salt of the electrolyte. This is an effect produced by the evaporation of water and the formation of crystals on the sides of the container. More of the solution rises between the crystals by capillary attraction and evaporates, leaving more crystals above, the action continuing until the crystals may even pass over the edge of the container.

The "creeping" may be prevented by dipping the tops of the
jar and the elements in varnish or melted paraffine to which
substances the solution does not adhere.

Despite its many objections, the Leclanché cell is widely used,
probably because it is easily set up and requires but little attention. It can stand for long periods unused with practically no
ill effects, except the local action due to the concentration change.
There is but little local action due to other causes.

DRY CELLS

Dry cells are cells prepared in such a manner that they are
portable and can be used in any position. Although they are
called dry cells, it should not be understood that they are really
dry, since if such were the case, no action would occur in the cell.
In fact, it is a very important consideration of the manufacturer
to so construct them that they will remain sufficiently wet.

Dry cells are essentially Leclanché cells, containing the same
elements as the Leclanché, but with their electrolyte and depolarizer in the form of a paste. It is rather difficult to state
any definite method of construction, since there are so many
different forms. In one form the zinc anode is in the form of a
cylinder, which serves as the outer containing vessel. The
carbon kathode occupies the interior of the cell, surrounded with
the depolarizer—a mixture of finely divided graphite or other
carbon and manganese dioxide, the latter being as pure as
possible. The kathode is suitably insulated from the bottom
of the cell to prevent internal short-circuiting. Between the
depolarizer and the zinc container is the electrolyte, which is
about a 25 per cent. solution of ammonium chloride, maintained
in a layer form against the zinc by being held in some porous
material as pasteboard, paper pulp, sawdust, charcoal, infusorial
earth, gypsum, magnesium oxide, zinc oxide, etc. The two
oxides just mentioned have the disadvantage of setting to a
hardened mass by the formation of an oxychloride compound
and thus increasing the cell resistance. Hygroscopic substances,
as zinc chloride, calcium chloride, and glycerine are sometimes
added to aid in preventing the cell from becoming too dry. The
contents of the cell are kept in place by suitable packing and the

whole is sealed in with bituminous material. One or two small openings are left in the container to allow the escape of ammonia gas produced during the working of the cell. When these openings become clogged the cell may burst because of the pressure of the retained gas.

Since dry cells are required to stand idle for long periods, it is very necessary in manufacturing them to see that no defects occur that would lead to local action. The materials used must be very pure. Salts of copper, lead, tin, iron, etc., are very bad, since zinc deposits these metals from their salts, and then they become kathodes for local action.

The voltage of the cells varies according to the brand or type, but may be considered to be equivalent to that of the Leclanché containing the fluid electrolyte; *i.e.*, from 1.0 to 1.8 volts.

The internal resistance increases during the working of the cell, this being due both to the formation of insoluble zinc compounds as explained on page 404, and to the evaporation of water through imperceptible openings in the container or seal. Sometimes the container may be ruptured by the pressure of the gas produced during working. The cell is then liable to fail because the moisture will evaporate and it will become too dry. Such cells may be made to produce several hours' more service by removing the outer pasteboard cover and placing them in water for an hour or two. The same treatment may be given cells in which no ruptures are apparent if small perforations are made in the case. During soaking a small opening should be left above the water for the escape of air. Deterioration due to drying out may take place even in unused cells, but if well-made, they will remain active for about 2 years.

References

Cooper: Primary Batteries, London, 1901.
Jones: Outlines of Electrochemistry, New York, 1901.
Lupke: Elements of Electrochemistry, London, 1903.
Thompson: Applied Electrochemistry, New York, 1911.
Allmand: Principles of Applied Electrochemistry, London, 1912.

CHAPTER XIX

SECONDARY CELLS

Secondary cells may be considered to be reversible primary cells. They consist of systems so arranged that when electrolysis is caused to occur in the cell, the electrodes assume such a condition that they may later exercise the same functions as the elements of a primary cell. Such systems are also called storage cells, because during electrolysis, chemical energy is stored up that may be converted into electrical energy at will. The process of storing up energy in the cell by electrolysis is known as charging, and the taking of current from the cell, as discharging. Since an understanding of electrolysis is of fundamental importance in a study of secondary cells, a discussion of it will be included here.

Electrolysis.—When a current passes through solutions of chemical compounds, or through compounds in the molten state, a chemical decomposition results. This decomposition is known as electrolysis, which means the setting free of combined substances by means of the electric current. When a current is passed through a solution containing a dissociated substance, the ions swarm to the electrodes and give up their charges. As was explained in the discussion of "Electrolytic Dissociation" in the preceding chapter, the ions exist in the solution before the current is passed, and begin their migration the instant the circuit is closed. The positive ions travel toward the kathode and the negative ions toward the anode, where they deliver their charges and thus serve to carry the current through the solution. When the ions have reached the electrodes and delivered their charges they may separate as elements, or undergo secondary reactions. When a current passes through an electrolyte dissolved in water, there may be two kinds of ions appearing at both the anode and kathode—ions from the dissolved substance and ions from the dissociated water—but only one kind may give up

408

its charge. For example, when a current is passed between insoluble electrodes, through a dilute solution of sulfuric acid, there will gather about the kathode, hydrogen ions from both the sulfuric acid and the dissociated water, and these give up their charges there. About the anode will gather both SO_4 and OH ions, but because the former give up their charges less readily than the latter, it is only the OH ions that discharge negative electricity to the anode. Two discharged hydroxyl ions unite to form a molecule of water and oxygen, which gas escapes from the solution. The SO_4 ions undergo no change at the anode; they merely carry negative electricity to this electrode, and then form sulfuric acid with the hydrogen ions of the water. Precisely similar actions may occur in the electrolysis of bases and salts.[1]

However, it is not always hydrogen and oxygen only that are evolved. Sometimes the ions of the dissolved substance give up their charges more readily than the ions from the dissociated water. For example, copper and silver discharge more readily than hydrogen, and chlorine more readily than the hydroxyl ion.

Also, it sometimes happens that the anode dissolves under the conditions of electrolysis. For example, if a copper anode is used in dilute sulfuric acid, copper ions will enter the solution and produce copper sulfate with the SO_4 ions that collect there. No oxygen will escape as when the insoluble platinum electrodes were used. If a solution of copper sulfate is used in place of the dilute sulfuric acid, the conditions will then be suitable for electroplating. Copper ions will enter the solution at the anode and deposit at the kathode.

Kinds of Secondary Cells.—Theoretically, there are a great many combinations that may serve as storage cells, and much painstaking effort has been expended in attempts to develop various systems into successfully working cells, but practical difficulties have caused all but two to be abandoned. These two are the well-known lead accumulator, produced by Planté about 1860, and modified by Faure about 1880; and the iron-nickel-alkali cell rather recently developed by Edison.

[1] For a more complete discussion of these cases, see JONES: Outlines of Electrochemistry, p. 40.

THE LEAD ACCUMULATOR

Construction of the Lead Accumulator.—When the accumulator is in a charged condition, that is, ready to deliver current, it consists essentially of a plate of lead and a plate of lead peroxide, dipping into a solution of sulfuric acid that has been prepared to contain 24 per cent. of sulfuric acid by weight. This solution has a specific gravity of 1.2. The plate containing the lead peroxide is generally known as the positive plate, and the lead plate as the negative plate. However, these terms are strictly applicable only half the time, for during charging, the current enters at the peroxide electrode, thus making it the anode, and flows through the electrolyte to the lead electrode, now the kathode. On discharging, the current flows through the cell in the opposite direction, leaving the cell by the peroxide plate, thus making it the kathode, and enters the cell by the lead electrode now the anode.

Types of Plates.—The plates of the lead accumulator are of two main types: first, those on which the active lead and lead peroxide are formed from the substance of the plate itself, by direct chemical or electrochemical action; and second, those on which the active material is formed of substances applied to the plate. The former are of the type originally produced by Planté, and are known as Planté plates, although the modern plate has been so highly improved, that it scarcely resembles the original. The latter type are as modified by Faure, and are generally known as Faure, mass, or pasted plates.

In making the *Planté type*, the surface of the plate is increased as much as possible by scoring with knives or else by casting it in a suitably ribbed or corrugated form. Plates mechanically prepared in this manner are then given a coat of lead sponge or lead peroxide by a process known as *"forming."* In the earlier methods, the first step in the forming process was to produce on both the positive and negative plate a coating of lead peroxide, this being accomplished by connecting them as anodes in an electrolyte of dilute sulfuric acid. The lead sponge plate was then produced from a lead peroxide plate, by connecting it as a kathode and reducing it. To secure a sufficient amount of active material the process of oxidation and reduction had to be repeated many times, so that forming was exceedingly slow, being completed in only about a year's time. Now the plates are prepared in about ½ day by a process known as "*rapid forming.*" To the dilute sulfuric acid is added some reagent whose negative ion by reacting at the anode, tends to produce a lead salt that is at once converted into lead sulfate. In this way a layer of lead sulfate is produced instead of lead peroxide; and since the sulfate layer is more porous than the peroxide, the action proceeds quite rapidly. The sulfate can then be converted into lead sponge or lead peroxide as desired by electrolyzing in a bath of dilute sulfuric acid. The substances used as "*forming agents*" are such salts as the acetates, nitrates, chlorates, chlorides, and their corresponding acids.

In making the *Faure or pasted plates*, grids containing grooves or pockets are first prepared and filled with a paste containing lead oxide and sulfuric acid, and often other substances. This paste soon sets to a hard mass, and then it is changed to lead sponge and lead peroxide by the reactions of electrolysis.

There are two or three noteworthy modifications of the preceding types. These are the plates of the so-called "chloride accumulator," the "box negatives" and the "iron-clad" vehicle plates. The chloride negative was formerly made by the Electric Storage Battery Co. of Philadelphia, in the following manner. Lead and zinc chlorides were melted together and poured into a supporting lead grid. The zinc chloride was then dissolved out, leaving a porous mass of lead chloride, which was later reduced to lead sponge by electrolytic action. The positive plate of this battery was made by the Planté process. Although this chloride method is no longer used,[1] the name is retained. The peroxide plates of the so-called chloride accumulator are now made by inserting in the grid, buttons produced by winding up corrugated lead ribbons in the form of a spiral. These buttons project from the grid a slight distance as shown in Fig. 66. The active material is then formed on the surface of this lead spiral.

In the box negative, the active material is not actually "pasted," but is held loosely in cavities in a grid, being kept in place by a perforated cover of sheet lead on each side. Because

[1] THOMPSON: Applied Electrochemistry, p. 154.

the cover keeps it in place the active material can be allowed to be very loose and porous—a very desirable feature.

The "iron-clad" vehicle plate is peculiar, in that it depends for support upon an insulating grid or envelope of some substance as rubber, celluloid, porcelain biscuit ware, wood, etc. The envelope of material of this sort surrounds the active material and prevents shedding, electrical contact being made with a central lead strip or wire.

Fig. 66.—"Chloride accumulator."   Fig. 67.—Showing construction of "Iron-clad Exide" cell.

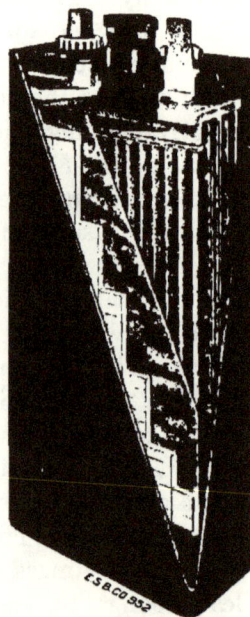

In the preparation of the Faure type plates the metal grid used is often made of a lead-antimony alloy, containing about 4 to 6 per cent. of antimony.[1]   This alloy is much harder and more rigid than pure lead, and is designed to resist "buckling"—a warping effect produced in the plate by the volume changes that occur in the active material during charging and discharging. Planté plates are made of soft lead.[2]   In assembling the plates

[1] ALLMAND: Principles of Applied Electrochemistry, p. 223.

[2] WATSON: Storage Batteries, p. 21.   Quoted by THOMPSON: Applied Electrochemistry, p. 153.

in the battery, peroxide plates are not used as end plates because the volume change in the peroxide plate is much greater than in the lead sponge plate, and if active on only one side there would be a great tendency toward warping.

Comparison of Types.—Planté and Faure plates each have characteristic advantages and disadvantages. The latter have less weight in proportion to their active mass, and are, therefore, especially suitable where low weight is required. However, they are mechanically weaker than Planté plates and the active material tends to fall out in time, particularly if the cell is charged and discharged at a rapid rate so that the volume changes are rapid. These detached pieces may lodge between the plates and short-circuit the battery. The Planté plates are more rigid and are used where ability to withstand heavy service is necessary.

Reactions during Charge and Discharge.—The following explanation is based on the theory developed by Gladstone and Tribe, which is now generally known as the sulfate theory. According to this theory sulfuric acid reacts with both plates during discharge, producing lead sulfate. During charge, the lead sulfate is converted into lead sponge on the negative plate and into lead peroxide on the positive. The following equations, indicating purely chemical transformations, are used to explain the reactions that occur *during charge.*

At the peroxide electrode:

$$PbSO_4 + SO_4 + 2H_2O \rightarrow PbO_2 + 2H_2SO_4$$

At the lead sponge electrode:

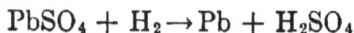

$$PbSO_4 + H_2 \rightarrow Pb + H_2SO_4$$

Sulfuric acid is set free at both electrodes and the density of the electrolyte increases.

The direction of the current through the electrolyte during charging is from the peroxide plate to the lead, but the direction is reversed during discharge, and the hydrogen ions are set free at the peroxide electrode and sulfate ions at the lead sponge electrode. The following equations show the reactions that occur *during discharge.*

At the peroxide electrode:

$$PbO_2 + H_2 + H_2SO_4 \rightarrow PbSO_4 + 2H_2O$$

At the lead sponge electrode:

$$Pb + SO_4 \rightarrow PbSO_4$$

Water is liberated and sulfuric acid is consumed, so that the density of the electrolyte grows less.

The *ionic nature* of the reactions are much more clearly set forth in the following explanation of Le Blanc's theory[1] as given by Allmand.[2] The reactions indicated are those that take place *during discharge*. They proceed in the reverse direction during charge.

At the positive electrode:

1. PbO_2 solid→PbO_2 dissolved
   $$\rightarrow \overset{++}{Pb} + 2\overset{--}{O}$$
2. $4\overset{+}{H} + 2\overset{--}{O} \rightarrow 2H_2O$
3. $\overset{++}{Pb} \rightarrow \overset{++++}{Pb} + 2 \oplus$
4. $\overset{++}{Pb} + \overset{--}{SO_4} \rightarrow PbSO_4$ undissociated
   $$\rightarrow PbSO_4 \text{ solid.}$$

The total of the preceding reactions is equivalent to the following:

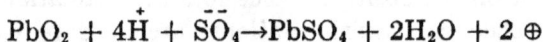

$$PbO_2 + 4\overset{+}{H} + \overset{--}{SO_4} \rightarrow PbSO_4 + 2H_2O + 2 \oplus$$

At the negative electrode:

1. $Pb + 2 \oplus \rightarrow \overset{++}{Pb}$
2. $\overset{++}{Pb} + \overset{--}{SO_4} \rightarrow PbSO_4$ undissociated
   $$\rightarrow PbSO_4 \text{ solid.}$$

The total of these reactions is:

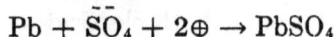

$$Pb + \overset{--}{SO_4} + 2\oplus \rightarrow PbSO_4$$

Polarization.—Because of the concentration changes that accompany the reactions of charge and discharge, the voltage of the cell is affected. During discharge the electromotive force decreases since the strength of the acid is lessened. The amount of acid actually used is but a small fraction of the total amount in the electrolyte, but since it is all abstracted from the solution

[1] For other plausible theories see DOLEZALEK: Theory of the Lead Accumulator, translated by CARL VON ENDE.

[2] *Loc. cit.*, p. 232.

contained in the pores of the active material, the effect is very pronounced. On standing, this change in concentration is corrected by diffusion between the acid of the pores and that of the main body of the electrolyte. In this way, cells that have been discharged below their normal voltage "recover" upon standing. With a high rate of discharge, the density of the acid within the pores becomes exceedingly low, because of lack of time for diffusion; but with slow discharge rates, diffusion may keep pace with the acid consumption.

Self-discharge.—When a charged cell is allowed to stand idle, it slowly discharges spontaneously. This is known as self-discharge, and even for a cell in good condition amounts to 1 or 2 per cent. a day. Self-discharge is due to local action. It occurs especially between the peroxide and the lead support of the positive plate, lead sulfate being formed at the surface of contact. This action is more noticeable in the Planté positives than in those of the paste type.[1] On the negative plate the difference of potential between the sponge lead and the solid supporting lead is so small that the self-discharge due to this factor is very slight. But in the lead sponge of many of the negative paste plates are such substances as graphite which are designed to increase porosity and cause the active mass to conduct more readily. With graphite present the negative plate may discharge itself by local action quite as rapidly as the positive.

Effect of Impurities in the Electrolyte.[2]—It is of greatest importance that the electrolyte be free from impurities, since they may serve to greatly accelerate self-discharge. The impurities that are the most objectionable in the electrolyte may be divided into three classes: first, small amounts of "forming agents," such as the nitric and hydrochloric acids, and other substances mentioned on page 411; second, metals having low solution tensions, as platinum and silver; and third, soluble salts of such metals as iron, whose ions are capable of being oxidized to a higher or reduced to a lower valence.

The "forming agents" are objectionable because each time the cell is charged they become active in converting more of the lead of the positive plate into lead peroxide, so that in time the lead

[1] MORSE: Storage Batteries, p. 207.
[2] See LYNDON: Storage Battery Engineering, p. 52.

upon which the plate depends for support may become peroxi-
dized to such an extent that the plate may fall to pieces.

Metals with low solution pressures that may assume the
kathodic position to lead are very objectionable because they set
up local action that produces self-discharge. Such action is
especially likely to occur at the lead plate, because these impurities are present as dissolved salts, and when the metallic
ion comes into contact with the lead, it replaces the ion in solution,
and the foreign metal is deposited on the surface of the plate.
Lead peroxide is not able to throw the metallic ions out of solution in this manner so they do not deposit on this plate. Platinum, gold, silver, and copper are objectionable in the order
named. Beside being kathodic to lead, it is relatively easy for
hydrogen to discharge to these metals, and this is a great aid to
local action.

Sulfuric acid may acquire traces of platinum during the process
of manufacture, by being concentrated in platinum retorts.
Thompson[1] says that one part of platinum in a million of acid may
produce a rapid self-discharge of the lead plate. The presence of
platinum and the similar metals may be recognized sometimes
by the continued evolution of gas at the lead plate after the
charging current has been interrupted.

The presence of ions that can exist in more than one stage of
oxidation, such as those of iron and manganese, are also aids in
bringing about self-discharge. These ions are oxidized to their
high valence at the peroxide electrode, and then diffuse to the
lead electrode, where they are reduced, thus bringing about the
discharge of both plates.

After commenting on the various factors that bring about self-
discharge, Allmand says[2] that a charged cell with a bad electrolyte
may lose as much as 50 per cent. of its energy in a day through
local action.

Sulfating.—When a cell is allowed to stand for some time in
the discharged state, the lead sulfate becomes transformed
into a dense layer that is very resistant to the action of the
charging current. This inert condition is the result of the
formation of large-sized lead sulfate crystals, which are both

[1] Applied Electrochemistry, p. 170.
[2] *Loc. cit.*, p. 227.

poor conductors and inactive chemically, because but little surface is exposed in proportion to their masses. In a normal discharged plate, the lead sulfate is finely divided, and beside, is mixed with lead sponge or lead peroxide, which are relatively good conductors. The production of the large crystals from the finer ones is due to slight temperature changes. As the cell becomes warmer the smaller of the crystals dissolve, and as it cools the dissolved lead sulfate separates again. The larger crystals, which have remained undissolved, act as nuclei and the separating lead sulfate precipitates upon them, thus increasing their size. If this so-called sulfating has not proceeded too far, the cell may be recharged by a long-continued application of the current, but the cell will probably not attain its normal value. If the cell has been allowed to stand discharged for several weeks, the plates may become so badly affected that it is advisable to replace them with new ones. It is very evident, therefore, that when a battery is allowed to stand idle, it should be kept fully charged.

The use of the word *sulfating* in describing this inert condition due to crystal growth is somewhat inapt, since the plates of a discharged cell in the normal condition are also covered with lead sulfate.

Charging and Discharging.—The normal electromotive force of a discharged cell is about 2 volts. As the cell is charged the voltage rises gradually. Oxygen gas is evolved at the peroxide plate at 2.2 volts. The completion of the charge is denoted by the evolution of hydrogen gas at the lead plate, this commencing at 2.3 volts; but it is good practice to continue the charging current for some time longer.

Batteries should not be charged too slowly, generally not below the 16-hr. rate. If the period required is too long, or in other words, if the current density is too low, lead sulfate instead of lead peroxide is formed on the positive plate.

During charge, a part of the energy of the current is transformed into heat, because of the resistance of the system. Care must be exercised that the temperature does not rise too high.

In discharging, the general rule is to interrupt the current when the drop in voltage has amounted to one-tenth the original voltage of the cell; for example, from 2 to 1.8 volts. But this is

27

not invariable. Morse says[1] that when discharged at low rates, as in telephone or train-lighting service, the lower limit is about 1.8 volts; in regulating power-plant loads, and in much of the other regular battery work, it is 1.7 volts; at very high rates, as when an emergency battery is called upon to take the entire load of a large station, it may be carried as low as 1 volt. In discharging at the higher rates less total current is available, since but little of the active mass is used, due to the fact that the acid has not sufficient time to diffuse into it, thus bringing about concentration polarization as has been explained.

The effect of variation in charge and discharge rates is illustrated in the following curves and discussion quoted from Morse.[2]

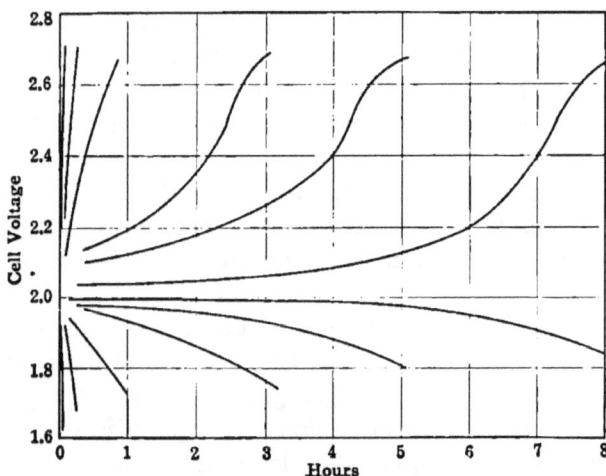

Fig. 68.—Curves of operation of Planté plates at various rates.[3]

"The rates for the curves of Fig. 68 are:

	Amp.
For 8 hrs. of charge or discharge	1.0
For 5 hrs. of charge or discharge	1.4
For 3 hrs. of charge or discharge	2.0
For 1 hr. of charge or discharge	4.0
For 20 min. of charge or discharge	8.0
For 5 min. of charge or discharge	16.0

These are the rates usually specified in practice.

[1] Storage Batteries, p. 116.
[2] Pp. 112 and 113.
[3] Reproduced by permission from Storage Batteries by Morse.

"The capacities corresponding to these rates are:

	Amp.-hrs.
For 8-hr. charge or discharge	8.00
For 5-hr. charge or discharge	7.00
For 3-hr. charge or discharge	6.00
For 1-hr. charge or discharge	4.00
For 20-min. charge or discharge	2.67
For 5-min. charge or discharge	1.33

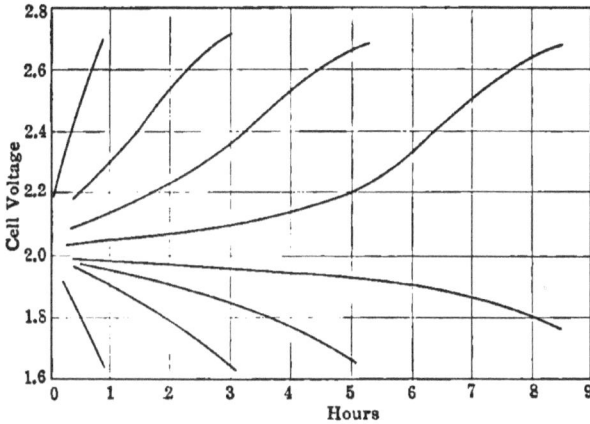

FIG. 69.—Curves of operation of mass plates at various rates.[1]

Morse says further:

"It makes a great difference in the discharge curve of a cell, whether the cell has been charged at a high or a low rate, and just as great a difference in the charging curve, whether the previous discharge has been fast or slow. Take a single case. Suppose a fully charged cell has been discharged at the 5-min. rate. It is evident from the figure that only 1.3 amp.-hrs. have been drawn from it. We only need to return a little more than this to the cell to charge it completely. In the same way, if our cell has been completely discharged at a low rate, and then charged at the 5-min. rate, we can only get about 1.3 amp.-hrs. into it. It may be fully charged for a 5-min. discharge, but it is by no means charged for a 3-hr. discharge."

The following general rules for the operation of batteries are also quoted from Morse.[2]

"For Planté plates:

1. Keep the battery charged.

[1] Reproduced by permission from Storage Batteries by Morse.
[2] Pp. 223 and 224.

2. Charge at a fairly high rate. Usually this means at the 8-hr. rate or a little higher.

3. Inspect frequently and remove all possible short-circuits immediately.

4. Keep acid density at the proper point.

5. Keep the acid above the top of the plates.

6. If plates buckle, straighten them out as soon as possible.

7. Do not let the temperature reach too high a point (100°F. is a safe limit)."

"Discharge at almost any rate does not harm good Planté plates, provided they are charged immediately after the charge is finished."

"For paste plates:

1. Charge at a low rate, 12-hr. or lower.

2. Overcharge occasionally by 10 per cent. or so. Once a week is often enough for the overcharge if the battery is in daily service.

3. Use an ampere-hour meter and regulate charge and discharge by that.

4. Try to give a nearly complete discharge before recharging. If the discharge is extended over 2 or 3 days no harm is done.

5. Watch temperature carefully. High temperature is much more destructive to paste plates than to Planté types.

6. Test each cell frequently and inspect at the least sign of trouble."

"The most usual trouble arises from continued net undercharge, especially in private installations."

The following further observations have been derived from Lyndon's Storage Battery Engineering..

If an instrument must be put into the cell for any purpose, never use metal. Use glass, hard rubber or wood.

If batteries are to be put out of commission for some time—say 6 weeks or more—they should not be allowed to stand in the electrolyte unless a small charge can be given them at least once in 2 weeks. When they are to be unused for a considerable period, the elements must be removed from the liquid and dried. But this must be done only after the plates have been specially prepared and they must be dried with the observance of certain precautions. For the methods of procedure for such dismantling, see Lyndon, page 263.

THE EDISON STORAGE BATTERY

In the cells of this battery in the charged condition, the active material on the positive plate consists of hydrated nickelic oxide and peroxide, while that of the negative plate is made up of finely divided iron. The electrolyte in new cells consists of about a 21 per cent. solution of potassium hydroxide to which has been added 50 grams of lithium hydroxide per liter. During discharge the active material is converted into the nickelous and ferrous hydroxides respectively.

Construction.[1]—For the active material of the positive plate nickelous hydroxide is prepared, and placed in layers in $\frac{1}{4}$-in. perforated tubes, $4\frac{1}{2}$ in. long, made of spirally-wound, nickel-plated, steel ribbon. The seam is made spiral to resist expansion. The layers of nickelous hydroxide are about $\frac{1}{100}$ in. thick, and are separated by electrochemically-prepared nickel flake. The arrangement is shown in Fig. 70. Each $4\frac{1}{2}$-in. tube contains a total of about 630 layers of nickel flake and nickel hydroxide, pressed closely into contact by a pressure of about 1 ton per square inch.

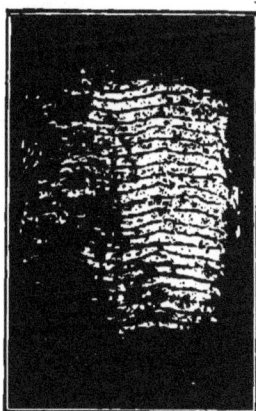

FIG. 70.—A magnified cross-section of a positive tube of an Edison cell.

The layers of nickel flake are in metallic contact with the sides of the tube and thus good electrical connection is insured between the active material and the supporting grid in which the tubes are clamped. The tubes are reinforced by eight seamless steel rings.

For the active material of the negative plate, iron oxide is prepared. This is mixed with a little oxide of mercury (subsequently reduced) to increase the conductivity and capacity, and the mixture is packed into pockets, about 3 in. long by $\frac{1}{2}$ in. wide by $\frac{1}{8}$ in. deep, made of perforated, nickel-plated steel. When filled, the pockets are placed in their steel, supporting grid

[1] From *bulletins* of the Edison company.

and pressed into close electrical contact with it by hydraulic pressure. The dies between which the plates are pressed are corrugated, the corrugations being imparted to the walls of the pockets, which are thus given greater mechanical strength.

The manner of assembling the plates and the necessary insulating parts, together with their positions in the container, is shown in Fig. 71. The lid of the container is permanently welded in place, since it is never necessary to renew the parts.

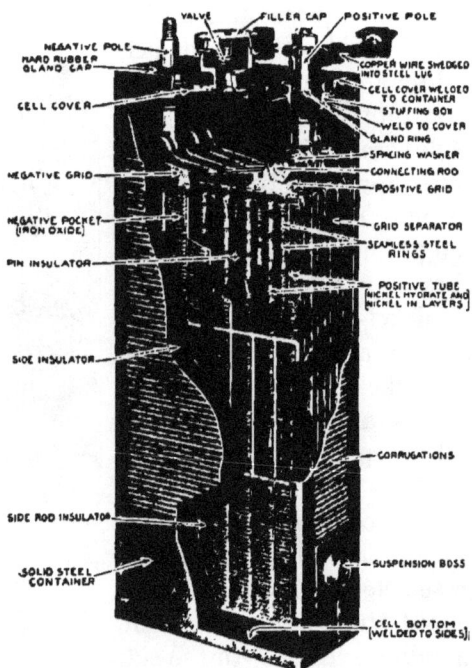

Fig. 71.—Showing parts of the Edison storage cell.

The lid is provided with a valve, which is designed to allow the escape of the gases produced during charge but to prevent the entrance of gases from the air. It is necessary to exclude carbon dioxide and oxygen, since the former reacts with the electrolyte, and the latter converts the ferrous hydroxide into the ferric form, which cannot be completely reduced again. The valve serves also as a convenient opening for filling.

As is obvious from the manner of construction, the Edison cell is of a very rugged type, and well-suited for service where an ability to withstand rough handling is essential.

Forming.—The plates are "formed" by repeated cycles of charge and discharge.

Types.—There are two general types of Edison storage cells designed for ordinary service—type A and type B. These differ in respect to the size of the plates. The *type A* positive contains two rows of 15 tubes each, and has a rated ampere-hour capacity of 37.5 hrs. The *type B* positive has one row of 15 tubes, and half the capacity of an A plate. The number of *positive* plates in a cell is indicated by the numeral following the letter indicating the type. Thus, an A5 cell contains five positive plates and has a rated capacity of 187.5 (5 × 37.5) amp.-hrs. A cell marked B4 contains four type B positives, and has a rated capacity of 75 $\left(4 \times \dfrac{37.5}{2}\right)$ amp.-hrs., etc. The number of negative plates in any cell is always one greater than the number of positives.

Both type A and type B cells are made in a "high" form, being designated by the letter H, as B6H. The high forms are identical with the others in all respects, except that the container is deeper. This permits an excess of electrolyte to be carried, so that water need not be added so often.

Reactions during Charge and Discharge.[1]—The active material in the tubes of the positive plate when it is in the discharged condition, is the green nickelous hydroixde. During charging, this is converted partly to nickelic oxide and partly to nickel peroxide. It has been found that when nickelous hydroxide is oxidized chemically, it is always first changed into nickel peroxide, and it is reasonable to assume that the same result is brought about by the oxygen liberated electrolytically at a nickelous hydroxide anode. Moreover, this assumption is borne out by the facts. Thompson says[2] that a freshly charged nickel plate contains more oxygen than is required by the formula, Ni_2O_3, the amount of the excess being sufficient to correspond to at

[1] The explanation given here is based on ALLMAND's and THOMPSON's discussion of FOERSTER's researches. See ALLMAND: Principles of Applied Electrochemistry, p. 239; and THOMPSON: Applied Electrochemistry, p. 178.

[2] P. 178.

least 8 per cent. of nickel peroxide. The nickel peroxide formed in this way, then reacts chemically with the nickelous hydroxide to produce hydrated nickelic oxide.

It has been shown[1] that the colloidal hydroxides of nickel and iron are not definite compounds, but rather oxides containing more or less occluded water; therefore, in writing the equations for the reactions in this cell, these substances are not always shown as definite hydrates, or as containing integral molecules of water. The following equations show the reactions that take place *during charge*.

At the positive electrode:

$$1. \qquad NiO \cdot H_2O + 2\bar{O}H \rightarrow NiO_2 + 2H_2O + 2 \ominus$$

By a secondary non-electrochemical reaction, this peroxide is changed as follows:

$$2. \qquad NiO_2 + NiO \cdot H_2O \rightarrow Ni_2O_3 \text{ hydrated.}$$

Reaction (2) takes place only during the first part of the charge, when much nickelous hydroxide is present. Toward the latter part of the charge, when the nickelous hydroxide has been largely used up, the peroxide, being unstable, decomposes spontaneously, as:

$$3. \qquad 4NiO_2 \rightarrow 2Ni_2O_3 + O_2$$

Because the peroxide is dissolved in the nickelic oxide, reaction (3) does not proceed as rapidly as it otherwise would, but it runs to completion on standing. Also, during the same period of the charge when reaction (3) takes place, some oxygen is liberated directly from the electrolyte, thus:

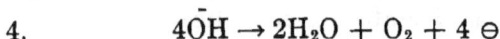

$$4. \qquad 4\bar{O}H \rightarrow 2H_2O + O_2 + 4 \ominus$$

Reaction (4) is due to the same condition that brought about reaction (3); that is, the gradual using up of hydrated nickelous oxide, which, when it was present in sufficient quantity, united with all the discharged oxygen. By reaction (4), current is used up to no purpose, thereby decreasing the efficiency of the cell.

At the negative electrode:

In the pockets of the negative plate when the cell is in the dis-

---

[1] ALLMAND: *Loc. cit.*, pp. 239 and 240.

charged state, the active material is ferrous hydroxide. This is reduced to metallic iron as is shown by the following equation, which represents the reactions *during charge:*

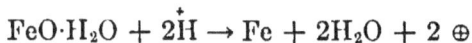

$$FeO \cdot H_2O + 2\overset{+}{H} \rightarrow Fe + 2H_2O + 2 \oplus$$

Water is thrown off at the negative plate, as well as at the positive (equation 1), so that the electrolyte becomes more dilute on charging. During discharge, the process is reversed and the electrolyte becomes more concentrated. The reduction of the iron oxide does not proceed readily, hydrogen being given off in the gaseous state from almost the beginning of the charge, the amount gradually increasing as the charge proceeds. In this way considerable current value is lost, even much more than at the positive electrode. It is perhaps needless to remark that since hydrogen and oxygen are escaping simultaneously during charging, lighted matches or similar sources of ignition must be kept away from the cells.

On *discharging,* the reactions are the reverse of charging and may be represented by the following equations.

At the positive plate:

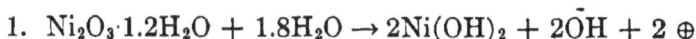

1. $Ni_2O_3 \cdot 1.2H_2O + 1.8H_2O \rightarrow 2Ni(OH)_2 + 2\overset{-}{O}H + 2 \oplus$

But if the discharge takes place soon after charging, the nickel peroxide present is first reduced, thus:

2. $\quad 2NiO_2 + H_2O \rightarrow Ni_2O_3 + 2\overset{-}{O}H + 2 \oplus$

At the negative plate:

$$Fe + 2\oplus + 2\overset{-}{O}H \rightarrow 2Fe(OH)_2$$

Charging.—The following curve and data relating to charge and discharge are taken from bulletins issued by the Edison company.

The length of the charge is determined by the extent of the previous discharge. If totally discharged, the cells are charged for 7 hrs. at the normal rates shown in Fig. 72; if only half discharged, for only half of 7 hrs., etc. If the extent of the previous discharge is unknown, they are charged until the voltmeter shows a constant reading for 30 min. of about 1.8 volts per cell with normal current flowing. In charging it is essential that the rate of charge be not less than the specified normal rate, or, if a tapering charge is given, that the *average* rate be not less than normal. The capacity of any cell is determined by the number of plates and can be calculated as shown on page 423. The temperature during charge should not be allowed to exceed 46°C.

Discharging.—The normal discharge rates for the various cells are the same as the charging rates, as shown in Fig. 72. The

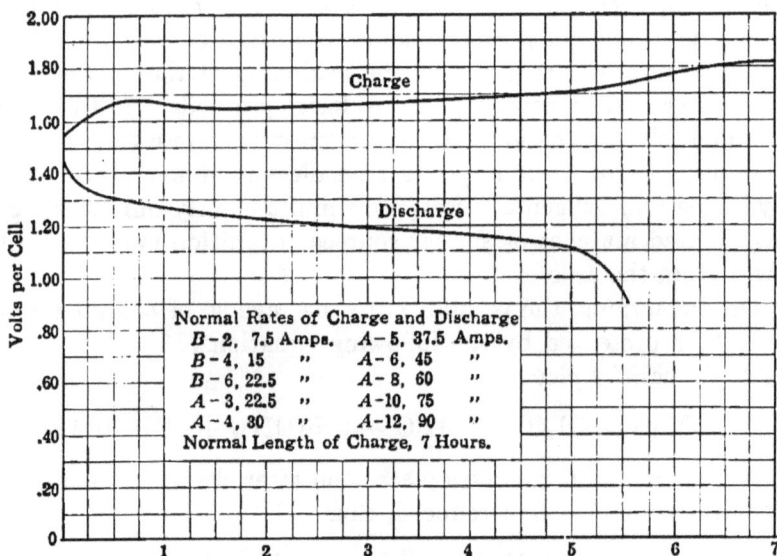

Normal Rates of Charge and Discharge
$B-2$, 7.5 Amps. $A-5$, 37.5 Amps.
$B-4$, 15   " $A-6$, 45   "
$B-6$, 22.5   " $A-8$, 60   "
$A-3$, 22.5   " $A-10$, 75   "
$A-4$, 30   " $A-12$, 90   "
Normal Length of Charge, 7 Hours.

Fig. 72.—Curves showing a discharge at the 5-hr. rate and a normal 7-hr. charge.   A and B type Edison cells.[1]

discharge voltage for any type of Edison cell is about 1.2. The discharge may be considered complete when this has fallen to about 0.9 volt.

Self-discharge.—On standing, the Edison cell undergoes a certain amount of self-discharge, it being most pronounced during the first 24 hrs., when it may amount to as much as 10 per cent. This is due to the evolution of oxygen by the decomposing nickel peroxide, thus:

$$4NiO_2 \rightarrow 2Ni_2O_3 + O_2$$

After the peroxide has decomposed, the self-discharge proceeds

[1] From bulletin of the Edison Storage Battery Co.

much more slowly, so that it reaches 15 per cent. only after several weeks.[1]

A noteworthy feature of the Edison cell is that it may stand idle indefinitely, either charged or discharged, without injury. This is in marked distinction to the lead accumulator, which must never be allowed to stand discharged.

Addition of Water.—During each charging of the cell, water is lost through electrolytic decomposition, as was explained in discussing the reactions that take place during charge. Since the total amount of the electrolyte is not large in proportion to the size of the cell, this loss must be replaced by the addition of distilled water. Enough water only must be added to bring the electrolyte to its normal level as shown in Fig. 73. In the bulletins of instruction issued by the Edison company, the proper heights of the solution above the plate tops are indicated as follows:

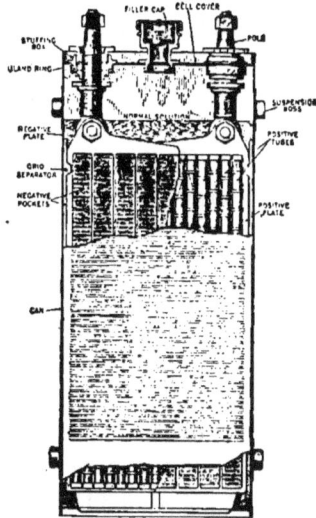

Fig. 73.—Showing normal level of electrolyte in Edison cell.

Type A3, A4, A5, A6, A8, A10, and A12 . . . . ½ in. above plate tops.
Type A3H, A4H, A5H, A6H, A8H, A10H, and A12H 3 in. above plate tops.
Type B2, B4, and B6 . . . . ½ in. above plate tops.
Type B1H, B2H, B4H, and B6H . . . . 2¼ in. above plate tops.

The height of the solution may be determined by inserting a glass tube at least ¼ in. outside diameter (it must be sufficiently large to prevent a false indication due to capillary attraction) and with straight-cut end, until it rests upon the plate tops. Then the upper end is closed tightly with the finger and the tube is withdrawn. The height of the liquid in the tube indicates the height of the electrolyte above the plate tops. The cells should not be filled above the levels indicated, since the bulk of the electrolyte increases during charging, due both to the liberation of water at the plates as shown by the equations,

[1] ALLMAND: *Loc. cit.*, p. 238.

and to the expansion caused by the rise in temperature. If the cell is too full, the electrolyte will flow out.

Renewing Electrolyte.—It is necessary to renew the electrolyte periodically, the frequency of renewal depending upon the extent to which the battery has been used. During use, the density of the electrolyte is reduced gradually by the slight amounts of potassium hydroxide thrown out by the escaping gases during charge. The time between renewals varies from 1 to 3 yrs. or even longer, and is determined by the specific gravity. If the gravity falls below 1.160 with the battery fully charged and the surface of the electrolyte at the proper height, it is an indication that the solution should be renewed.[1] The cell should be completely discharged before removing the electrolyte; and it must not be allowed to stand empty, but must be filled at once with the fresh solution. If the moist negative electrode is allowed to stand in contact with the air, the iron is oxidized to the ferric state and this cannot readily be completely reduced. The renewal solution consists of about a 25 per cent. solution of potassium hydroxide to which has been added 15 grams of lithium hydroxide per liter, thus differing from the electrolyte carried by a new cell, which is a 21 per cent. solution of potassium hydroxide with an addition of 50 grams of lithium hydroxide per liter.[2]

References

DOLAZELEK: The Theory of the Lead Accumulator, translated by VON ENDE, New York, 1904.

NIBLETT: Storage Batteries, Chicago, 1911.

LYNDON: Storage Battery Engineering, New York, 1911.

ALLMAND: Principles of Applied Electrochemistry, London, 1912.

MORSE: Storage Batteries, New York, 1912.

[1] Recommended by the Edison company.

[2] TURNOCK: Active Materials and Electrolyte of the Alkaline Storage Battery, *Met. Chem. Eng.*, September, 1916.

CHAPTER XX

HYDROMETRY

The Principle of a Hydrometer.—When a body floats in a fluid, the weight of the fluid displaced is equal to the total weight of the floating body. Upon this principle hydrometers are constructed. These instruments are usually of glass so designed and made that when floating, the long axis is in the vertical position, the required displacement being secured by a bulbous middle portion beneath which a weight is fixed. The upper portion consists of a slender stem, which in the most common forms is graduated.

General Classes.—Hydrometers are of two general classes:

(a) *Those of constant volume.* In using an instrument of this class, the amount of fluid displaced is the same for each determination, since the hydrometer is weighted until a fixed mark on the stem sinks to the level of the surface of the liquid. From the weight required the specific gravity is determined.

(b) *Those of variable volume.* In using instruments of this class the determination is made by noting to what point on the graduated scale the instrument sinks when placed in the fluid being tested. It is the hydrometers of this sort that are used almost entirely for liquids, there being two kinds in general use in the United States.

1. *The Baumé* (bō may') *hydrometers.* This hydrometer is used for the coarser varieties of work, as in industrial operations, and for specifications in mercantile transactions.

2. *The direct-reading specific-gravity hydrometer.* This is used most generally for the more accurate scientific work, but it is used in those cases mentioned in the preceding paragraph as well.

The Baumé Scale.—The scale of the Baumé hydrometer cannot be called logical. It is, indeed, very arbitrary. There are two varieties of the instrument.

For liquids heavier than water, the scale begins at zero. This zero point was determined by floating the instrument in pure water at 60°F. and marking the stem at the surface of the water for 0°Bé. A solution of common salt, NaCl, containing 15 parts of the pure salt and 85 parts of pure water, was then prepared. The instrument was floated in this solution at 60°F., and marked 15°Bé. at the surface of the liquid. The space between the two marks thus determined was divided into 15

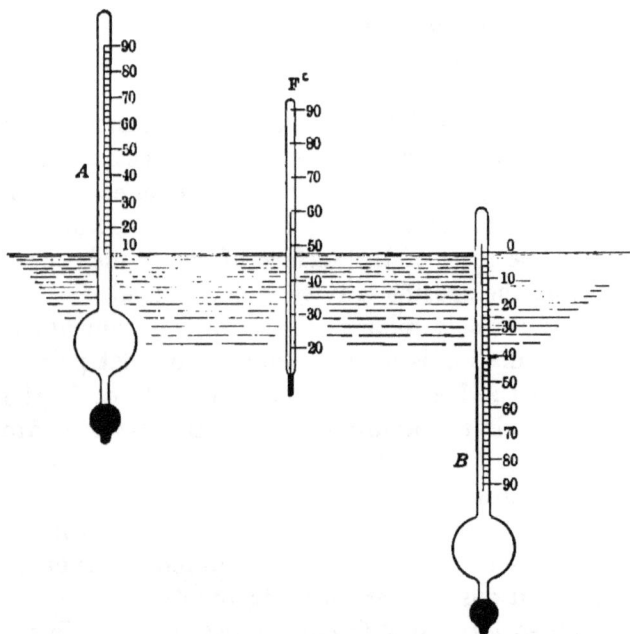

Fig. 74.—Showing (A) Baumé hydrometer for light liquids, and (B) Baumé hydrometer for heavy liquids in pure water at 60°F.

equal parts, and the scale was completed by continuing these divisions along the stem of the hydrometer.

For liquids lighter than water, the scale begins at 10. It was constructed according to the same arbitrary plan as the preceding. The point of the scale level with the surface of the liquid when the instrument was floated in a 10 per cent. solution of common salt, NaCl, at 60°F., was marked zero. The point to which it sank in pure water at 60°F. was marked 10. The intervening space on the stem was then divided into ten equal parts, and these graduations were extended along the scale.

It will be observed that the numbers from 10 upward appear on both Baumé scales, and when speaking of an unfamiliar liquid it should be stated whether it is lighter or heavier than water.

The Direct-reading Scale.—The scale on the direct-reading specific-gravity hydrometer begins at *one* for liquids both heavier and lighter than water. The instrument reads 1.000 when floated in pure water at 15.5°C. (60°F.). For liquids heavier than water the scale reads one plus a decimal; and for liquids lighter than water, a decimal only. The decimals are usually shown to the third place.

The Use of Hydrometers.—The density of a liquid is most conveniently measured when it is contained in a tall, narrow, glass cylinder, known as a hydrometer jar. The best type is that which bears an enlarged top as shown in Fig. 75. This enlargement prevents the upper part of the hydrometer stem from floating near the edge of the jar, for if this should happen, capillarity would render an accurate reading impossible.

When making a reading on any hydrometer, the eye must be on a level with the surface of the liquid, as shown in Fig. 75, and the scale must be read where it touches the lower curve of the meniscus. If the eye is not held in the proper position, the readings will be incorrect as shown by the dotted lines.

Since the specific gravity of a liquid changes rapidly with changes of temperature, if accurate determinations are desired, the temperature of the liquid being tested should be brought to exactly the temperature for which the density is being ascertained (generally 60°F.), and the temperature should be uniform throughout. Sometimes, rather than to actually modify the temperature of the liquid, *corrections* in the readings are made for temperature differences. On the Baumé scale the correction generally used is 1°Bé. for every 10°F. In using the instrument for liquids lighter than water in a liquid that is too warm, *i.e.*, too much expanded, the instrument will sink too far and read too much, because the larger numbers are at the top of the scale and signify the lighter liquids. Consequently, the correction must be subtracted in this case. With the instrument for liquids heavier than water, the larger numbers are at the bottom of the scale, hence in an over-warm liquid the instrument will read too little, and the correction must be added. With liquids that are too cold, the reverse is true in each case. However, since the degree of expansion varies with the kind of liquid, and is not the same per degree at all tempratures, no sufficiently accurate, generally applicable correction can be assigned for the direct-reading scale, where small decimal differences are important, although corrections may be worked out for a given liquid within specified ranges of temperature.

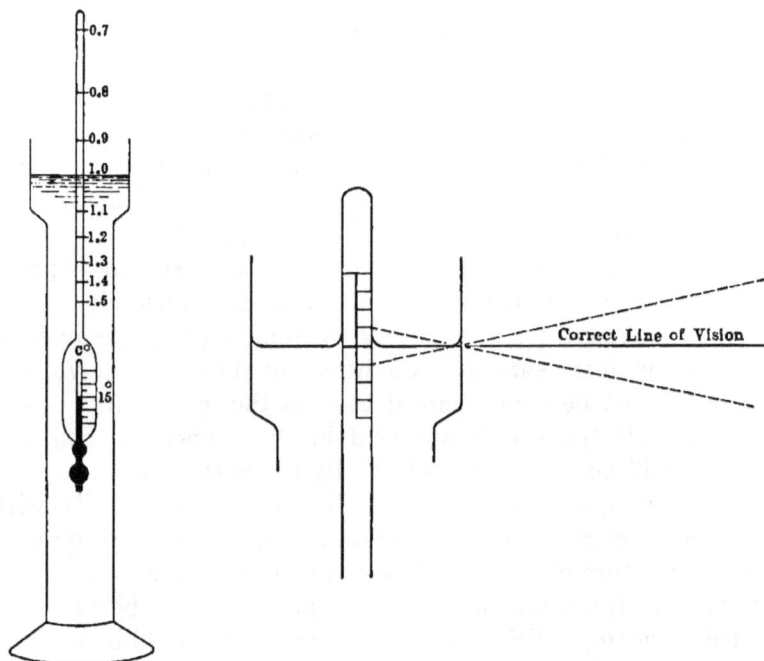

FIG. 75.—Direct-reading hydrometer in pure water at 15.5°C. (60°F.). Enlarged section indicating correct method of reading scale.

The hydrometer should always be allowed to remain in the liquid several minutes ·before the reading is made—as much as 15 minutes in viscous liquids, as in some oils for instance. It should be noted that the accuracy of the hydrometer increases with the thinness of its stem or spindle.

I. *Instruments in Liquids Lighter than Water*

The Baumé scale begins at 10.
The direct-reading scale begins at 1.

To convert the values of one scale into the corresponding values of the other. The known value is substituted in the following formulas and the expression is solved algebraically.

For liquids lighter than water:

$$\text{Bé.} = \frac{140}{\text{Sp. gr.}} - 130$$

$$\text{Sp. gr.} = \frac{140}{130 + \text{Bé.}^\circ}$$

II. *Instruments in Liquids Heavier than Water*

The Baumé scale begins at 0.

The direct-reading scale begins at 1.

To convert for liquids heavier than water:

$$\text{Bé.} = 145 - \frac{145}{\text{Sp. gr.}}$$

$$\text{Sp. gr.} = \frac{145}{145 - \text{Bé.}^\circ}$$

INDEX

www.ingramcontent.com/pod-product-compliance
Lightning Source LLC
Chambersburg PA
CBHW020905210326
41598CB00018B/1782